PRAISE FOR *BARREN LANDS*

"Not since John McPhee has a writer made me care so much about the men who care for rocks. There is an intense joy in watching a writer successfully work such a vast multitude of threads. *Barren Lands* is bursting with life . . . breathtaking."
— Marc Zabludoff, former editor-in-chief, *Discover* magazine

"A true adventure story, with secrets, spies, attempted murder, mysterious fires, plane crashes, and feats of daring and courage. Fascinating, fast-paced, and enormously enjoyable."
— Richard A. F. Grieve, chief geoscientist, Natural Resources Canada

"The writing sparkles." —*National Post* (Ottawa)

"*Barren Lands* is a brilliant, lucid, and, in parts, sensational investigation."
— Edward Jay Epstein, author of *Dossier: The Secret History of Armand Hammer* and *The Rise and Fall of Diamonds*

"This tale of avarice and ambition—every word of it is true—has never been told so dramatically, or with such scrupulous attention to detail." —Stefan Kanfer, author of *The Last Empire: De Beers, Diamonds and the World*

"Krajick, a well-informed journalist, vividly relates the saga."
— *The Washington Post*

"Armchair adventure at its very best." —*Rock & Gem*

BARREN LANDS

AN EPIC
SEARCH
FOR DIAMONDS
IN THE
NORTH AMERICAN
ARCTIC

KEVIN KRAJICK

A W.H. FREEMAN / OWL BOOK
HENRY HOLT AND COMPANY
NEW YORK

Henry Holt and Company, LLC
Publishers since 1866
115 West 18th Street
New York, New York 10011

Henry Holt® is a registered trademark of Henry Holt and Company, LLC.

Library of Congress Cataloging-in-Publication Data
Krajick, Kevin
　　Barren lands: An Epic Search for Diamonds in the North American Arctic / Kevin Krajick
　　　　p. cm.
　　Includes bibliographical references and index.
　　ISBN 0-8050-7185-7 (pbk.)
　　1. Diamonds—Northwest Territories—Gras, Lac de, Region. I. Title.
　　TN994.C2K48 2001
　　338.4'7622382'09712—dc21　　　　　　　　　　　　　　　　　　　　　200100815

Henry Holt books are available for special promotions and premiums.
For details contact: Director, Special Markets.

First published in hardcover in 2001 by Times Books

First Owl Books Edition 2002

A W. H. Freeman / Owl Book

Designed by Michael Minchillo

Printed in the United States of America
10　9　8　7　6　5　4　3　2　1

FOR
RUBY AND STELLA

Man puts an end to darkness, and searches every recess for ore in the darkness and the shadow of death. He breaks open a shaft away from people, in places where there is no foothold, and hangs suspended far from mankind. That earth from which bread comes is ravaged underground by fire. Down there, the rocks are set with sapphires, full of spangles of gold. Down there is a path unknown to birds of prey, unseen by the eye of any vulture; a path not trodden by the lordly beasts, where no lion ever walked. Man attacks its flinty sides, upturning mountains by their roots, driving tunnels through the rocks, on the watch for anything precious. He explores the sources of rivers, and brings to daylight secrets that were hidden. But tell me, where does wisdom come from? Where is understanding to be found?

JOB 28: 3-12

It is the ambition of every prospector to unroll the map northwards.

BERTRAM BARKER, NORTH OF '53

CONTENTS

PART III

PART IV

ACKNOWLEDGMENTS

Thanks to the memories of diamond seekers and Arctic adventurers past and present, nothing here is fictionalized. As far as I know, this is the way it happened. From southern Africa to the Arctic Ocean, many shared details of their lives, dug out their memoirs, and physically retraced their steps. Several guarded my well-being in some dangerous places.

Nearly every living character helped in the research. Their names are obvious in the text, but special thanks are due to Chuck Fipke, Stewart Blusson, Hugo Dummett, John Gurney, Tom McCandless, Chris Jennings, Brent Carr, Paul Derkson, Mike Waldman, Ray Ashley, Fred Sangris, and Moise Rabesca. Deep thanks also to their various parents, spouses, children, brothers, and sisters, who gave much. People at BHP Diamonds helped often, especially Rory Moore and Chris Hanks. De Beers Consolidated Mines has a fierce reputation for secrecy, yet current and ex–De Beers scientists welcomed my interest and volunteered many previously hidden facts about their work. They include Barry Hawthorne, Mousseau Tremblay, Joe Brunet, and Craig Smith. Three most important northern guides hardly appear in the text, but their spirits are there: Steve Matthews and Tom Andrews of Yellowknife and Allen Niptanatiak of Kugluktuk. I bless them all and trust none will get into trouble for what they revealed.

A large amount of history is contained in these pages. Whenever possible I used primary sources: maps, diaries, newspapers, scientific articles. The Geological Survey of Canada (GSC), whose work quietly helped set the stage for the northern diamond rush, provided piles of material via Walter Nassichuk, Ron DiLabio, Bruce Kjarsgaard, and many others. Unique historical documents came also from generous colleagues at the state geologic surveys of Wyoming, Wisconsin, Arkansas, Colorado, Maine, Illinois, Indiana, Kentucky, Michigan, Montana, New York, Ohio, Virginia, Georgia, and North Carolina. Rare memoirs of the Barren Lands were provided by the Archives of the Northwest Territories and the wonderful Yellowknife Public Library. The New York Public Library and the geology library of Columbia University were incomparable sources of old books and articles on diamond prospecting.

My first opportunity to journey to the Barrens came in the summer of 1994, when I reported on the diamond rush for *Discover* magazine; later pieces for *Newsweek*, *Audubon*, and *Natural History* furthered my research. My agent, Regula Noetzli, spent two years finding a publisher for this book. W. H. Freeman's Holly Hodder finally accepted the project, and it was greatly strengthened by editor Erika Goldman and the hard work of Mary Louise Byrd, Bob Podrasky, Michele Kornegay, Susan Goldstein, and T. J. Fitzgerald. Ex-GSC man William Shilts and W. Dan Hausel of the Wyoming Geological Survey served as expert advisers, preventing many mistakes.

Three people deserve the most thanks of all: my parents, Katherine Distin Krajick and Rudolph Adam Krajick, who always encouraged me to go wherever curiosity led; and my wife, Ruby, who stood behind me no matter what. When I disappeared for weeks into the tundra and then for months into my office, it caused her worry and hardship, I know, but she was there always.

I narrate here the deaths of some fellow human beings. To their loved ones: I tried to do this with the greatest respect. I also mean no disrespect to the people of the north by locating the Barren Lands wholly within Canada's Northwest Territories. In 1999 the northern half of the Territories, and thus roughly half the Barrens, became Nunavut, homeland of the Inuit. All events described here took place before then.

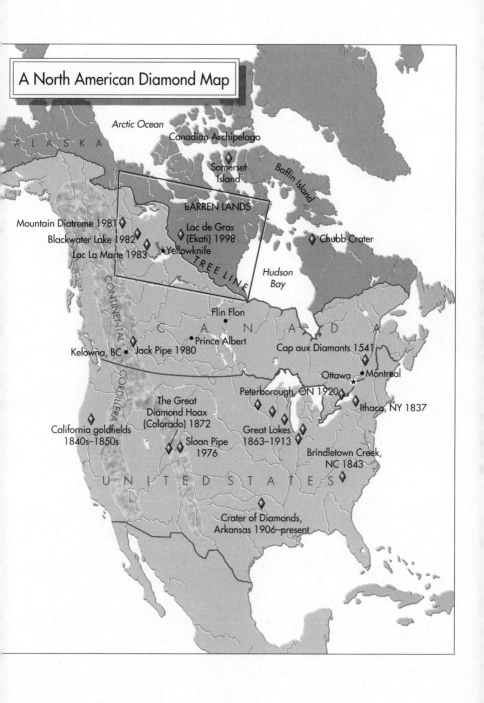

A North American Diamond Map

Arctic Ocean

Canadian Archipelago

ALASKA

Somerset Island

Baffin Island

BARREN LANDS

Mountain Diatreme 1981

Blackwater Lake 1982

Lac La Marte 1983

Lac de Gras (Ekati) 1998

Yellowknife

Chubb Crater

TREE LINE

Hudson Bay

CONTINENTAL

CANADA

Flin Flon

Prince Albert

Kelowna, BC

Jack Pipe 1980

Cap aux Diamants 1541

Ottawa

Montreal

CORDILLERA

Peterborough, ON 1920

Ithaca, NY 1837

The Great Diamond Hoax (Colorado) 1872

California goldfields 1840s–1850s

Sloan Pipe 1976

Great Lakes 1863–1913

Brindletown Creek, NC 1843

UNITED STATES

Crater of Diamonds, Arkansas 1906–present

The Barren Lands and Surrounding Regions

North American Coast

Mackenzie River

ARCTIC CIRCLE

TREE LINE

Bloody Fall • Coppermine (Kugluktuk)

BARREN

Norman Wells

Great Bear Lake

Contwoyto Lake

Mountain Diatreme

Acasta Gneisses

Coppermine River

Yamba Lake

Lac d Sauva

Main Esker

Norm's Camp

MACKENZIE MOUNTAINS

De Beers camp

Blackwater Lake

Desteffany Lake

GSC Mapping 1950s–1970s

Blackwater River

Lac de Gras (Ekati)

Misery Poir The Narrov

Willowlake River

Horn Plateau

Lac la Martre

Yellowknife

Fort Simpson

Glacial Lake McConnell

Mackenzie River

Great Slave Lake

BOREA

NORTHWEST TERRITORIES

100 km

N

0   100 Miles

To Edmonton

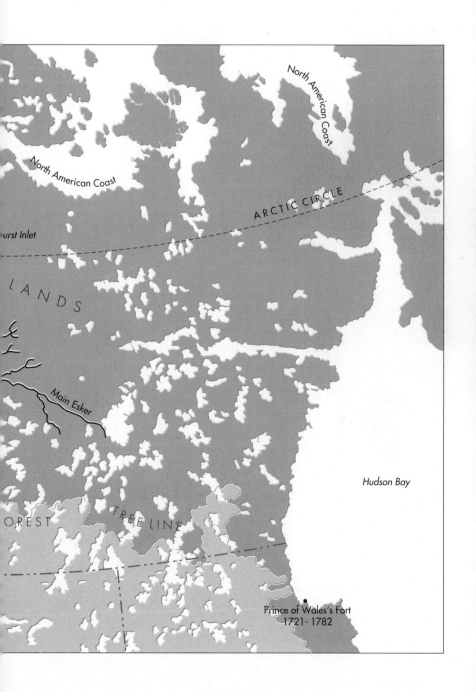

BARREN LANDS

Prologue: Heading North

Having neglected to bring a pickax on this particular trip, Charles E. Fipke was nearing the bottom of a seven-and-a-half-foot hole in the snow and ice by tearing at some rocks with the small pick-end of a geologist's hand hammer. His son Mark was at the distant top, shouting down curses about the cold, the wind, the risk of dying, and the uselessness of it all.

Fipke ignored him. It was his own turn to dig. It had taken them five hours to get down this far, and, as usual, he was not going to stop until he got what he wanted: a twenty-pound bag of sand and gravel from the frozen earth at the bottom. Twelve years into this mad prospecting enterprise there seemed to be no end in sight. That is, unless you considered the empty bank account, the crystals of wind-driven snow now eroding their faces, the cold progressing up their limbs, and the fact that they were a several-week walk from town in the middle of the tundra.

Fipke had picked the shoreline of an unnamed little lake to dig, based on a glimpse of it from the air the day before. With their last dollars, they had chartered a $600-an-hour helicopter and flown here to follow up the tantalizing clues of earlier trips and at long last to stake out their claims. Millions of frozen tundra lakes, covered with snow and pocked with lichen-blackened boulders around the shores, looked the same from the air—except this one, sitting in its own craterlike depression, a bit more circular than the others. On both the north and south shorelines, dark, unusual cliffs dropped straight to the ice, as if someone had drilled a hole in the bedrock. In camp that night a dozen miles off, Fipke could not sleep. Every time he closed his eyes, he would see the lake with the two dark cliffs on opposite shores, plunging to the ice.

In the morning they had flown back and landed near where the shore sloped more gradually to what seemed like a beach. It was hard to tell; snowdrifts filled the low ground. Under the snow were apparent outlines of boulders, sticking up like pumpkins under a bedsheet. The digging began where Fipke presumed waves might wash up in summer, leaving sand. Beneath six feet of hard snow he hit a foot and a half of ice laced with large stones. At the bottom, Fipke cleared a path, tossing dislodged rocks, ice chunks, and powdered ice up to Mark, until he hit the apparent terminus. Here, he discovered something extremely unfortunate: Underneath the rocks was not sand, just more rocks. In the parlance of the northern prospector: loon shit. Useless junk. This would not work. You needed sand, not rocks, to figure this out.

This was not their first hole today. Fipke climbed out and surveyed beyond a deranged semicircle of other fruitless excavations they already had dug. Farther out, low snow- and stone-covered hills rose and fell like great waves on the choppy high seas, the long rooflets of thin snowdrifts on them honed sharp as knives by wind. Across the lake he spied a low ridge cutting through all, running in both directions to the horizon. On its flanks was some high ground where wind had driven the snow off a patch of coarse gravel. Maybe at least they could climb up on there and hack out gravel. Once they got home—if they got home—they would sift through it and take a look under the microscope. Fipke told the pilot where he wanted to go, and Mark began cursing again. The date was April 14, 1990.

In early July 1994, I traveled to the tundra of Canada's Northwest Territories to report on the discovery of a huge field of potential diamond mines there. My destination was that vast region known as the Barren Lands, the uninhabited edge of North America that lies beyond the regions where trees exist. Here the small-time prospector Charles Fipke and several partners had tracked down the remains of a single ancient volcano under the surface of a tiny, unnamed circular lake. Deep in the rocks were eighty-one tiny diamonds.

This discovery sparked one of the greatest mining rushes in history. The only road was hundreds of miles south, within the tree line, but hundreds of companies homed in with aircraft, staking claims larger than whole countries. The claims ringed the discovery site, near the shore of the huge Lac de Gras, or Ekati, as it is called by the Dene Indians. In English both names mean Fat Lake. Some aboriginal elders say it refers to the abundant smell of caribou fat burning in campfires during long-ago fall hunts there. Others say it is for the white quartz veins lacing the crystalline rocks along the shore, which resemble veins of that same fat.

Lac de Gras rattled the diamond world hard. Diamond mines are so difficult to uncover that decades may pass between the finding of one and another. A single discovery can be worth $70 billion, so the few who know the arcane methods of diamond prospecting guard them well. The South African De Beers cartel has run most of these mines since the nineteenth century, and the cartel was here like everyone else, of course, frantically deploying drills, instruments, computers, and geologists. So far they hadn't found a thing: Fipke had beat De Beers to the diamonds, as the Fipkes of the world almost always do. It is only afterward that the cartel usually takes over. This time it was not clear if it could.

Fipke was an unlikely diamond magnate: nearsighted, goateed, short and tough as an oak stump, his speech jolted by a stutter and a hair-trigger laugh. He often forgot to tie his shoes and had other absentminded-professor mannerisms, though without benefit of the fancy degrees supposed to attend absentminded professorship. His main assets were steady industry, endless optimism, and his family. His main partner, an ascetic loner named Stewart Blusson, was equally unlikely. Blusson, recently of the elite Geological Survey of Canada (GSC), did have a Ph.D., as well as a license to fly helicopters—both great assets in the north, where there is much rock and no roads. Wiry, weathered, quiet, and intense, Blusson was a wilderness survivor par excellence, but always brushing just a little too close to death. Among other scars, he carried a grizzly-claw mark on his left arm from his years in the Yukon. When these two found something, it threatened to change everything—the diamond market, the Barren Lands, themselves.

When I first heard of the strike, I was fascinated. There was a small stack of books about the cartel's supposed dark manipulations of the diamond market, but nothing on how prospectors actually locate diamonds. There were whole shelves on polar explorers, but little on the obscure Barren Lands, where the North American Arctic begins. Finally, among old papers, I discovered a secret saga: Prospectors had sought a North American diamond mine for 450 years, from Arkansas to the islands of the Arctic Ocean. Fipke and Blusson had spiritual ancestors, all working in the same direction, whether they knew it or not. So, I decided to write the story of how mineral hunters—eventually these diamond hunters—arrived at the literal edge of the earth. It takes shape in the south over many years and miles. Then, like an old prospector seeking new frontiers, it turns north.

I first met Charles Fipke at his lab in Kelowna, British Columbia, a small city that lies about 200 miles northeast of Seattle, Washington. No one in Kelowna calls him Charles, not even his parents. He's just Chuck. Friendly and unassuming, Chuck told me bits and pieces of how he and his crew had explored snaky brushlands; edges of glaciers; tops of high, crumbly precipices; and other places normal people don't go. Then he offered to show off his big find.

A few days later we were on a commercial jet to Yellowknife, Northwest Territories. This is the jump-off point to the Barren Lands, a lone frontier town of 13,000 souls hammered into the bare rocks of the forest at the gravel dead end of the South and North American road systems. A hundred miles further, trees stop and tundra begins. The name Yellowknife comes from the Yellowknives, a tiny group of Dene said to have once possessed a mysterious mine far out on the tundra. Aside from Indians and government officials, its population is mostly immigrant prospectors and related folk. It lies about 1,800 miles northeast—mostly north—of Seattle.

On the flight up, we passed over hours of wilderness: the north-south continental cordillera, which runs from Guatemala nearly to the northern ocean, its ranges upon ranges of barren peaks frozen forever under glaciers and icy tarns; then the swampy, undernourished boreal forests that form the deep buffer between the treeless Arctic and the populated south. The slender road to Yellowknife, intermittently visible far below, was the sole sign of human penetration. Finally we flew over an enormous wind-whipped water, Great Slave Lake, to whose north shore Yellowknife clings.

When we arrived it was near midnight, but the sun still hung on the horizon in that peculiar, spooky twilight of the far northern summer. After a few hours' sleep we went down to Air Tindi, a charter outfit with a lakeside float-plane dock, and squeezed into a de Havilland Twin Otter alongside a cargo of diesel drums. In a land one-third covered with lakes, ponds, creeks, and marshes, float planes and helicopters are the only way to travel. In winter the airplane pilots trade floats for skis. The Otter revved up, wafted off, and we turned north once more.

Soon the spruces below looked ever hungrier, like rejected Christmas trees. They began to spread out and keep distance from one another. Big bare spots opened, where whalebacks of naked bedrock reared up and plunged into ponds. Then along a jagged, invisible edge, we passed the tree line. Below was a mosaic of stone, splashed with uncountable lakes and parallel watery striations stretching to the horizon. The Barrens form an immense triangle bounded south by the trees, north by the continental coast, east by Hudson Bay, and west by the woods at the foot of the high cordillera—roughly 500,000 square miles. No feature rises more than a few hundred feet; it is made of the world's oldest rocks, and they are simply wearing out.

Around 16,000 B.C., the last continental ice sheet receded from its southern limit near Exit 10 on the New Jersey Turnpike (Metuchen), but did not finish melting up here until about 4000 B.C. Jumbles of car-size boulders still lie where ice dropped them, and raw dikes of surviving igneous bedrock jut like rows of fairy castles. There are eskers—great, sinuous ridges of sand and gravel left by watercourses that once ran within the ice. Hundreds of feet high, hundreds of miles long, the eskers are leveled on top with beachy sand and look

eerily man-made, like interstate highways. The land is still frozen
from the top to 1,500 feet down. There is little rain or snow, so it is
technically desert; water lies everywhere only because permafrost,
low relief, and lack of river systems defeat its movement. There is a
term for this: deranged drainage. With nothing to block wind, wind
blows most of the time. Winters hit 90 degrees below zero Fahrenheit. Lakes freeze six feet thick, rendering the land "more a place of
physics than biology," as one bush pilot put it. Deep rock cracks disguised by snow can swallow human or animal in an instant. When
the thaw comes during brief summer, caribou migrate up from the
trees, grizzlies wake from an eight-month torpor, and the land
hatches a blaze of blueberries and enough biting insects to eat the
planet. In this beautiful season, a bolt of lightning may come from
the vast sky and strike you dead. All this is why it is called the Barren Lands.

By some definitions, the Arctic Circle marks the start of true
north. At 66 degrees, 32 minutes, it is the latitude above which the
sun never sets on at least one summer day and on one winter day
never rises. But it is the irregular tree line, which waves both above
and below the invisible Circle, that is the actual boundary. That is
where weather patterns change abruptly and familiar creatures like
coyotes and amphibians disappear. Plants and human history bend
into strange shapes. The world as most of us know it ends.

Like some other sojourners of the twentieth and twenty-first centuries, I have always been deeply saddened by the journals of explorers like Meriwether Lewis. In the early 1800s, Lewis crossed east
over the treeless Great Plains of the American West, looking for the
Pacific Ocean. He wrote in September 1804: "Vast herds of Buffaloe
deer Elk and Antilopes were seen feeding in every direction as far as
the eye of the observer could reach." Of wolves, he noted, "We
scarcely see a gang of buffaloe without observing a parsel of these
faithfull shepherds on their skirts in readiness." Along the dry bed of
the Medicine River Lewis glimpsed what seemed a mythical combination of wolf and wildcat—probably a wolverine, a real beast now
long extinguished almost everywhere in North America. Meriwether
Lewis, I thought, was among the last to witness such miracles, now
hunted down, plowed up, paved over.

I believed this until shortly after the Twin Otter landed. Four days later after midnight, with the moon sitting low on one horizon and the sun on the other, I crept up the flanks of a great esker, peered over, and saw the migratory Bathurst caribou herd moving back to the trees, newborn calves in tow. The adults' great velvety horns swarmed like giant puppets in the air ten feet from a boulder behind which I cowered. Then they flowed around me on both sides, stretching as far as the eye could see to all points of God's compass. Except for the popping of tens of thousands of leg joints and the occasional clatter of a hoof knocking over a stone, the land was silent. Out of the twilight a single brown wolf trotted calmly over a rise and headed for an old bull.

The Barrens have never really been lived upon, only traveled through; the wolverine, an eater of the dead, is almost its only year-round resident. For millennia Dene living in the shelter of the trees followed the caribou into the tundra to hunt them, but they always retreated with their quarry. The Inuit, or Eskimos, living off the resources of the sea on the tundra's opposite end also entered to hunt the caribou, then retreated. The center, remote from sea or trees, was no-man's land. Now nomadic peoples have retreated for good, and Europeans have never replaced them. Only the stone tent rings and spearheads of the ancients are left, scattered along the eskers where they traveled, camped, and hunted. Among the Dene, only a few dwindling elders remember *Hosi*—the treeless land—and the names of lakes, hills, and "dreaming places," where terrible visions will appear to anyone who dares sleep on them. Aerial photography has sketched in blank areas, but cartographers have attached names to few features, and there is little real knowledge of what is there. When I arrived biologists were still studying the journal of young Samuel Hearne, the first European—and still one of the only ones—to cross the Barrens. The years of his journey were 1769–1772.

Hearne was the original Barren Lands prospector; and for reasons that soon will be apparent, most of his European successors in this region have been prospectors, too, hunting gold, uranium, and other treasures. Hearne was seeking a fabulous copper mine, but his trail, laced with bloody footprints and stalked by massacre, never led to it. Others have failed out here as well. That is why, unlike most aboriginal people, the Dene have never come up with a word for "white persons." They refer to outsiders as *kwet'i*—rock people—after the inexplicable thing they all seem to be hunting.

However, until Fipke and Blusson showed up, the search for diamonds never penetrated here. It was instead going on far south, and with great gusto. Canadians and Americans have long been enthralled by diamonds. Americans became the world's leading importers in the mid-1800s, a fact that has never changed since. Almost every year Americans set a new record, and they now buy one-third of the world's diamonds—far more than anyone else. Canada, with one-tenth the population, would hold this honor if only it had more people. Eight of ten Canadian women own one or more pieces of diamond jewelry, making Canadians the world's top per capita diamond consumers.

In search of a domestic supply, hundreds of individuals and companies in both countries have vied for the big strike. They often have been so overanxious that there is a long rap sheet of places where land promoters, stock sellers, and assorted swindlers have made fortunes on nonexistent deposits. Maps are pocked with the resulting place-names: Diamond Canyon, Diamond Butte, and Diamond Peak, Arizona; Diamond Field Draw, Colorado; Diamond Mountain, Utah; Diamond Hill, Rhode Island; Diamond Crater, Oregon; the Diamond Range and Diamond Valley, Nevada; the Diamond Mountains, California; Diamond Basin, Idaho; Cap aux Diamants, Quebec; assorted Diamond gulches, creeks, islands, and springs; and a dozen towns with the word Diamond. There are quartz "Herkimer diamonds" from upstate New York, "Rock Springs diamonds" from Illinois, and "Cape May diamonds" from the Jersey shore.

The search is stoked by more than mirage: Genuine diamonds have in fact turned up all over North America, and many strange stories are connected with them.

In the summer of 1886, twelve-year-old Willie Christie of Dysartville, in rural northeastern North Carolina, was sent to a farm spring to fetch a pail of water. As he sat down on a box by the spring, he saw a translucent stone shining up at him from two feet away. Thinking it "a pretty trick," he took it and showed it around at the local grocery store. Eventually it was mailed to the great American gem expert George Frederick Kunz, a self-taught mineralogist from Hoboken, New Jersey, who later played a pivotal role in the search. Kunz confirmed it was a diamond, "a distorted and twinned hexoctahedron, of $4^{1}/_{3}$ carats, transparent, with a grayish-green tint." The facets formed a tiny spiderweb pattern. Kunz bought it for $150, and for decades it was prominently displayed at the American Museum of Natural History. Its whereabouts are now unknown; like so many other old American stones, it is lost. But that is another story.

In April 1928, twelve-year-old William P. "Punch" Jones of Peterstown, West Virginia, was playing horseshoes with his dad, Grover, in the yard. Punch's toss hit the stake and kicked a glassy fragment from the dirt. "See, I found a diamond!" he joked. Later, Punch became briefly famous, but not for this. His mother, Grace, set the world's record for consecutive male births—Punch was eldest of seventeen kids—and U.S. President Franklin Roosevelt hosted Grover Jones Family Day at the 1939 New York World's Fair. Punch kept the fragment in a cigar box until June 1943, when he mailed it to a Virginia geology professor. At 34.46 carats, it was the second largest known North American diamond; the Smithsonian Institution displayed it. Shortly after, Punch joined the U.S. Army and was killed in combat in Germany. Afterward his mother grew sick of hearing about the famous "Punch Jones Diamond." "I wish they'd a threw it in the New River sometimes," she said. In the 1960s the family asked the Smithsonian to return the stone, then stored it in a safe deposit box at First Valley National Bank. In 1984, Sotheby's of New York sold it for them to an anonymous investor in Asia for $74,250.

In the summer of 1990, Darlene Dennis was jogging on a gravel road just after dawn, a quarter-mile from her home outside Craig, Montana. In the gravel she spied what appeared to be a clear, melted plastic coat button. A jeweler in Great Falls identified it: 14 carats, nearly flawless. The jeweler mentioned it to New York antiques dealer Alexander Acevedo, who knocked on Dennis's door and wrote her an $80,000 check for the stone, which he dubbed the "Lewis and Clark Diamond." It was four times the normal price, but it was American, and thus, he felt, extra valuable. Acevedo extracted the find location from her, then secretly crawled for hours on the road, in a nearby riverbed, and in a gravel pit looking for more. He received extreme sunburn and, some time later, a bad skin cancer on his nose. I visited Acevedo one afternoon in 1999 at his Madison Avenue gallery. He sat me down on an expensive old couch and let me heft the rough stone in my hand. It is real.

Real diamonds of more than 2 carats also have been found in the states of Alabama, Arkansas, Virginia, North Carolina, Georgia, Tennessee, Texas, California, Oregon, Idaho, Colorado, Wyoming, Washington, Ohio, Wisconsin, Minnesota, Michigan, Illinois, and Indiana and in the province of Ontario. Smaller stones, plus unverified finds, are reported from Alaska, Maryland, New York, South Carolina, Kentucky, New Mexico, Arizona, Alberta, Saskatchewan, Quebec, and British Columbia.

Each new random find sparks prospecting ventures, but all face the same problem: Systematic, scientific searches in the same places rarely yield more stones—nor clues to their origins. In fact, as these stories suggest, many discoverers are children, the world's luckiest gem prospectors: full of energy, eyes built close to the ground, happy to pick up any piece of dirty junk, not looking for anything in particular. Only one consistent clue has emerged. A definite swath has turned up in terminal moraines left by the last glaciation—the outer edges of the northern ice.

Now, if there is one thing Americans and Canadians both love, it is the idea of the one-mule prospector heading out to find the mother lode. The California forty-niners; the Klondike; unbathed men with names like Swiftwater Bill, Skookum Jim, and Lucky Swede; tales of lost mines and headless skeletons found sitting in lonesome miners' cabins: In these two nations whose economies and outlooks are founded on minerals, yarns about prospectors are the quintessential folklore. But these are almost always stories about gold and silver. When was the last time you heard one about diamond hunters?

Diamonds are infinitely rarer than gold or silver. They are, technically, a mineral—a naturally occurring solid with a more or less predictable chemical formula and crystalline structure. Quartz and mica are minerals; most rocks, agglomerations of minerals. Diamond, like the mineral graphite, is made of the single element carbon but is rearranged by heat and pressure into a lattice that is totally different—supremely lustrous, hard, and heavy. Meteorite craters in Arizona and Siberia contain diamonds crystallized from the high-pressure inferno of the impact; some diamonds may even have arrived inside the meteorites, from space. Below earth, they may form when tectonic plates collide and one block of crust is shoved under another. The submerged block gets squeezed, then later refolded up and up, like a seed squishing out of a peach, until it is at the top of a mountain. There are diamondiferous peaks in eastern China, the Tibetan Himalayas, and northern Kazakhstan. Diamonds may travel from unimaginable depths; geologists recently found some on the southwest Pacific island of Malaita in an eruptive rock called alnöite, thought to have come up 500 miles. No one knows the depth limit, if any.

However, diamonds made in these ways are generally sparse, ugly, and microscopic. The ones prospectors want—big, clear, and concentrated in minable quantities—are the ones in our story. We should not pretend we know exactly how they form; ideas change from time to time. But lab scientists have analyzed minerals found alongside diamonds, broken down separate minerals encased inside (flaws, or "inclusions"), and synthesized diamonds themselves. As a result, we believe most gem-quality stones come from below the nuclei of continents, in earth's most ancient, stable regions, called cratons. It is said they form 75–125 miles down, in a "diamond stability zone" of 1,700–2,300 degrees Fahrenheit, at 45,000–60,000 atmospheres, in two rock types, eclogite and peridotite. The carbon source is probably primordial fluids, gases, or semisolids—nothing as pedestrian as the proverbial hunk of coal. Crystallization may take a second or 10 million years; we do not know. Crystals may be stored for eons; some have been dated at 3.3 billion years, more than three-quarters of earth's history.

Diamonds form mainly under cratons because under much of the rest of earth the geothermal gradient is too high; that is, as depth increases, temperature goes up too steeply. Most of the near-surface has been ravaged repeatedly by diamond-unfriendly melting events—rifts, plumes, volcanoes, continental collisions. As a result the lithosphere—the stiff, congealed stuff of tectonic plates floating over earth's hot, gooey asthenosphere—averages only sixty miles deep. However, within the unmoving cratons, things have long been insulated from upset; here cooled, solidified lithospheric roots may penetrate the hot interior as much as 250 miles, looking in cross section like the teeth in your gums. Toward the bottoms, diamonds form. The very oldest cratons are called archons, for the Archean eon, which ended 2.5 billion years ago. They underlie parts of Africa, Siberia, India, Brazil, Greenland, Scandinavia, Canada, and the United States—all places where diamonds may be found.

For most of history, no one knew where diamonds came from; they were found scattered in a few riverbeds. Then, around 1870, miners in South Africa discovered the means by which they emerge: small, deep-seated structures called kimberlites. These are semimolten, gassy eruptions, often starting within or below the diamond stability field and tearing up through mazes of faults, fissures, and weaknesses. They come in worldwide waves. No one knows why, nor has anyone seen one erupt; the last of seven known waves ended 50 million years ago. We surmise one is like a tornado, traveling upward at twenty or thirty miles an hour, smashing through innumerable layers, ripping things out on the way and creating a supercharged cereal of liquids and solids. If by chance it intersects diamonds, they may come along; but considering the vast assemblage of things a kimberlite may pick up, diamonds form a minor part, if any at all. And, if the journey takes too long, gets too hot, or too much oxygen gets in, the gems will burn; after all, they are carbon, subject to the same laws as charcoal briquets. This all makes diamonds even rarer.

Most kimberlites probably never hit the surface; they sit far below like blind, unblinking eyes. But if one gets near enough, it is progressively liberated from the pressure of overlying rock. At top, it expands and speeds, breaks through, and rockets into the sky to blow out a deep, carrot-shaped crater. Lava rains back in like a soufflé that has exploded and settles into itself to solidify into a kimberlite pipe. Each one is usually just a few acres on top, but often they emerge from their complex, tapered plumbing in clusters from six to forty. Eons later, humans dreaming of clear, tiny stones go looking for them.

All the diamonds found in rivers during most of human history probably would not fill more than a wheelbarrow. But in the ten years after pipes were discovered, diamond production multiplied ten times over. Since 1880 it has multiplied forty times again, and the pace of discovery and mining has increased ever faster. Humanity has now mined over 500 tons of diamonds—a third of them in the 1990s.

In some ways, diamond hunting is like any other prospecting. Pick an area that, for whatever reason, appears prospective. Then dig gravel, soil, or rock in many spots and search it for footprints of an orebody. Ores of zinc, lead, or copper may erode from a mass and leave a trail of intact rocks ("float"), an invisible chemical trail, or pure particles of the substance itself, which may be followed. Gold is, in fact, where you find it. Like diamond and most other valuable minerals, gold is heavier and more durable than your average substance. Fragments of such heavy minerals may survive long journeys in streambeds or glaciers. Swirl a panful of sand with some water, and light, usually worthless, grains like quartz float off; in the middle, gold flakes and other heavies settle.

Once you find footprints, it is time to track them to the source. Following up rivers is standard, for they conveniently concentrate debris from identifiable watersheds. Somewhere upstream the material may suddenly peter out, which means you have just passed the first-order tributary feeding it in. Go back, pick up that tributary, and follow until the same thing happens at a second-order tributary. And so on. The closer you get, the more and bigger the float, mineral grains, or other signs. At the spot itself, you may see metals inside bedrock leaching out to oxidize in the air, creating a bright rusty stain, or gossan. Chunks of pure, or so-called native, copper may lie about. Visible gold may sparkle in a quartz vein. Dig. You have hit the mother lode; you and your descendants will live in splendor for many generations.

That is the way it is supposed to work, anyway; it rarely does. Mountain ranges rise and fall, and rivers change their courses. Oceans, lakes, volcanoes, and ice sheets may smear the trail or bury it altogether. You have to be smart and well informed to make the right adjustments—and, in the end, lucky. Keep an eye out for followers, bandits, backstabbers, and claim-jumpers; the scientific complications of prospecting are rarely shown in old prospector movies like *Treasure of the Sierra Madre*, but the treachery and madness are real.

Diamond prospecting is the worst. Not only are diamonds rarer than gold and kimberlite more obscured than gold ore, but most kimberlite contains no diamonds; and when it does, they are vanishingly rare. Gold ore grades are measured in ounces per ton, up to a pound per ton. A diamond carat, based on the weight of a tiny dried locust tree seed, equals one-fifth of a gram, and rich deposits are measured in carats per *100* tons—that is, parts per million. Even Russia's fabulously rich Mir Pipe has only about 60 carats per 100 tons—three-quarters heavily flawed industrial stones. (Gem-quality diamonds are so much more valuable that they typically earn 90 percent of a mine's income.) Given this, only about 30 of the world's 6,000-some known kimberlites have ever become major mines. Even when companies think they have a mine, they may test-dig for months or years to know for sure.

Clearly, you cannot find diamond deposits by looking for diamonds; they are too rare. Instead, you must look for surrogates, secret signs. The most useful are other rare heavy minerals inhabiting kimberlite in quantities 100,000 times greater—variously called diamond indicators, pathfinders, satellites, captives, or slaves. Bleeding from kimberlite in sand- to BB-size grains, the best indicators are certain rare blood-red pyrope garnets, pea-green chrome diopsides, and shiny black ilmenites. They may travel dozens or hundreds of miles before settling in sandbars, eskers, or beaches. Four or five grains culled from among billions may start the trail.

The specialized job of finding, distilling, and tracking such heavy-mineral grains was Chuck Fipke's business. He and Stewart Blusson had come a long way to find their pipe, but things were still uncertain. When I met them in the diamond-rush summer of 1994, the question was whether there were enough diamonds to make mining in this exquisitely isolated and hostile region a feasible idea. The other question was whether it was a good idea.

PART I

CHAPTER 1

Misery Point

A thunderhead of mosquitoes attacked Chuck Fipke's entire body as he scrambled like a crab across the tundra looking for something. The rest of us were uselessly trying to wave off this plague of the forty-day warm season and sweating in the uncharacteristically still air. Chuck took no note of the insects thronging his wrinkled forehead, tangling his thinning hairline and goatee, and drilling each muscle of his thick, gnarly forearms. He was in his customary prospecting position: hands, knees, and eyes to the ground; rear end high in the air; wearing his usual blue jeans and white golf shirt. One running shoe trailed an untied shoelace. His nose was now meeting up with a gray chunk of rock protruding from some ankle-high blueberry bushes by the side of a small lake, and he had twisted his glasses crazily to one side of his head so he could peer directly at the rock with the little magnifying glass that customarily hung from his neck by a white cord.

"Nope, nope, that's not it," he muttered to Walter Nassichuk. Chuck did not bother rising; instead, he charged at a stoop toward the next rock and snatched off its crinkly moss covering with the pick-end of his geologist's hammer to see what was underneath. Nope. He charged again. Walter and I zigzagged behind his various courses, trying simultaneously to keep up, watch out for unusual stones, and guard against the possible surprise approach of grizzly bears.

We were up against a tiny shallow lake whose bottom and shoreline consisted solely of assorted boulders. In the low bushes near the shore, bare circles of coarse gravel three feet across had heaved sparse plant life aside—frost boils, places where ground-up glacial till is forced from the earth by perennial thawing and refreezing. In the north, a fine place to look for minerals, Chuck spread his short legs, doubled over at the waist, and picked through rocks lying around the boils with the hammer. He dropped some into a pile and flung the rest aside sidehanded. "Oh. Oh. Here, Walter, here."

"What is it, Chuck?"

Without answering, Chuck grabbed a gray cantaloupe-size rock with both hands, set it on a nearby boulder, and began whacking at it hard. Chips flew, so we covered our eyes with the backs of our hands. The thing split open. Chuck kept smashing away till he had four or five fist-size pieces. When it was safe, we stuck our heads down and looked close. "Is that it, Chuck?" asked Walter in a patient voice.

"That's it, hey," said Chuck. "Look." He jabbed a stubby forefinger at the rock's innards. Along the fresh surface where it had split were telltale flecks of bright green chrome diopside, standing out against the gray groundmass. They were the size of pinheads, and the color resembled that of emeralds. He turned another fragment over. "Oooh, look, Walter, look at all the garnets in here! These are pyropes, hey. Gee, look at them all!" When Chuck got excited, he spoke in a high, childlike voice, and he was often excited. His sentences tended to be peppered by the all-purpose "hey" where others might take a breath. The exposed inner surface was studded with a dozen or more tiny dark red flecks—pyrope garnets. They were peculiar, almost bluish—dark as the darkest blood. They glistened in the sun like the miniature jewels they were. "Oh, oh, look," cried Chuck, turning another chunk. "This one has both!" In this chunk was a tiny crescent of pyrope garnet; next to it, a chrome diopside, pea-size. Pure, gorgeous kimberlite float. Somewhere under the tiny lake was the ancient spout from which it came, filled with it. Chuck gaped silently. You could hear mosquitoes buzz.

"What's the orange stuff?" I ventured, pointing to a bright splotch powdering an edge of the rock. "That. Oh, uh, uh. Huh," answered Chuck. He put the magnifier to his eye. "That's, that's lichen. Like, hey, it's a lichen, growing on the rock. You know? An orange one. Lichen. Hey?" He paused for a long moment and contemplated the lichen under magnification.

Walter and I were getting the grand tour of "the property," as Chuck liked calling it—960,000 acres of mineral claims some 200 miles northeast of Yellowknife. Walter was an old buddy of Chuck's, a paleontologist with the Geological Survey of Canada (GSC). I, a reporter, had known Chuck just a few days, but I was fast becoming familiar with his operational mode: fast, often heading for the cliff's edge, never quite going over. Employees called him "Captain Chaos," for he was constantly stuttering, changing course in mid-sentence, losing wallets and tools, failing to work gadgets like fax machines. He seemed as naive as he was disorganized; Chuck was already telling me about the troubles with his wife, Marlene, as if I were an old friend, too. He, Stew Blusson, and various colleagues were also fabulously rich. Shares in their outfit, Dia Met Minerals, had gone from 12¢ to nearly $70 after an Australian mining conglomerate agreed to front massive exploration costs for a share of a possible mine on "the property." Crews already knew of a pipe under the lake we were at—the so-called Leslie Pipe—plus twenty-five others, and the stock market was betting on diamonds aplenty. Chuck, riding high, had an uproarious laugh and was always ready to buy drinks in town. Without a word, he handed me the rock with the chrome diopside and the crescent-shaped pyrope garnet embedded side by side—a gift.

Earlier he had burst from the helicopter that was our transport here and, before the rotor stopped turning, assumed the prospecting position. Now loaded with kimberlite float, we piled back in and shut the plastic-bubble doors against the bugs. It was a nice A-Star, a million-dollar machine with room for five. Technically it belonged to an offshoot of Dia Met, but in fact it was Chuck's personal taxi on the tundra. Our jumpsuited pilot lifted off, turned, and we tore over the land at eighty miles an hour. Some 500 feet below, a small herd of 400 or so caribou panicked when they saw us and began streaming pointlessly around, trying to get away, like a big amoeba on the move.

We circled a few minutes to take photos, then decided to quit harassing the animals.

Ten or twelve miles and hundreds of small lakes on, we reached a lake about 1,500 feet across. Underneath was another kimberlite pipe—the so-called Misery, for Misery Point, where it was situated. At the moment it was deserted and looked like all the other lakes, though if you looked close it seemed a little more circular and had a couple of unusual cliffs on opposite shorelines dropping straight to the cobalt-blue water. Chuck led another dash toward a flock of frost boils, pulled out his hammer, laid it on a rock, reassumed the prospecting position, and began a wandering narration of how he had found it four years before.

"The first day is always the shits for bugs, hey, because you're bathed. After, it's no problem. I figured the ice direction from the glacier was this way so I went around the other side, hey . . . The trouble is people have been coming here and; oh, here's one—a lot of them have been taken. We've had quite a few visitors, hey. Look at that pyrope, hey. Look at this pyrope, Walter. The right spot, they're just like all over the place." Lying in the frost boil were free, tiny pyrope garnet grains liberated from the rocks by erosion. Chuck began pinching them out with his fingertips. Some had stuck onto the heads of nearby mosses, blown there by wind like so much confetti. We knelt and looked at them under magnifiers.

"What happened is, we came in over in there and of course there was six feet of snow, hey, on the ice. So we dug through the ice. We could see the boulders in the ice, hey. We hacked them out but when we got to the bottom we couldn't get no fines, hey, just big rocks. And that took five hours, hey, and we got nothing. So then I decided to go over here and there were a few little bare spots with exposed till and we started hacking out frozen till."

Chuck pointed with the hammer-handle a half-mile off toward a behemoth esker. It was in fact the grandest esker of all, the Mississippi of eskers, running more or less straight 300 miles east to west, with hundreds of tributaries merging into the main channel like a dendritic river system. Its sandy sides reared up as high as 300 feet, and it was studded with ranks of boulders lined upright like sentinels. It had no particular name.

"So we were hacking, hacking up there, hey . . ." Chuck paused for a moment, then appeared to have forgotten what came next. We examined some more grains.

Presently he stood up and scanned the land. "That esker is going to be a road one day," he said.

We had lifted off and were flying back toward camp when Chuck pounded his knee hard and cursed. "Damn! Damn!"

"What is it, Chuck?" said Walter.

"I, I, hey . . . I, oh it's too late to go back. Forgot my hammer. It's back on that rock. I . . . Damn!"

That night we stayed next to another unnamed lake in a huddle of prefab trailers. Next door was a giant diesel-powered generator and piles of fifty-five-gallon fuel drums for heat and electric. Everything was surrounded by a new "bear fence," a high-voltage affair meant to keep out bears and wolverines who chewed sneakers geologists left outside—and the occasional entire building.

A lounge had a big-screen TV and dozens of videos. "Oh, get rid of that bloody *Fatal Attraction*," hollered Chuck when he saw the pile. "Throw that thing down the toilet. Bloody thing. My wife bought that movie, hey? Where's my Madonna tape?" He giggled.

In the mess hall a genuine French chef named Pierre had made fresh pastries, and the refrigerator was stocked with unlimited chocolate milk and ice cream for drillers coming off their twelve-hour shifts. For supper, Pierre charred great, bloody steaks and fish on a grill outside. Yards away, drillers were catching the fish, trout up to thirty pounds, which grow so slowly in the frigid, nutrient-poor waters that they were probably fifty or one hundred years old. Good food is essential to morale in the bush. The only thing missing was alcohol: People tend to get too crazy out here. "Anything that's fun is prohibited!" groused Chuck. "I used to encourage people to bring beer and pornography out—then I would confiscate it for myself!" He burst into laughter.

Perhaps the most unusual amenity was the special toilet, which kept us from polluting the lake being emptied of the centenarian trout. It was OK to pee as usual, but for any further business, you had to lift the lid off the bowl and line it with a white plastic bag from a dispenser.

Then you did your business into the bag. Afterward, you closed the lid and pressed a foot pedal to drop the bag into a hidden chamber. You then pressed a big metal button to unleash a long, powerful blast of electricity, which cremated the bag and its contents with a loud drone and an odd smell. Important tundra survival tip: Never sit on the bowl when you hear the drone.

The next day we went to a mine shaft, angled from yet another lakeshore underneath the lake bed, into one of the new pipes. The Australians were fixing to drag out tons of ore to count the diamonds—the final test of the pipe's worth. We donned hard hats mounted with battery-powered lamps and entered a ten-foot-wide black hole with two giant yellow ventilation pipes protruding, as if from the maw of a monstrous insect. A bone-jarring half-mile tractor ride through the dark brought us to the mining face. The workers had just gotten past country rock into the pipe itself. High-powered lights on folding stands illuminated this innermost chamber, where water seeped through the walls and bits of garnet and other indicator minerals sparkled. The raincoated miners stepped back to let Chuck wander the chamber's edge. He poked a finger into a hole, pulled out a pebble, and eyeballed it with an intense scowl. Then he tossed it on the floor. Out of a little blue backpack, he pulled a borrowed replacement hammer and commenced to pound the wall, as if to mine the thing himself. He put some chunks in his backpack and handed out more free samples all around.

The grand finale was the plant that processed the ore. After emerging into the sun, we took off in the helicopter once more. Gradually something white and alien shimmered on the surface. Closer, it looked like a settlement on the surface of Venus. It was a tentlike structure covering many acres, with arched portals, side structures multiplying in every direction, and vents jutting through the roof like minarets—a sort of Taj Mahal of the tundra, airlifted in pieces and reassembled.

Low-slung trucks were hauling kimberlite along a road built of esker gravels. A frontloader shoved the ore into a deafening assembly of crushing wheels, screens, and sprayers. A fast-whirling cyclone floated off the lightweight, generally light-colored crustal materials like quartz and retained generally darker heavy minerals. End of the line was a padlocked mesh cage. Inside it, a vertical pipe was spitting a clattering stream of blackish, damp, nasty-smelling gravel from the cyclone into a yellow metal barrel—the heavies, or "diamond concentrate," largely ilmenites and chromites. There were no visible gems; it looked like sludge. The Australians' hard-hatted security guard in blue stood by the cage, hands behind his back, and a video camera in the ceiling watched the guard. Chuck asked if he could examine the contents of the barrel by hand. The answer was no. "Sir—please don't touch the cage," barked the guard when Chuck made a reach for it.

The barrel would be sealed and flown to Reno, Nevada, where the contents would be mechanically fed into a locked conveyor belt. At the end, particles would plunge off one by one and meet an X-ray beam. Anything that was a diamond would luminesce light blue, triggering an air jet to blow the particle into a locked box. The rest would plunge into a tailings heap. This machine, called a Sortex, was invented by the Russians.

If any precious stones were in the locked box, human hands would not touch them until they reached Antwerp, Belgium. There, hereditary appraisers whose ancestors have handled diamonds for many generations would sort the stones for size, shape, color, and clarity and put a price on each. If they were good and plentiful enough, Chuck, Stew, and their partners would have themselves a mine.

Chuck Fipke stood back respectfully from the cage, twisted his glasses along his head at that same crazy angle, and peered deep into the sludge.

CHAPTER 2

Cap aux Diamants

In the years 1534 and 1535, Jacques Cartier, navigator of Saint-Malo, Brittany, explorer of new lands, and emissary of the French King François I, sailed up the Saint Lawrence River, the waterway that now separates part of eastern Canada from upstate New York. He was seeking the Strait of Anian, or the Northwest Passage, as it came to be called—a mythical channel thought to lead through the newly found continent of North America to the Pacific Ocean, thence to the riches of Asia. Cartier's was the first real stab into the interior.

Not far up he met the locals—Huron Indians who circled his ships in canoes giving wild cries, then fled when the cannons were fired. Within a day or so, the two peoples had sniffed each other enough to trade furs and meat for hatchets and cloaks. Cartier promptly took advantage to kidnap three Hurons, a leader named Donnaconna and his sons Taignoagny and Damagaya.

Transported to France and tutored in French, Donnaconna and his sons communicated most astonishing news: Feeding the Saint Lawrence from the north was another river, so long that no one had ever seen its source. Up there was a mysterious kingdom called

Saguenay, where people mined immense amounts of pure copper, which the Hurons called *caignetdaze*, and, according to Cartier, "infinite quantities of gold, rubies and other gems."

The story was at least partly plausible. Aboriginal peoples had been making ornaments, tools, and weapons of native copper for millennia. Among other places, copper deposits lay along the shores of Lake Superior, to the west. On the other hand, the Huron tongue had no words for gold or gems. Nevertheless, noted Cartier, "This chief is an old man who has never ceased travelling about the country by river, stream and trail since his earliest recollection." Besides, Cartier had other witnesses: He had met Iroquois Indians farther upriver, who, he reported, "seized the chain of the Captain's whistle, which was made of silver, and a dagger handle of yellow copper-gilt like gold" and pointed northwest, up the Ottawa River. Up there you will find this, they said.

These stories were repeated under oath before a notary public and presented to King François. The king was so impressed that in 1541 he sent Cartier with a far larger expedition supplemented by fifty-some condemned convicts in chains to find not the Strait of Anian, but the Kingdom of Saguenay.

At the time, it seemed like a good idea. The Spanish *conquistadores* were just starting to loot the apparently endless wealth of the southern New World. In 1519 Hernán Cortéz had killed the Mexican Aztec emperor Montezuma after receiving imperial gifts, including a gold disc the size of a cartwheel and another of silver. In 1530 Francisco Pizarro captured the Peruvian Inca Atahuallpa and had the emperor buy his freedom by filling a twenty-two-by-twenty-seven-foot room with gold higher than a man could reach. When the room was full, Pizarro changed his mind and instead had Atahuallpa tied to a stake and strangled with a crossbow string. The treasures were so great that the stories seem like myths to us now, but they were real.

Cartier landed his northern New World prospecting expedition in August 1541, on a point of land where the Cap Rouge River enters the Saint Lawrence, nine miles upriver of the current city of Quebec. The Cap Rouge ran in the desired direction: "The mouth of the river is toward the South, and it windeth Northward like unto a snake," Cartier wrote in his journal. Its banks were lined with gigantic hemlocks, beautiful oaks groaning with acorns, grape-laden vines, and land "as good . . . to plow and mannure as a man should find or desire." He set a fort on the high promontory of reddish limestone cliffs that give Cap Rouge its name, planted a garden of turnips and lettuce, and took a walk. Saguenay was just over the hill.

Cartier wrote: "Upon that high cliffe wee found a faire fountaine very neere the sayd Fort: adioyning whereunto we found good store of stones, which we esteemed to be Diamants." In the red rocks on the other side of the mountain was "the best yron in the world," and at the end of a meadow of fine hemp plants, veins of minerals that shone like gold and silver. Cartier described the metals as "goodly," but it was the diamonds that transfixed him. To him, they were "the most faire, pollished and excellently cut that it is possible for a man to see. When the Sunne shineth upon them, they glister as it were of fire."

The discovery aroused huge excitement among the crew—including that substantial number given a reprieve from the gallows for the trip. Cartier, a cautious man, had the diamonds sealed in a barrelful of sand and mounted a guard over them. At first opportunity he sent back two ships with the barrels, plus samples of the iron, gold, and silver. He hoped they would speedily "bring newes out of France how the King accepted certain Diamants which were sent him." They were in such a hurry that when they encountered vessels of Cartier's colleague, the Sieur de Roberval, whom they were supposed to aid, they made only a brief rendezvous, then snuck off in the middle of the night.

It is not recorded who examined the Canadian diamonds, but few would have been qualified. The reason was simple: In the Europe of the time, almost no one had ever seen a diamond. Book 37 of the *Natural History* of the Roman Pliny the Elder was still the main scholarly work, and Pliny, who knew little enough himself, had been dead for 1,462 years. Pliny taught that diamond, or *adamas* in Greek, is "the most highly valued of human possessions . . . which for long was known only to kings, and to very few of them. . . . Here nature's grandeur is gathered together within the narrowest limits. [Its hardness is] indescribable, and so too that property whereby it conquers fire and never becomes heated. Hence it derives its name,

because . . . in Greek, it is the 'unconquerable force.'" He said it originated in mysterious mines somewhere in India.

The last bit was true: In India, where diamonds had been known
from at least 400 B.C., royals reserved the best stones for themselves.
After about 100 B.C., a few small, inferior specimens filtered out to
Rome and its neighbors, where people wore them as talismans. The
trade was short-lived; when Rome fell well after the death
of Pliny, so did the commerce. In succeeding centuries diamonds
became ever scarcer in the West, so that eventually European and
Arab philosophers, alchemists, rabbis, and doctors unburdened by
facts or observations of any kind used Pliny as a takeoff point to
invent the wildest possible theories regarding their origins, properties, and methods of procurement.

The chief tale regarded a place called Valley of the Diamond. Ver-
sions appeared in everything from the mythical voyages of Sinbad the
Sailor to the actual *Travels* of Marco Polo, who visited India at the end
of the thirteenth century. According to this, somewhere in India were
mountain ranges laced with deep valleys and sheer precipices. On the
valley floors, giant serpents slithered from caverns, their visages so
fearsome that anyone who gazed on them would die. These valley
floors lay strewn deep with diamonds.

Of course diamond seekers did not descend to meet the serpents.
Instead, it was said, they made the long, dangerous trip to the
precipices' edges, killed sheep or other animals, flayed the carcasses,
and tossed them into the depths. When the bloody meat hit bottom,
stones would adhere. Eagles who lived in aeries among the high
rocks flew down to bring it back up for devouring, and it was at that
moment that the "prospectors" had their chance. They scrambled up
through the rocks, battled the raptors off the meat, and skimmed the
stones into a leather bag. Marco Polo pointed out quite seriously that
if the eagles devoured the meat before you got to it, all was not lost;
you watched where they roosted and in the morning picked through
their dung.

Beliefs about diamond formation were equally imaginative. Many
considered diamonds distillations of divine forces—perhaps gathered-
up starlight or thunderbolts. One writer theorized that they pos-

sessed a celestial energy that changed air into water, then hardened it into the gem. Gemologist Jerome Cardan said they formed when juices dripped off gold and became distilled. Alchemists constantly tried making them; the English John Mandeville stated they could be cultivated. "I have oftentimes tried this experiment that if a man keep them with a little of the rock, and water them with May dew often, they shall grow every year," he wrote. Italian physician Joseph Gonelli believed they grew when particles of poisons floated through the air, settled onto a diamond kernel, and, unable to penetrate its dense mass, solidified. A related belief: The Valley of Diamond's serpents were pricked by the stones as they slithered, and their venomous blood nourished the gems. Some said diamonds were sentient beings, males and females. François Ruet, a doctor from Zurich and contemporary of Cartier, relates that "a lady worthy of credence" of the house of Luxemburg had two that made babies regularly.

As for the properties of diamonds, they were said to drive off fires, thieves, and floods; cure disease; imbue owners with courage; produce spiritual ecstasies; neutralize magnets; make wearers invisible; revive dead animals; and reveal whether a spouse was committing adultery. When Pope Clement VII was seized by illness in 1534, doctors dosed him with 40,000 ducats' worth of ground-up precious stones, including pure diamond dust. He quickly passed away. That was the trouble: Impure possessors could corrupt their powers. Not only might they fail to cure; they might also provoke sleepwalking or madness. Triangular stones were said to inspire quarrels; square ones, vague terrors. The rare diamond containing a tiny blood-red flaw—the indicator mineral pyrope garnet—was thought to bring death to anyone who dared wear it.

By Cartier's time, diamonds were switching identity from objects of magic to objects of pure lucre. This was partly the product of a resumed, growing trade. In the 1490s the Portuguese figured out how to sail around the Horn of Africa. In 1510 they occupied the Goa coast of India, and more stones began filtering in through Lisbon to whet the appetites of the European royals who bought them. Chief was King François, Cartier's patron. François loved wearing lavish baubles, and now, in his forties, he had somehow gotten possession of a necklace holding eleven large table-cut diamonds from the Indian trade. He prized this necklace over almost all else he owned. It is no wonder he was so excited about getting his own source in North America.

François's lapidaries examined the Canadian diamonds closely, for Pliny had warned that quartz could be mistaken for diamond. Pliny also warned of fakes; craftsmen in the southern Mesopotamian city of Ur had already concocted artificial gems of glass by 3000 B.C. He advised as a first test striking stones on an anvil, because diamond could not be broken. This was disastrous advice; diamond is by far the world's hardest substance, but it is easily shattered along its crystal planes by a direct hammer blow—probably the fate of many "tested" over the centuries. A scratch test is better. Pliny recommended testing also by taste and weight. Here he was right: Starting with Archimedes in his bathtub, the ancients knew how to measure specific gravity, or mass. Diamonds have a heavy specific gravity of 3.5, or three and one-half times the density of water; quartz, only about 2.7. Such measurements are described in ancient Indian texts and were possibly known by wise old European lapidaries. Likewise, diamond is an extraordinary heat conductor; press one on your tongue and it will suck away body heat, creating a perceptible chill. Hence the nickname "ice."

Found in nature, diamonds are often drab, dirty, dull-looking little things, but Europeans were spoon-fed gems cut and polished by faraway artisans to bring out brilliance. Thus English lapidary Thomas Nicols put down the last, not-so-dependable trial, the "glister" of fire Cartier saw in the Quebec sun: "The true diamond . . . will snatch colour and apply it and unite it to itself; and thus will it cast forth at a great distance its lively shining rayes. By this . . . do the most judicious of jewelers distinguish the true Diamond from those of bastard kinds."

The French lapidaries were beginners, but not fools. They had enough assays to identify Cartier's stones for what they were: the "bastard kinds," beautiful, transparent examples of quartz. Assays also showed that the gold and silver from the Cap Rouge hemp field were pyrite—fool's gold.

Considering the rarefied state of knowledge, Cartier could have been forgiven—but he was not. François never entrusted him with another long-range expedition. The country buzzed so loudly and long about the "discovery," it was recorded in a French proverb that has survived these 450-plus years: *"Voila un Diamant de Canada!"*—loosely, "Here's a Canadian diamond . . . sucker!" or "Fake as a Canadian diamond." It is equivalent to the American phrase "Phony as a three-dollar bill." One promontory along the Saint Lawrence is called Cap aux Diamants to this day.

"It is to the southward not the icy north, that everyone in search of a fortune must turn," concluded Peter Martyr, a writer of the day. But Cartier's map survived, with his notation alongside a river flowing from the far northwest: "By the people of Canada and Hochelaga it was said, That here is the land of Saguenay, which is rich and wealthy in precious stones."

North America turned out to be far more complex than anyone had imagined, with its endless labyrinths of bays, fjords, rivers, estuaries, sounds, gulfs, peninsulas, channels, islets, inlets, and straits—and that was just the east coast. Thirty-five years later, in 1576, a group of English merchants sent English ex-pirate Martin Frobisher far north up this coast to look again for the Strait of Anian—history's first documented Arctic expedition. When he hit the Arctic, Frobisher traveled through landscapes out of Greek myth: intense snows falling on the sea, freezing fog, 1,200-foot icebergs calved off glaciers, whirlpool currents that spun the ships around with waterfall-like roars, weeklong gales that tipped them bow to stern.

They made landfall off the northern Canadian mainland on a windswept, barren rock later identified as Baffin Island, and on a satellite, crewmen picked up a curious black stone with gold-colored flecks. They also glimpsed floating objects that looked like porpoises. When they got close enough, they saw the objects were Tartar-like men in beautifully made skin canoes—a first encounter with the Inuit.

Back in London, three assays were made on the black rock. Only one showed gold, but people got so excited, some were sure Frobisher had found Ophir, the lost mine where King Solomon is said to have obtained the gold for the Temple of Jerusalem. (Ophir, long a popular object of search, was also said to be in Arabia, Africa, or India. Frobisher's recently late colleague Christopher Columbus claimed it to be on the Caribbean island of Hispaniola.) Money was raised from rich investors, and Frobisher was sent back exactly like Cartier, with a royal charter and a bunch of convicts to look for treasure first, Anian second.

Of the 1,200 tons of rock he returned with, much was locked in the Tower of London. Then the truth came out. Further assays confirmed the results of two earlier ones, both ignored: The "gold" was pyrite. The venture's main backer went to debtors' prison, Frobisher returned to piracy, and the black rock was used to pave roads, among other projects. Four centuries later archaeologists are still finding it in English drainage ditches, houses, and garden walls.

Frobisher did, however, get a glimpse of the deeper northern interior when he sailed past Baffin partway up a west-leading strait, and he saw a powerful tide flowing through. This, many believed, signaled the entrance to the Pacific. In 1610 English explorer Henry Hudson followed it—and to his horror hit another coastline, much too soon for it to be China. He was trapped in Hudson Bay, a vast inland sea with a 7,600-mile shoreline. With winter weeks away, his ship was soon frozen in. After the ice released its grip nine months later, his starving, half-crazed crew mutinied and set him and his small son adrift in an open boat to die. A later engraving shows the heavily bearded, cloaked Hudson sitting with the boat's rudder in one hand, his boy's hand in the other, staring in despair away from the unexpected coast of ice and rock. Hudson had found the start of the true northern inland, the region that came to be called Les Terres Stériles—the Barren Lands.

A succession of other explorers charted Hudson Bay and surrounding coasts. Dozens of ships didn't return, but eventually sailors learned by trial and error how not to die of scurvy, ice, and sheer bewilderment. Once the Bay was somewhat better known, they also found something more dependable than minerals: furs. The huge fur trade, the original basis of the Canadian economy, began around the Saint Lawrence, then expanded via French and English companies to the windswept western Hudson Bay coast by the mid-1600s. French and English built rival forts in the lower reaches, below the tree line, where there was wood for fuel and building, then engaged in constant warfare, so some forts changed hands every few years. The main fur trader was the English Hudson Bay Company, chartered in 1670, the mightiest force in the north for centuries.

For decades the Europeans never tried penetrating the interior; the woodland Cree Indians were quite willing to bring pelts in exchange for European guns, kettles, and other coveted objects of iron. Farther up the coast past the tree line, sailors saw only rocks, moss, and curved-horn musk oxen, their great coats of hair dragging on the ground—no people. The Cree informed traders that the Dene Indian tribes, the unseen people said to roam these regions, were subhumans with pointy tails. James Knight, a gruff and tenacious early Hudson Bay Company commander, described the barren-ground travelers by hearsay, though it is not clear whether he meant the Dene or the "Esquimaux": "Them natives to the Norward are more Savage and brutelike than [woodland Indians] and will drink blood and eat raw flesh and fish and loves it."

Then, some time around 1708 a remarkable event swung the Europeans' gaze to the treeless north. The Cree showed up at a French-controlled fort on the Hayes River with ornaments and tools made of brilliant copper. The commander, Nicolas Jeremie, wrote they came from a rumored Dene tribe far to the northwest, the Dogrib. Along with the so-called Yellowknives (also called the Red-Knives or Copper Indians), they were said to have "in their country a mine of red copper so abundant and so pure that without putting it through the forge, just as they obtain it at the mine, they pound it between two stones and make anything they wish. . . . Our Indians constantly bring it from there when they go on war parties."

The French were driven off Hayes River a few years later, and Hudson Bay Company's James Knight took over again. Now gray-bearded and a thirty-year veteran of the north, Knight was fascinated by the copper. Could it come from the Kingdom of Saguenay, discredited 180 years ago? Shortly another puzzle piece fell in. A native woman captured by the Cree escaped and showed up at the fort half-starved. She was a Dene. Not only did she lack a pointy tail, but she was good-looking, and so good at both talking and fisticuffs that even some of the men deferred to her. In other words, she was a natural leader. Knight, whose own truculent streak helped keep control of his rough outpost, quickly came to respect this woman who had come so far. Her name was Thanadelthur, or, say some, Wetsi Weko—Hearts on Fire.

Knight took her under his protection, and she quickly learned English. He saw in her his link to the mine; in 1715 he sent Thanadelthur, trader William Stuart, and a small Cree escort on a year-long expedition to contact her people, some 1,000 miles to the west. Instructions to Stuart were to make peace between Dene and Cree and open a Dene fur trade. But "above all . . . you are to make a Strict Enquiry abt. there Mineralls . . . if you find any Mineralls amongst them You must seem indifferent not letting them know nor the Indians as goes with You as it is of any Value but to bring back some of Every Sort you see."

They traveled many weeks west through the high boreal forest. Then somewhere near the shores of Great Slave Lake southeast of current Yellowknife, Thanadelthur said she was going ahead alone. To their terror, ten days later she returned with 400 warriors at her back. Fortunately, she came in peace; she orated to both sides until she was hoarse, convinced everyone to smoke a peace pipe, and got a few Dene warriors to follow back to Hudson Bay, bearing samples of furs—and more copper. Knight convinced them to chalk a map to the place from where this metal came.

This map, a version of which still lies preserved in the Hudson Bay Company archives, showed a great, treeless landmass extending far beyond the known reaches of the Bay coast. From it flowed seventeen rivers northward into a great ocean. The sixteenth the Dene called Chanchandese, or "Metal River." At its mouth, so far north that in summer the sun did not set, was "ye Coppermine." In addition to copper the Indians spoke of "yellow mettle"—gold, Knight assumed—and portrayed the river as so broad and long that Knight imagined it to be the Strait of Anian. The Dene said it was not; you couldn't sail, there being no interior waterway and the northern ocean all being frozen. To walk there and back would take three years, they claimed. They urged the Europeans to forget the copper and to trade iron for furs. Knight, however, could not conceive of such a distance uninterrupted by navigable waters. Thanadelthur became his main liaison and adviser for the Dene fur trade, but Knight soon was pestering directors in London to let him look for the Coppermine River. He met delay, skepticism, and worse, when Thanadelthur took sick in 1717 and died—of smallpox, possibly. She may be seen as a Canadian Sacagawea, the Shoshone woman who guided Lewis and Clark and whose image graces the U.S. dollar coin.

A new fort was laid out farther north on the Bay coast at the trees' edge, closer to the Dene fur trade—and the supposed mine. Lacking enough wood for construction, the Hudson Bay Company commenced building a stone fortification forty feet thick with gunports, a sloping parapet, and a giant British flag whipping in the center to fend off the French. Named Prince of Wales's Fort, it took forty years to build, well beyond Knight's time. One later commander got a plow from Europe and uselessly tried growing oats outside; the Dene, eager for European iron, stole it and took it 1,000 miles west to near Yellowknife. Inside the fort, the great flagstone-floored chambers were fitted with huge fireplaces to keep everyone warm, but they were so drafty and firewood so scarce, men had to hack ice off the inside walls with axes. This is how the Dene, snug in their small caribou-skin tents, began to call their new partners *kwet'i*—the stone-house, or stone, people. The word *kwet'i* survived centuries, coming to signify any outsiders, but it acquired a new connotation after enough prospectors came through: people obsessed with stones.

Knight sailed to England in 1718, when he was nearly eighty—incredibly, not to retire but to persuade the Hudson Bay Company to let him personally find the tundra mine. Now in possession of a more refined map, he told skeptical directors that he "knew the way to the place as well as to his bedside." The committee debated his proposal all winter; Knight threatened to find other investors if they refused. Some thought he was crazy, while others backed him; the price of copper was so high back then, it was essentially a precious metal. In the end, the truculent old man won. He and about fifty others departed Gravesend in June 1719 with a ship and a sloop loaded with large iron-bound chests to stow the treasure. No one ever saw them again.

CHAPTER 3

The Coppermine

For a while people figured Knight had sailed through Anian and into the Pacific and would return some day like a boomerang. The Dene kept bringing furs—that was the main thing. In succeeding decades a few captains sailed desultorily up the Bay coast looking for the copper, then lead and silver. According to one chronicle, a 1744 expedition by captains Thomas Mitchell and John Longland picked up minerals at the mouth of the Little Whale River "that they thought might be diamonds." Since there is no further mention, it seems likely that the captains expeditiously, and wisely, dropped the idea. This story had been heard before.

Forty-eight years after Knight's disappearance, one Hudson Bay Company man arrived who would cross the treeless land in search of the treasure. This was a blond-haired, blue-eyed ex-navy sailor in his twenties with no apparent interest in prospecting: Samuel Hearne.

Hearne got started in 1767, when he sailed up the Bay's west coast to meet the Inuit (trading was now established with some bands) and sent back a startling report. Along one desolate island were guns, bricks, a smith's anvil, a ship's hulls, and an identifiable ship's figurehead—the remains of Knight's expedition. Hearne wrote: "[N]either stick nor stump was to be seen [here] near sixteen miles from the main land. Indeed the main land is little better, being a jumble of barren hills and rocks [and] the woods are several hundreds of miles from the sea-side." Hearne encountered some elderly Inuit who actually remembered Knight, and through an interpreter collected their pitiful account. After the vessels broke up near shore, the crews had landed, and the Inuit gave them what food they could spare. But it was no use; over two winters the visitors dropped of starvation and sickness, partly because the well-meant donations of raw seal meat and whale oil played fatal havoc with their bowels. Finally only two were left. The Inuit saw they "went to the top of an adjacent rock, and earnestly looked to the South and East" for rescue. Then "they sat down close together and wept bitterly." When one died, they said, the other dropped while digging his grave. It was true: Young Hearne climbed the rocks and found the two skulls and some large bones, side by side. The last man, the friendly old folks told him, was the one who made metal implements—the smith.

So ended Knight's prospecting trip—and began Hearne's. Unknown to him, some at the stone fort had maintained their unhealthy fascination with the mine. Some said a boy named Richard Norton had found it, then lost it, while traveling with the Indians in the 1720s. On an island one captain found a "Sepulchre" with 500 corpses—many holding metal-tipped spears. Norton later became fort commander, then was succeeded by his son, Moses Norton. Moses renewed the search, sending two Dene scouts into the tundra. The scouts, Matonabbee and Idotliazee, disappeared for five years. Suddenly, upon Hearne's arrival, they rematerialized. They carried a chunk of copper—and a deerskin map.

This map, like Knight's, showed a broad river mouth emptying into an unknown ocean at the tundra's far northern end. Matonabbee reported he had reached the place. He claimed there were three mines "so rich and valuable that if a [fort] were built at the river, a ship might be ballasted with the oar. . . . [T]he hills were entirely composed of that metal, all in handy lumps, like a heap of pebbles." Moses Norton immediately took the map, went to London, and obtained orders for a white man to go settle once and for all if this place really existed—and, as the usual afterthought, to see if it were the Strait of Anian. Next thing Hearne knew, "I was pitched on as a proper person to conduct the expedition."

Aside from running messages between forts on snowshoes and finding the bones of his predecessors, the twenty-four-year-old had no training in wilderness travel or prospecting; but the science of the day would have helped little anyway. Things like glacial erratics—boulders plucked from bedrock and transported by ice—were thought to be products of Noah's flood. Metal prospecting depended on spotting obvious glitter or, in some cases, witchery. Some still used forked divining rods supposedly attracted to metals—hazel for silver, pitch pine for lead, ash for copper—along with certain rings and crystals. However, Hearne had proved himself a strong and observant man; as a sailor he knew navigating and mapmaking, and he loved a good adventure.

Hearne started out for the Coppermine in 1769. Twice, poor guides bogged him down around the tree line for months, and he had to turn back. On his second return in September 1770, the cold was so intense that he was freezing for lack of skin clothing. His dog, not quite as tough as he, stiffened and died one night. Just as things looked truly bad, on the trail he ran into a handsome, older, six-foot-tall man with brooding eyes and a hawk nose. It was Matonabbee, maker of the deerskin map, returning from a fur-trading trip. He gave Hearne a warm otterskin suit and saved his life.

Son of a Cree woman and a Dene hunter, the personable, multilingual Matonabbee seemed the perfect guide. Like Thanadelthur, he was a diplomat. He had calmed tribal wars threatening to wreck the fur trade and was now middleman to the mysterious Yellowknives, who lived so far west they were still only a rumor. He had apparently found the mine with their help, and had there run into the Yellowknives' deadly competitors, the Copper Inuit—a group whose range in the north-central Barrens included the mines as well. Such Inuit-Indian encounters usually occasioned slaughter, but Matonabbee had actually made friends with the Inuit and had even given them a few precious pieces of iron from the fort.

Matonabbee asked Hearne if he wanted to see the mine. Hearne assured him, "I was determined to complete the discovery, even at the risque of life itself." Here was "the most sociable, kind and sensible Indian I had ever met, [with] the vivacity of a Frenchman . . . the sincerity of an Englishman [and] the gravity and nobleness of a Turk." Matonabbee, Hearne noted, could recite the story of Jesus Christ in much better detail than most Christians—and didn't believe a word of it.

On December 7, 1770, they walked off with the chief's extended family and a few hunters armed with bows, spears, and a few guns. Hearne, the lone European, lugged an outmoded quadrant discarded from a ship to plot positions. They would not return for two summers. He was about to enter a world no outsider had ever seen, and which few would see again for over 200 years.

This land was formed in a most unimaginable cauldron of fire, and it evolved energetically right until the prospectors passed through.

Even as earth's crust cooled and the first protocontinents congealed on the surface like scum, buried masses of stone were solidifying miles below. These are now known as the Acasta gneisses—at 4.05 billion years, the oldest dated earthly rocks. Heavy elements like chromium and magnesium settled in the deeps, while light ones, such as calcium, floated up to reside in the crust. As protocontinents formed, a few glommed onto each other. Then others; these stitched-together provinces eventually formed North America. One core piece, now called by geologists the Slave Province, formed around what is now Lac de Gras. It is a small archon—maybe 110,000 square miles. Its long history could never be recited even if we knew it all. But by about 2.6 billion years ago, the giant igneous events still broiling most of the world halted here. No more continental collisions, volcanic island arcs, or subducting plates. Here in the geologic heartland, things cooled, and deep, solid roots with pressure and temperature conditions hospitable to the formation of diamonds developed. As a last gasp, four great underground flows welled up from the side, cracking the shallow subsurface with fingers of magma 100 feet wide and many miles long—dikes. Ding a windshield with a rock, and such shallow, radiating cracks will weaken the glass's surface. Ding an archon, and they may help small eruptions, like kimberlites, to worm their way up later.

As the archon's keel hardened, the land buoyed on it into a vast plateau. Erosion split the plateau into giant mountains, then took them to their bottoms. When it all got low enough, seas came and went, dropping debris to form sedimentary rocks. These grew so deep that heat and pressure metamorphosed them to new rocks. Folds, fractures, and faults worked the strata. Seawater circulated through them, dissolving hot metals below, then jetting back near seafloors to precipitate gold into quartz veins. Native copper lodged in leaves, slabs, and nuggets. Troves of zinc, silver, nickel, uranium, palladium, platinum, tungsten, beryllium, bismuth, niobium, and cobalt formed by various mechanisms. On and on—you understand. Just about every rock or mineral ever made anywhere was made in the Slave Province. Prospectors love such old rocks; the more experience rocks have, the greater chance that some interesting deposit or other has formed in them. Kimberlites included: Possibly using weaknesses made by the dikes, they cracked up through the uncountable layers 540 million years ago; again at 450 million, 170 million, 84 million, and 47 million. Hundreds, maybe thousands, planted themselves at various levels.

Finally came ice sheets. In the last 2 million years at least four have bulldozed the continent from the north, then receded. We know little of the first three, since each wrecks traces of the last. But the final mass started building in two domes a few miles high on either side of Hudson Bay about 75,000 years ago. The ice then moved in a downsloping manner west across what is now the Barrens toward the cordillera and south through what is now boreal forest to the Great Lakes and New York City, tearing 50 to 300 feet off the surface and entraining the mess in its lower parts.

After all the erosion, nearly all the former seas and mountains of the Slave Province were now gone or pulverized, leaving outcrops of multibillion-year-old bedrock once two or four miles down—ancient granites, the dikes, the Acasta gneisses. Interspersed in it were our kimberlite pipes, like carrots in the garden. Since the surrounding rocks were hard and the kimberlite crumbly, the ice scooped out the

pipes' tops and smeared the contents all over—diamonds, pyrope garnets, ilmenites, chrome diopsides. The pipes' tops looked like craters,
but mostly they lay intact below. When the ice melted, the craters
became lakes, as did millions of other depressions of all sorts. Today,
most look alike.

As the glaciers melted, rivers formed within them to create eskers.
From the Irish *eiscir*, or ridge, eskers occur also from Scandinavia to
North Dakota, but nowhere else so spectacularly as in the Barrens.
They seem to defy gravity, running up one side of a bedrock ridge
and down the other, so many were probably made by water flowing
under pressure in deep ice tunnels. Debris in the ice was liberated by
this water and pushed through the tunnels, filling them with sediment. When the flow slowed, it dropped a rainbow compendium of
the region's history—gravels and sands made of pulverized dead seas,
magma flows, metamorphoses—and dabs of kimberlitic minerals.
Tunnel waters issued from receding glacial snouts, and when the glaciers were gone, there lay wormlike casts of the vanished tunnels—
eskers. Overriding the deranged topography of the Barrens, their
headwaters radiate from the ice's high point near Hudson Bay. The
central esker, the biggest one of all, runs right through Misery Point,
east to west.

About 6,000 years ago trees moved up beyond their present line,
during a time warmer than ours. Great herds of caribou, and the Indians who lived on them, followed. Soon both were migrating yearly
into the treeless land during the brief summers. About 4,000 years ago,
the Siberian forerunners of a competing race, the Inuit, reached the
tundra's opposite edge—the northern coast and the vast archipelago
above it. They, too, ventured into the tundra, to meet the caribou from
the other direction. Both groups had a fine eye for the Barrens' geologic resources—soft soapstone for pots, hard graywacke for scrapers,
blood-red rhyolite for axes, white chert for arrowheads, flint for fires.
Most prized were plates of native copper lying in streambeds and
sticking out of ice-scoured stones to provide needles, knives, fishhooks, bracelets, and spears.

To both Indian and Inuit, the tundra was a separate world, from
which everyone retreated to their separate corners for shelter, fuel,
and food when the caribou left. For Indians, who otherwise lived

among the trees, it was taboo to remove certain objects, for that would disastrously upset the order of things. These included the berries that grow so abundantly on sandy, well-drained esker tops, and any leaves, for which they diligently searched their belongings before leaving. As for the Inuit, the rich food web roiling under and on the arctic sea ice provided a steadier living—fish, whales, polar bears, seals. Caribou fat is a poor substitute for seal blubber used to fuel lamps; dry, powdery Barrens snow, constantly remilled by wind, is useless for igloos built of the packable coastal variety. The Inuit forbade hunters to cook caribou on sea ice.

As climate swings pushed the tree line back and forth, the two peoples pushed on one another. When it was warm, Indians advanced farther north; when cold, Inuit farther south. The tundra's shifting center was always lurking with the possibility of starvation or deadly ambush by the other people.

At the outset, Matonabbee jovially informed Hearne that his previous guides had failed because they hadn't taken along any women to tote dried meat, pitch tents, mend clothes, and keep everyone warm at night. Wives were made for labor, he said, easily hauling twice as much as a man with little more upkeep than the licking of their fingers. He should have known: He had a floating cast of six to eight, dragging 150 pounds each—more in winter, when they could use sleds. Hearne suspected they stayed strong because they helped themselves to the meat, which they carried, whenever the men were not looking.

In the treeless land, Matonabbee warned, people and animals moved through unpredictably. You had to depend on meetings of the two occurring at the right moment to stay alive. Thus now, in winter, he planned to keep to the tree line, though it meant many detours along its zigzagging course away from the mine. Here caribou nibble lichens off low trees while the tundra plants are frozen under the snow. In spring, said Matonabbee, they would turn and follow the caribou straight north, hit the mine at summer's height, then return with the deer.

They first struck across the thin snow west-northwest. Hearne thought the weather relatively "mild" for December, but one should note the world was then in the tail end of a long cold snap, the Little Ice Age. New York harbor frequently froze solid; soldiers in George Washington's 1777–1778 winter camp in Valley Forge, Pennsylvania, froze and died.

Soon they entered regions composed of thin woods separated by long stretches where the next trees were far over the horizon. In places like this, Hearne penned his primal description of the tundra: "On the barren grounds, whether hills or vallies, there is a total want of herbage except moss, on which [caribou] feed; a few dwarf willows . . . and a little grass may be seen here and there, but the latter is scarcely sufficient to serve the geese and other birds of passage during their short stay in those parts, though they are always in a state of migration. [These regions are] incapable of affording support to any number of the human race even during the short time they are passing through them in the capacity of migrants." The Dene called this *Hosi*, or *dechinule*, "land of little sticks."

In the trees you could make a fire and pitch a tent fortified with boughs and piled snow. Where trees were lacking, misery was instant. "We thought ourselves well off if we could scrape together as many shrubs as would make a fire; but it was scarcely ever in our power to make any other defence against the weather, than by digging a hole in the snow down to the moss, wrapping ourselves in our clothing, and lying down in it, with our sledges set up edgeways to windward," wrote Hearne.

The caribou were mysteriously absent. Just a week out, the travelers grew hungry. Come Christmas, Hearne had not tasted anything for three days except a bowl of tobacco and a drink of snow water. He felt depressed and wished himself back home in England having holiday dinner. Then he saw the good humor with which the Indians starved. A few days made them "merry and jocose on the subject, as if they had voluntarily imposed it on themselves; and would ask each other in the plainest terms, and in the merriest mood, If they had any inclination for an intrigue with a strange woman?" When things got a little more pressing, they casually stopped to examine their wardrobes to see which parts could be spared for gnawing: perhaps some deerskin rotting off a tunic, or a spare shoe. "I must acknowledge that examples of this kind were of infinite service to me, as they tended to keep up my spirits," Hearne remembered.

By early February they shunted southwest along the trees, but the barren ground continued to stretch in plain sight to their right. Travel was rapid; some days they made fifteen miles over the stones; on flat frozen lakes, that far in a couple of hours—one reason they had set out in winter to begin with, since in summer those lakes would all be in the way. The Indians navigated by the stars and sun and seemed to carry in their heads a map of features in all directions. Hearne realized that the vast overland distances they spoke of were not figments of their imagination. With their relentless pace and lack of horses or navigable rivers in most places, he wrote, "Their annual haunts . . . following the lead of the deer . . . is so remote from any European settlement, as to render them the greatest travellers in the known world."

In mid-April, they were heading west-southwest when one of Matonabbee's wives went into a fifty-two-hour labor, occasioning an extraordinary stop for the whole thing. The moment she gave birth, though, the signal was made to move. With things now starting to melt along the tree line, she hoisted the infant and her pack and waded through miles of knee-deep slush like everyone else. Hearne had never liked this particular woman, but as he heard her sigh repeatedly in pain, at that moment he felt more for her than he had for any woman in his life.

When caribou appeared, the Dene killed them all and ate just the best parts. Hearne was appalled at the waste, but the Dene had no concept of it; food either came or did not, and when it did, there was no point in parsimony. In addition to the staples of caribou jerky or fresh caribou roasted or boiled, the party feasted on raw caribou brains; the genitals of unborn caribou, both male and female; squirming insect larvae from under caribou skin; and the big specialty, a haggis of smoked caribou entrails stuffed with blood, fat partially chewed up by young boys, and part-digested herbage from the animal's own stomach. When caribou or cooking fuel ran out, anything would do: semiraw fish, mucus and blood running from one's own nose, and head lice, which people generously shared.

Halfway across the lower edge of the barren ground, the tree line veers northward sharply. They hit this turn in mid-April, and it shaped their course. At one of the last stunted groves before pure tundra began, they came to a traditional spot for making summer birch-bark canoes and stopped to cut trees. The frames, explained Matonabbee, would be broken up for snowshoes on the return. Here they were joined by upward of 200 strange Indians, and everybody worked together on canoes in preparation for following the caribou.

It was here that things first started going wrong. When Matonabbee heard from someone that the former husband of his favorite wife was present, he coolly took a long knife, went to the man's tent, and stabbed him in the back three times without saying a word. Others pulled him off before he killed the fellow. Matonabbee called for water to wash the blood, easefully smoked his pipe, and asked Hearne rhetorically if he had not done the right thing. Hearne, who had never suspected his friend could do such a thing, did not record his reply.

On May 22, everyone took off into the open Barrens, including all the hundreds of new arrivals. By evening on the 23rd, they were thirty miles clear of trees, but kept finding themselves surrounded with ghost forests of withered stumps blown over by wind— reminders of a once-higher tree line. In this blasted landscape, Matonabbee awoke one morning to find his favorite wife gone— eloped, it seemed, with her wounded former lover. He was impossible to console. Next, an extremely fierce newcomer bullied Matonabbee out of another wife. Matonabbee then came to Hearne and told him he was too ashamed to continue; it was better to retreat to the trees, where people were civil. Hearne saw the shipwreck of his expedition looming. He entreated all day, using every argument he could muster, including a promise that Matonabbee and his descendants would always be honored at the stone fort if only he would show the way to the mine. By late afternoon Matonabbee straightened up and said they would have to hurry if they were going to get there in time. He added that most of the supplies, and all the women and children, had to stay behind.

On May 31 they set out, still dogged by the joined-up male extras, the piteous crying and yelling of wives and children following them across the rolling land until distance and low hills cut it off. In that moment the dark truth struck at Hearne: The hundreds of extra men were all shaping thick wooden shields and talking enthusiastically about war. They were not just following along in the same general direction to hunt caribou; they had attached themselves to him and Matonabbee and intended to follow them to the Coppermine so they could plunder the copper and kill the Copper Inuit. This was apparently not Matonabbee's idea; it may secretly even have been the reason he wanted so badly to turn back.

Hearne tried to dissuade them; he hadn't planned to kill anyone. The Dene angrily called him a coward, obliging him to change his tone for his own safety. He made a speech, telling them he "did not care if they rendered the . . . Esquimaux extinct; adding at the same time, that though I was no enemy to the Esquimaux, and did not see the necessity of attacking them without cause, yet if I should find it necessary to do it, for the protection of any one of my company . . . far from being afraid of a poor defenceless Esquimaux . . . nothing should be wanting on my part." This twisty bit of diplomacy brought back smiles.

Lightened now of most everything but weapons, they sped ever faster over the still-melting land. The caribou were moving north in such great numbers that their antlers looked like moving forests on the ridgetops. The Dene slaughtered more and more, taking only marrow and tongues. The tundra itself magnified the killing in a mighty way. The migrating deer meander north over numerous deep, braided trails, but in a land where mazes of lakes block every turn, all trails converge in rare spots where big water bodies pinch to a few hundred feet, and the crossing can be swum. At these narrows, deer clumped in masses to paddle across, and the Dene waited in canoes or on shore to kill them. Narrows were considered holy; people returned to certain ones generation after generation; tribes warred over them. By now the tundra also was filling with other migratory wildlife, including millions of birds. The Indians could not spot the smallest nest without uselessly slaying the young or breaking the eggs.

Hearne stopped daily to add to his map, but it was a limited affair, confined to the narrow thread of land he saw. Given poor instruments and pressing circumstances, the map was not very good; positions were often off, and many native names he noted were later lost even to the Dene, unattached to any known location. However, we know that around now he passed somewhere near the northeast corner of Lac de Gras. Hearne does not mention seeing the lake, but the Dene told him it was the source of the copper river. The northeast corner contained the greatest caribou crossing of all—a spot through which the tundra's central esker runs, separating the more northerly Lac du Sauvage from Lac de Gras. At one point the esker breaks, and water flows through a narrow gap year-round into Lac de Gras—the caribou narrows. On the west side of the narrows was Misery Point, where Chuck Fipke would one day dig a desperate hole in the snow. The Yellowknives had long warred with the Dogribs over it. Hearne's translation of the lake's name was different from the modern one, but eerily prescient: Large White Stone Lake.

They had another 300 miles to go, northwesterly. Around June 15, ice masses in the biggest lakes started breaking up with a weeklong round-the-clock barrage of crackling and thunder, accompanied by continuing snow and sleet. On June 21, rain and fog obscured the way. Then at 10 P.M. it cleared and the sun shone bright. It did not set that night. Hearne, the trained mariner, knew he had passed the Arctic Circle.

The next day—it is not clear whether by plan or accident—they ran into the Yellowknives, camped on the far side of a small river. Never having seen a white person, they flocked and gently examined Hearne like scientists classifying a new species. They pronounced him a perfect human being, except perhaps for his golden locks, which reminded them of a urine-stained animal tail. They were excited at the prospect of trading directly with Europeans. Upon hearing of his mission, they enthused over the Coppermine, confirmed that ships could get upriver, and offered to help. Hearne did not have much choice. After another week of caribou hunting, this now-great army set off.

It was early July, but snow and sleet pelted them until they were forced to hide in the lees and crevices of great stones standing over the land—an unusual but not unheard-of summer event. Hearne almost suffocated in a snowdrift covering his rock; some of the Indians turned back.

As they traveled amid the intermittent snow squalls, the land turned to a "confused heap of stones," a mass of boulders stretching many miles. The Yellowknives led, often forced to crawl on hands and knees; but in a few spots there was a path wide and well worn as any English country byway, indicating they had been here many times. Out in the middle were several huge table stones covered with many thousands of small pebbles. The Yellowknives told Maton-abbee and Hearne that by custom, everyone heading to the mines had to lay down a stone here. So, says Hearne, "each of us took up a small stone in order to increase the number, for good luck."

When they emerged from the boulders it finally cleared and warmed—which brought insufferable hordes of mosquitoes. Now Hearne began to think he was losing his mind. They were in a region of open marshes with strange little islands of loam sticking out. Some contained long furrows with enormous rocks shoved out, as if giants had been plowing a garden. Hearne was sure they were lightning strikes. The Yellowknives assured him they were made by grizzlies digging up sik-siks, squirrels living underground.

On July 13, near the top of a chain of long hills, the Indians announced the river was at hand. They reached the crest—and it was not there. Scouts were sent out in various directions, and Hearne tried to doze a few hours in a haze of insects. The scouts came back and pointed west-northwest. Ten miles on, writes Hearne, he arrived at "that long wished-for spot, the Copper-mine River."

Hearne was aghast. Instead of the navigable Strait of Anian, he saw a northward-rushing stream a few hundred feet wide, everywhere full of shoals, rocks, and with no less than three water-falls in sight from where he stood. It could barely be crossed in a canoe. Deranged drainage.

Spies were instantly dispatched to look for the Eskimos while the main mass moved along the river toward the northern ocean, now just forty miles off. Hearne dutifully plotted the watercourse as it pinched in between looming cliffs seventy-five feet apart and ripped over more falls. The opposite banks were of solid rock, correspon-ding exactly with each other, which made him imagine that the chan-nel had been made by some terrible convulsion of nature.

At noon on July 16 the spies returned. The Inuit were camped eight miles from the river mouth near a great rapid. Hearne's survey was dropped. The Dene painted their shields with strange creatures, then began advancing swiftly and silently over a broad plain through knee-deep swamps to avoid being spotted on hills. Hearne wondered what miracle might save the Inuit. At the lead was none other than his friend Matonabbee—put in the place of honor at the insistence of a powerful Yellowknife elder.

By the dusky light of midnight, they came upon a high, dark volcanic dike cutting the tundra, its sheer sides looming monstrously out of the landscape. The river roared through it and curved around a quarter-mile out of sight. They climbed up its height, and below could see the water broaden and calm. Not 600 feet off, at the foot, were five skin tents. The Indians stripped nearly naked despite the bugs, and at 1 A.M. rushed down. Hearne, afraid the Inuit might kill him if they caught him unarmed, borrowed a spear and followed.

When they were almost there, someone in the tents woke up. About twenty men, women, and children rushed out naked and half-asleep. Their backs to the river and weaponless, they shrieked and groaned as the Indians stabbed at all parts of them.

A girl of about eighteen streaked past Hearne, pursued by two Dene. One stuck his spear in her side. She fell at Hearne's feet and grasped his lower legs, screaming. Hearne reflexively tried shaking her off, simultaneously begging the attackers to stop. In answer, they raised their spears high and transfixed her to the ground. Then they looked Hearne sternly in the face and asked if he would like an Eskimo wife—all the while the still-living girl gasping and twining like an eel at his ankles. Hearne froze; if he made a wrong move now, he might be next. He meekly asked them to have the mercy to finish her. One warrior pulled his spear out and aimed a thrust for her heart. Years later, Hearne wrote: "The love of life . . . even in this most miserable state, was so predominant . . . it seemed to be unwelcome, for though much exhausted by pain and loss of blood, she made several efforts to ward off the friendly blow."

When the young woman was dead, Hearne stood in a daze of terror and grief, while the Dene ran around performing obscene examinations of the bodies and hunting more victims. At the foot of the thunderous rapid, a very old woman was spearing salmon with a pitchfork-like spear. It was so loud she hadn't heard a thing. They were within a few paces before she turned and looked; her half-blind eyes were red as blood. She tried to run, but the Indians took turns stabbing her arms and legs, then poked out her eyes before killing her. Then they tested out her tool; with masses of spawning salmon battling upstream through the great rapid, the fishing was fantastic. It was hard to come up with less than two at every stroke.

The Dene went to the tents and ransacked them of copious copper: bayonets; knives; thick, beveled hatchets with heads five or six inches long; hide scrapers shaped like aces of spades with long caribou-antler handles. In one tent lay the two pieces of precious iron from the stone fort, given by Matonabbee as gifts, fixed into ivory handles as knives. Everything nonmetal was smashed or thrown into the river. The Dene piled the loot atop a hill, clustered around it, and clashed their spears with many shouts of victory. To mock any survivors hiding within earshot, they yelled, *"Tima! Tima!"*—in the Eskimos' language, "Hello friend. How are you?"

A feast of fresh raw fish was pulled from the rapid, and by 5 A.M., when everyone had digested, the Indians informed Hearne they were ready to resume his survey. He gave the rapid a name still used today—Bloody Fall—and turned his back on it.

The Coppermine ran a few more miles between some low hills and then over an unnavigable sandbar into the Northern Ocean, such as it is in this region—more channel than ocean, choked with pack ice, islands, and seals. No big waves, no nice beach. It was 1 A.M. on July 18, 1771. There was no Strait of Anian. That left the mine.

They walked southeast about twenty-five miles to "ye Coppermine." Here is Hearne's entire description: "This mine, if it deserve that appellation, is no more than an entire jumble of rocks and gravel, which has been rent many ways by an earthquake. Through these ruins there runs a small river; but no part of it, at the time I was there, was more than knee-deep." They rooted but found only one good copper hunk of about four pounds, gleaming like a penny under its green tarnish. The Indians debated what animal the crescentlike hunk resembled—a tradition accorded all finds—and concluded Arctic hare. Then they told Hearne, belatedly, that the site had a problem. Its discoverer was a woman kidnapped by the Inuit, blindfolded and taken onto the sea ice. Escaping back to land, she stumbled across this mine, then home. She conducted her men back to the spot, but instead of thanking her they overpowered and raped her on the spot before loading up with metal. She refused to go back to the trees ever again; as revenge, she said, she was going to sink, along with this copper. Sure enough, year after year they returned to see her sunk further, until she was gone—along with most of the treasure. Hearne saw another explanation: In the hills and marshes leading to the site was a labyrinth of trails beaten over generations. The site was worked out.

The Indians, anxious to get out before winter or Inuit overtook them, allowed only four hours at the "mine." The next day they made a grueling forty-two miles, then repeated the performance after a brief rest. Even Hearne could not take this. A week at this pace made his legs swell and ankles stiffen. He tired so badly that he lost control of his steps and presently was banging his feet against the stones. He scraped the skin off the tops. Eventually his toenails festered and fell off. Sand and gravel got between his toes, removing the skin there. Somewhere back near the eastern shore of Lac de Gras—it is not clear if he ever did see the lake—he looked behind and saw he was leaving bloody footprints.

He did not die only because the band ran into the wives and children—a cause for celebration, and enough pause to salve his feet. As soon as they moved again, he witnessed what could have been his own fate. One woman developed a bad cough and grew too weak to walk. The family fixed her up with food and water, then turned their backs and walked away crying—the only choice the world's greatest travelers could have. To Hearne's horror, the woman unexpectedly caught up and tried to follow them—twice. The third time, she did not reappear. The Dene did not usually bury their dead, but left them for the wolverines, said to have extraordinary powers. These included the ability to hypnotize people into sleep and to capture and raise children for food. It was taboo to kill them.

On June 23, 1772, Matonabbee and Hearne reentered the stone fort, having walked 3,500 miles.

In 1776 Hearne was given command of Prince of Wales's Fort. Matonabbee, still active, brought in vast quantities of furs for his old friend, and Hearne honored him there as he had promised.

Then, in spring 1782, everything ended. One week while Maton-abbee was gone the French arrived with three heavily armed ships and demanded the fort. Hearne had seen combat with the British Navy and knew when he was outgunned. He surrendered without a shot, and the French allowed him to sail home unharmed. In half a day, they made off with the furs, torched everything else, and blew up the sixty-year-old fort. When word reached Matonabbee, he knew the place of his livelihood and honor was gone forever; Hearne, he believed, had been taken to the deep sea to be drowned. Some time afterward, with no warning and when no one was around, Matonabbee did something unusual, if not unprecedented, for an Indian: He hung himself. The following winter six of his remaining wives and four of his children starved to death.

The Indians now began killing each other off with ever greater vigor in various wars, including a continuation of the one between the Yellowknives and the Dogrib over the Lac de Gras caribou cross-ing. Some say the worst slaughter was over some bits of iron from the stone fort. The first of many epidemics of European smallpox and other diseases swept outward from the other forts and through the northern tribes. It was the beginning of the end.

Back in England, Hearne wrote a book about the search for the Coppermine, *A Journey from Prince of Wales's Fort in Hudson's Bay to the Northern Ocean in the Years 1769, 1770, 1771, 1772.* It came com-plete with scientific descriptions of Barren Lands wildlife and native customs. However, he could never leave his journey behind. He often asked restaurant waiters to serve his fish semiraw. He confessed he could never think about Bloody Fall without beginning to cry. Unused to handling money, he lost most of the book's proceeds. His health declined, and in 1792 he died of dropsy, an illness usually asso-ciated with cirrhosis of the liver. He was forty-seven. His Arctic hare copper chunk now resides in the British Museum of Natural History. Hearne faded into obscurity, as did the land in which he had searched for brightly colored stones.

CHAPTER 4

"A Deathly Stillness"

From that day almost to this, few outsiders entered the tundra. After Hearne, no known European crossed again for fifty years. The next incursion came with the conclusion of the Napoleonic wars. This suddenly gave the British Navy a great many underemployed seamen, and it decided to send some exploring.

Hearne had not found the mine, but that did not disprove its existence. He had found no water passage through the tundra either, but that did not prove there was not one north of the mainland—the so-called Northwest Passage, an alternative to the Strait of Anian that was to the early nineteenth century what space travel is to the early twenty-first. British Naval Lieutenant John Franklin was given a copy of *A Journey to the Northern Ocean* and told to follow the Coppermine to its mouth to chart the unknown northern coast. Dr. John Richardson, surgeon and naturalist, was assigned to study rocks on the way. The twenty-person expedition came well equipped for the 1819–1822 foray with the latest navigational instruments, canoes rowed by over a dozen hired Métis (trappers of mixed European and Indian blood), and two Hudson Bay Inuit interpreters to contact the Copper Inuit.

Initially it was easy. Instead of the killing walk along the tree line, they paddled now-developed fur-trading water routes within the trees all the way to Great Slave Lake. Near the present-day site of Yellowknife, the Métis sought out the Yellowknives, who lived there seasonally, and Franklin hired them as tundra guides. The tribe was now led by the violent, much-feared warrior chief Akaitcho, or Big Foot. The *kwet'i* paid him well, and he promised to travel in peace.

Paddling up through a chain of lakes and streams, the party reached the edge of the tree line and paused to winter about seventy-five miles southwest of Lac de Gras. Here, as they prepared to leave in spring 1821, Dr. Richardson saw the snow melting under a blinding spring sun. He noted: "[I]t looks like frost silver . . . and every little rising is studded with innumerable polished facets, as if sprinkled with diamonds. The intensity of all this splendour soon becomes painful to the eye."

When they reached the barren ground Richardson collected numerous rocks. Aided by the now-developing science of modern geology, he charted the strikes and dips of rounded formations of graywacke and gneiss and recognized them as remainders of some former world. However, he puzzled over huge pink granite boulders lying everywhere. He never figured out that the bedrock from which they came underlay much of the Barrens and that so much of the pink granite was lying around in pieces because it had been ripped out and transported by moving glaciers. Such a radical idea had not yet taken hold.

Hearne had passed just east of Lac de Gras; now Franklin and Richardson missed it by passing just west. A little farther on, the Yellowknives brought them to its outlet, the Coppermine, and they rode it north. In many places nearer its headwaters it was barely identifiable as a river; it spread out into endless mazes of lakes and chaotic channels that, without guides, would be impossible to trace. When the watercourse grew steeper and more definite, they had to make frequent portages. Richardson observed many narrow, dark valleys bounded by perpendicular mural precipices of greenstone, and here he picked up small plates of native copper and rocks laced with copper- and iron-bearing minerals. He also spotted veins of sparkly galena, known to carry lead. It was said the Indians knew how to smelt musket balls from it. Richardson tried this trick but couldn't make it work. All the copper was in loose chunks, obviously detached from somewhere else. "We did not observe the vein in its original repository, nor does it appear that the Indians have found it," he wrote.

Some weeks in, they reached the northerly spot where Hearne had hit the river and followed down to Bloody Fall. If anyone doubted Hearne's account, the massacre site was clear: The victims' skulls and bones still moldered in the grass. It was also still a great fishing spot—the Inuit called it Kugluktuk, or "Place Where the Water Falls"—and a number of Inuit were there when the expedition came up. Naturally the Inuit fled, except for one immobile old man, who tried stabbing the Europeans with a copper-tipped spear until the interpreters calmed him. The Yellowknives stood by, impassive, but they were spooked. Akaitcho said it was time to turn back; he feared ambush and oncoming winter. Franklin could not get him to stay, so they parted on friendly terms at the river mouth. Akaitcho told Franklin to retrace up the Coppermine when he was done with his coastal survey, and they would all meet back at the spring jump-off point. It was July 18, 1821, exactly fifty years almost to the hour after Hearne had been here.

The party set out eastward in the bark canoes to make their chart, but the tortuous bare sandstone and basalt sea-edge was nearly impossible to follow. It wound around and around senselessly, sometimes heading backward. Giant waves, lightning storms, and ice floes threatened to smash the canoes. Occasional distant Inuit appeared on the shore, then flitted like ghosts. Supplies ran low, frost appeared, and migratory birds sailed by, heading south. By late August, 200 miles east of the Coppermine, Franklin saw he could go neither forward nor back by water, and made a hard decision: Instead of retracing they had to turn inland and walk to Akaitcho in a straight line using their seaman's compasses—a calculated 600 miles. No one had any idea what was out there. Richardson's rocks were to be lugged along.

Two and half months of starvation, freezing, and death followed. Dark fogs enveloped them. They staggered against constant wind, floundered through deepening snowdrifts, and several times nearly dropped into profound chasms that sliced the way. The caribou had long fled, so they ate lichens growing on rocks, deer spines dropped by retreating wolves, and their own shoes. One day all Franklin's maps and records were swept away in a rapid.

By September 21, their expert navigation brought them back twenty-five miles above Lac de Gras—which they never did get to see. Now they were barely able to walk. Richardson dumped out his precious rock collection, which is probably still lying out there for someone to find. They recrossed the Coppermine and, strung out in a long line, began to die or just disappear. One day Richardson huddled alongside his dear friend Robert Hood, then staggered off for lichens. He returned to find a bullet in Hood's forehead. No one knew who fired the shot, but it was not suicide. Later Richardson began eating bits of "wolf" meat someone brought to camp—until he realized it was probably human. Richardson now suspected the meat-bringer as Hood's killer, and shot him in the head.

Of twenty men, only Franklin, Richardson, and seven others reached the Yellowknives. The Indians were so horrified at their condition that they fed, clothed, and nursed them like babies. Akaitcho himself cooked, a task he never performed even for himself, and every single member of the tribe from youngest to oldest came to look in on them.

Postscript: Sir John Franklin later became famous—but not for this long-forgotten trip. In 1845, back at sea, he sailed his ships *Erebus* and *Terror* westward into the arctic archipelago, again seeking the Northwest Passage. There he disappeared with all his 128 men—the worst arctic exploration disaster in history. Searchers spent decades looking for them. Franklin songs, Franklin books, and overly hopeful Franklin sightings became so routine, Sir John became the Elvis Presley of the 1800s—always rumored by the faithful to be alive somewhere. There were twentieth-century reports of blond Inuit inhabiting islands above the Coppermine—a sort of lost Franklin tribe. We know what happened, though: Inuit later testified they saw the ships seized by ice, which undoubtedly crushed them. The crews fled. From the 1850s to the year 2000, their bones have turned up on lonely islets and in the upper Barrens, where they tried walking out; many bear butchering marks from cannibalism.

Only in 1944 did a schooner, sailed by Royal Canadian Mounted Police Sergeant Henry A. Larsen, make the first single-season trip from the Atlantic to the Pacific through the labyrinthine so-called Northwest Passage. It has never been a practical travel route.

After Franklin all was quiet for a long time. French Catholic missionaries and fur traders in one-room cabin posts infiltrated as far as the forests around present-day Yellowknife, and here they stayed; only a few briefly followed the Dene to open land. One was a Frère Gossot, who traveled up with two Yellowknives in the 1840s. Somewhere south of Lac de Gras—again, he too missed the big lake itself—they came to a small stream surrounded by yellow hills. Gossot was about to drink when the Indians stopped him; the water was poisoned, they said. He realized it was stained dark blue-green; in the stream was a gigantic boulder full of copper, and metal laced the hills around—the place where the Yellowknives got their knives, they told him. A hundred years later prospectors were still trying to relocate this spot—one more lost mine.

There was hardly anyone else for another forty years; then an odd procession began. With the approach of the twentieth century the world was getting perhaps too civilized. Even at this date the American Wild West had been reduced, literally, to a circus act starring Buffalo Bill Cody and Geronimo. Diamonds discovered above the Cape of Good Hope were now a vast industrial enterprise, so even darkest Africa was losing its cachet. Victorian gentlemen with unlimited money and time for adventure had to look elsewhere. For this now-extinct breed of traveler, the barren ground was perfect: a nice, blank spot on the map containing large beasts for shooting, dusky-skinned savages to show the way, and the supreme bonus—the unkillable rumor of mineral treasure. From the last roadheads around Edmonton, Alberta, 700 miles south of Great Slave's northern shore, flatboats were now beginning to make their way through tortuous forest waterways to the traders and missionaries, and a dozen or so gentleman explorers took them. One made it to Lac de Gras—the only recorded outsider to visit before the time of living memory—and he managed to tramp right over the diamonds.

This was Warburton Pike, a manly, mustachioed Englishman from Wareham, Dorset. Pike arrived on Great Slave Lake in fall 1889 with his trusty 50-95 Winchester Express and a great stock of ammunition. There he hired a Métis guide by the name of King Beaulieu to take him musk-ox hunting on the tundra. Beaulieu was just the man: Some sixty years earlier his father had been one of the Métis who helped Sir John Franklin hook up with Akaitcho. Beaulieu introduced Pike to Akaitcho's descendants; Pike employed a number, and together they all canoed north up a chain of unmapped lakes and streams. At long last they entered that lost center of the land where Lac de Gras lay.

As they passed the protection of the trees, they had to lean into the wind that suddenly broke open from the north. Shallower waters were already freezing. Game and fuel were scarce, so they ate raw musk-ox corpses found floating at one lake edge. Several days on they came to Lac de Gras itself.

The rocky shoreline of Lac de Gras, its edges laced with quartz veins and its waters whipped white by wind, was too great to see or canoe across. They skirted on foot. When they reached the northeast end several days later, there was what Pike called a "peculiar ridge composed of fine gravel and sand, resembling at a distance a high railway embankment." It was the great trans-Barrens esker. They picked it up and followed, and presently came to the great caribou crossing, where water rushes through the esker from Lac du Sauvage into Lac de Gras. People customarily kept canoes there for anyone who wanted to cross, so they took advantage of these and made the other side across the strong current. The narrows was holy land—even camping nearby was considered irreverent—so they moved south a few miles along the shore of Lac de Gras and made a wretched camp on a windy peninsula. Lack of wood and game meant no fire, no supper. Pike got little sleep that night and decided he hated this place. He dubbed it Le Point de Misére—Misery Point. He was within walking distance of Chuck Fipke's little lake.

Pike walked north off Misery Point, eyes peeled for big game. A few days later he got his first musk ox—"my first shot settled him," he proudly stated. It was the first of dozens he killed, along with wolverines, caribou, and even loons. The tireless hunter cut off the giant, shaggy ox heads for trophies, but when he wasn't looking the Yellowknives, who did not feel like carrying them back, threw them away. King Beaulieu informed Pike that most of Franklin's men had died within a day's walk of where they were right now, and just about this time of year. He pointed around. Snowdrifts were deepening, the wind picking up. They should turn back. But Pike did not want to go quite yet; there was more shooting to be had. Beaulieu, his beard hung with icicles of blood from butchering the last kill, scanned the barren hills for more game but saw nothing. Then he unleashed a volley of blasphemies and told Pike it was time to leave whether he liked it or not. He dragged the hunter back for another night at Misery Point, and in the morning they recrossed the narrows and fled to the trees.

Pike was furious. He concluded that the northern native "is not a pleasant companion, nor a man to be relied upon in case of emergency. Nobody has yet discovered the right way to manage him. His mind runs on different principles from that of a white man."

Before turning back he climbed to the top of a steep hill to look around the wintry future diamond fields. The view, he wrote, was "the most complete desolation that exists upon the face of the earth. There is nothing striking or grand in the scenery, no big mountains or waterfalls, but a monotonous snow-covered waste, without tree or scrub, rarely trodden by the foot of the wandering Indian. A deathly stillness hangs over all, and the oppressive loneliness weighs upon the spectator till he is glad to shout aloud to break the awful spell."

Other visitors of Pike's ilk included the English gentleman Ernest Thompson Seton, who marveled at the masses of caribou and summer wildflowers in the lower tundra in 1907 and upbraided others for calling it "the Barren Lands." He dubbed it "the Arctic Prairies." Seton went on to help found an organization called the League of Woodcraft Indians, later renamed the Boy Scouts. There was also a wealthy American mining engineer, George Douglas. One day in Mexico, Douglas was visiting the private railcar of his rich cousin James, for whom the copper town of Douglas, Arizona, is named. James had read Samuel Hearne's *A Journey to the Northern Ocean* and half-jokingly suggested that George go check the Coppermine. To his great credit, in 1912 George and two companions did, somehow managing to follow old maps to Bloody Fall. At the rapid, spearing fish as usual, were the Copper Inuit—the very last North American natives untouched by outsiders. Shouting "*Tima! Tima!*" and communicating through signs, Douglas enticed them into trade. He gave them chocolate and other unknown items. They bequeathed him copper-tipped bone arrowheads and an ivory object used to carry dead marmots by their noses.

Around the same time New York sportsman Harry Radford entered the Coppermine area with a friend, George Street, to collect zoological specimens for the Smithsonian Institution. The Inuit instantly killed them. They also killed a couple of Catholic missionaries who arrived to save souls. They opened the priests' bodies and ate the livers to defend themselves against the spirits of the southern invaders. Some years later there was dilettante prospector and trapper John Hornby, son of a rich English cricketer, who boasted he could be dropped naked into the Barrens and survive. Northeast of Great Slave, just within the tree line, he wintered in a crude cabin with a nephew and a gentleman friend. They killed a wolverine for its hide but missed the caribou migration, on which they had proposed to dine. After eating the wolverine's hide, they dug up its heart and ate that. All expired in early 1927, their agony heightened by hair clogging their intestines. They were found some time later by a real prospector, who summoned the Royal Canadian Mounted Police to help bury them under some stones.

Prospectors and geologists were beginning to head north now in growing numbers. This was due in part to the rise of one of the world's most remarkable scientific outfits: the Geological Survey of Canada (GSC). Chartered in 1841 to map the 3.9-million-square-mile country's vast mineral deposits from coal to gold, the GSC sent a steady trickle of geologists to unknown forests, prairies, mountains, and islands. Their fellow civil servants, the Royal Canadian Mounted Police, might have been more famous and better attired, but never were they more dashing. GSC men, whose ranks were reserved for Ph.D.s, disappeared alone for months or years to travel by dogsled, canoe, horse, or foot. In addition to their main assignment of studying rocks, many did respectable jobs at topographic mapping, botany, zoology, paleontology, photography, learning aboriginal languages, and surviving every conceivable wilderness predicament. The GSC's first foray into the Barrens was 1893, and the expedition came back with seminal information.

The small party was led by the badly nearsighted but able Joseph Tyrrell. They canoed a river system in the tundra's southeastern corner, starting near the tree line and coming out on the Hudson Bay coast. By this time modern geology had established that apparently foreign rocks found down as far as the central United States had been pushed there by ice sheets sweeping somewhere from the far north. However, no one had yet quite figured out where the ice originated, nor just what routes it took. It was Tyrrell who assembled powerful evidence that one main mass had spread out from the southeastern Barrens from a central dome. He saw the signs in striations carved across naked bedrock and in erratic boulders moved great distances. Old Hudson Bay beachlines were strangely risen 500 feet above present sea level—a sign that the ice had been so deep here, it had pushed the earth down, and it was now bobbing back up. Clearly the ice path went not only south but west, across the still largely uncharted Barrens, into which Tyrrell saw eskers winding. He named the author of these works and mother of all glaciers the Keewatin—Cree for "North Wind."

Tyrrell found no diamonds; there is no record he looked for them or even thought about them. But 1,000 miles south, even as he was making his way along the Doobaunt, the Kazan, and other remote rivers, geologists in Wisconsin and other Great Lakes states were abuzz. Diamonds had been found in cornfields, streambeds, rock quarries, and the bottom of someone's well. All were found in glacial material. As soon as Tyrrell came home, his maps were torn apart for evidence of where that material came from.

The first northern prospectors were not propelled by diamonds; the clues were still too crude. They came looking for gold, for lack of anywhere else to go. The great nineteenth-century gold rushes, starting in 1848 with California and washing back into the interior West, were over; within decades the mines were worked out.

Panners turned north, working mainly up the continental mountain spine through Oregon and Washington, across the Canadian border into British Columbia, then up to coastal Alaska and east into the Yukon. Here, in the giant mountains west of the Barrens, was the last great gold rush, the wild 1896–1899 Klondike. When a panner named Skookum Jim found gold on Klondike Creek amid the deep Yukon peaks, 100,000-plus argonauts sailed to neighboring Alaska and tried hiking in. On ice fields and in high passes, many froze, disappeared in avalanches, or dropped straight into crevasses. More than half turned back. About 750 tried an even longer, more improbable route: from Edmonton through Great Slave Lake and on to the Yukon—some 2,500 miles altogether.

Only a handful made it. Many ended marooned on Great Slave's shores, homemade rafts broken by wind and waves. Some thought they saw sparkles in the sand and decided to poke around right here. Most was fool's gold as usual, but in 1898 near present Yellowknife, one man found the real thing: a quartz vein assaying 2.2 ounces per ton. Two GSC geologists were dispatched, but they reported the grade was too low: "The remoteness of any metal market even when the borders of civilization are reached . . . would place mining operations out of the question, except in the case of extraordinarily rich deposits." This only egged on prospectors to find just such deposits. As more arrived, claims were staked, holes blasted. Missionaries who previously had accepted tributes of furs from migratory Dene now encouraged donations of any shining stones they might see. Some priests staked their own claims. Lakeside galena veins were checked for precious metal. Most searchers had only vague knowledge; one group threw some galena into the campfire, and when bright metal melted out, shouted in unison: "Silver!" It was pure lead.

In the 1920s a jobless English barber and World War I veteran named Bertram Barker made his way to the edge of the Barrens, partly on the advice of an American hobo with whom he once bummed a train ride. The hobo had told him: "Follow me, buddie, and you'll wear diamonds!" Barker was not sure what this American idiom was supposed to mean, but he headed north anyway. Years later he wrote, "Work by day, talk round the fire at night. . . . Always of nuggets large as a pigeon's egg, a thing, I may say now, that I have never seen in all my life, nor met anyone else who has." After eleven years, he went home to cut hair again. Wavelets of *kwet'i* like this washed up on the woodland shores of the tundra for thirty-some years as native people looked on in puzzlement. Scattered in the bush, there were never enough outsiders to start a town.

Then, in 1934, a group of diggers spotted a quartz vein yielding 13.6 ounces of gold per ton. Big financiers perked up, bought the claims, and rafted heavy equipment north. With the new muscle, shafts were sunk hundreds of feet, and it became clear the surface showings were only signs of truly rich, deep gold reefs. Unfortunately, the reefs laced the Great Slave Lake peninsula where many Yellowknives had returned to fish and hunt seasonally for centuries. The government gave the mining companies title, the Indians were evicted, and the city of Yellowknife was born.

Open tundra was 100 miles north; the nearest roadhead now advanced to Grimshaw, Alberta, 450 miles south. But far-out Yellowknife soon had 1,000 residents, a beauty parlor, a radio station, four steady prostitutes, and a golf course laid out on the bare rocks. Along with their clubs, golfers carried rifles to snipe at gigantic ravens that kept stealing balls. Many veins didn't last long, and so mines like the Negus, the Discovery, and the Akaitcho—no relation to the Indians, who received no payment—came and went fast as a northern summer. Yellowknife always hovered somewhere between boom and bust. Within a few years 4,000 small exploratory claims stretched around it, soon tapping the very edges of the tundra. The GSC sent summer parties to draw maps of the bedrock in hopes of aiding development.

With all this activity, people did pick up some minor gems—bits of sapphire, jade, lazulite—and more gold. Farther north and west within the tree line, over by Great Bear Lake, mineralogist Gilbert LaBine found a fabulously valuable radium mine. It was a long, expensive journey, but radium, then in demand as an experimental cancer cure, brought $75,000 an ounce. Marie Curie herself imported some of the first produced. This mine, dubbed the Eldorado, also yielded a large amount of uranium, which was discarded until a use was found some years later: development of the atomic bombs dropped on Hiroshima and Nagasaki.

Prospectors were already probing the tundra with another radical invention, the airplane. The first northern bush pilots cut their teeth dueling with the Germans in World War I biplanes, then started civilian careers in the boreal trees—hence the appellation "bush pilot"—hauling prospectors, mail, and supplies. It was said that one old native Yellowknife lady, newly converted to Catholicism, saw her first airplane come over the horizon and thought it was God. She folded her hands over her head to pray, kept her eye on the plane until it passed overhead, and fell backward into a snowbank.

As planes grew sturdier and could take bigger loads longer distances, flights above the tree line became possible. The first were in summer 1929, when Toronto financier/prospector Colonel C. D. H. MacAlpine convinced investors to fund an outfit, Northern Aerial Mineral Exploration. Ski/float craft were flown in stages from New York City to an inlet on Hudson Bay, where pilots were armed with vague maps torn from school atlases and the journals of Richardson and Tyrrell. Then they took off into territories where charted features were as much as 500 miles apart. They navigated by dead reckoning: flying in a straight line by compass, checking their watches against their groundspeed so they could look out for the next known spot, usually a major lake or river mouth. Along the way they sketched features below or just filed landscapes in their heads like Indians.

This was a joy on good days, when the tundra was dappled with sun and great herds of caribou could be seen moving across the land. But that was good days. Pilots quickly learned that colliding arctic weather systems conjured extreme storms seemingly from nowhere. Many lakes apparently good for float landings actually contained hard-to-see rocks just under the water. Snow or fog on the mostly featureless surface made air, land, and water all converge, inviting crashes. It took bush-pilot savvy to survive. One man, just before landing, made a prospector climb out, hang onto the wing, and press with his feet to steady a ski gone askew. Another by the name of Elmer Fullerton broke a propeller upon landing. He created a new one out of carved oak sled boards and moosehide glue, flew home, and lived until 1968 to tell about it. Survivorship only went so far; there were four fatal crashes in 1930, and since then everyone has lost count.

Geologists in these planes quickly turned the tundra's complete openness to their advantage. With no more trees in the way, they could prospect from 500 feet, scanning for quartz veins zigging through bedrock or the bright metalliferous stains caused by gossans, out here ranging in size from desktop to shopping mall. Orangey-red iron oxide blooms might signal gold or base metals; yellow or pink, silver or uranium; blue-green to turquoise, copper. After noting nice spots from the air and drawing crude maps, they would have the pilot land two or three men with a canoe. The men hiked and paddled from one spot to the next, and a few weeks later got picked up at a prearranged spot. The problem was, once the plane buzzed away you were back in the stone age. Weather, water, and grizzlies presented the usual dangers. Radios were useless because of distance. In 1929 two geologists just forty miles from a Hudson Bay trading post panicked when snow swept in early and their plane could not take them out. They set out overland, but too many detours around lakes left them bewildered. On the second day one died of exposure. A week later, the other reached the post with his legs frozen and wound up a double amputee.

Over the northern reaches of the now-legendary Coppermine, fliers saw a view no one had ever seen. In the tortured country of waterfalls, canyons, and cliffs were strings of gossans in all colors, some stretching dozens of miles—chalcocites, sulfides, amygdaloids, shining native metal indicating copper, zinc, gold, silver. Geologists risking their lives on the ground, though, found the showings too poor to pay. Some might have paid—a thousand miles south, near a road. In the end, companies recognized that metals and mine equipment could not be taken in and out of this trackless place like 150-pound geologists. The only way to mine the Barrens would be in the case of "extraordinarily rich deposits." No one ever checked out Lac de Gras, for there were no signs of minerals. Against the scale of the land, these few planes were like sparrows landing on the moon and pecking.

In 1932 the GSC sent its first reconnaissance mission to the central Barrens: geologist Clifford Stockwell and an assistant, who spent a summer paddling around Lac de Gras and the upper

Coppermine, and reported nothing of economic significance. Then, during the late 1930s and 1940s planes from the Royal Canadian Air Force (RCAF) took systematic aerial photos—the start of rough topographic mapping. When GSC scientists saw the first results, they were fascinated: They knew of the esker running through Lac de Gras, but its scale had never been apparent until it jumped out of the photos. In July 1947, a Single Otter float plane dropped off a young GSC man, Robert Folinsbee, and a few geology students to make a geologic map of Lac de Gras. For navigation they had the huge-scale RCAF photos, badly distorted at the edges because they were taken with fisheye lenses. The latest aeronautical chart showed the giant lake's outlines and, just beyond, a great white space reading UNMAPPED.

Things had not yet completely melted. The GSC mappers plunged chest-deep into rushing streams still tinkling with candle ice and lined their freighter canoes with ropes. Insects were already out, so they powdered the insides of their tents with DDT. DDT worked: Dead blackflies coated tent floors two inches deep.

Folinsbee found the rocks dizzyingly complex and, as far as he could see, barren. In quartz veins along shoreline he found a few flakes of gold—very few—and in other rocks minor bits of lithium and tantalum. The main feature seemed to be dreadful boulder fields. Whenever Folinsbee got into one and was teetering precariously from one tippy boulder to the next, small packs of wolves would show up to watch from a few hundred yards. He figured they were waiting for him to break a leg. He looked down in the deep crevices between boulders and saw the bones of many caribou, and perhaps other creatures.

One day Folinsbee traversed Misery Point. Its name, adopted by Ottawa cartographers, was in fact the only name on the map except for Lac de Gras itself. Here Folinsbee mapped the visible bedrock— coarse granite and gray metamorphic schist with layers of sparkly mica interspersed. Dug into this hard solid stone, and surrounded by mounds of glacial till, were countless little lakes, all looking the same. However, he was impressed with the trunk esker. He hiked up its side and followed its flat top west. A while on he came to a giant S curve and, on the esker's south flank, a small lake with a beautiful sandy beach nestled into the curve. Here the fifty-foot ridge cut the north wind, so only modest waves lapped the protected cove. It was a fine campsite and perfect for landing float planes, so he set up. He then discovered something else: Along the curving strand line was a weird band of fine black and purple grains. Two feet wide and about 300 feet long, it appeared to be heavy minerals of some sort. The little waves, he noted, contained just enough energy to wash away common light-colored quartz grains, leaving weighty ones distilled. He shoveled twenty pounds into a canvas bag to take home.

Back home, under a microscope he saw mainly two things: grains of shiny black ilmenite and a peculiar, deep blood-red garnet. Folinsbee was no expert on garnets, but he knew different garnet types came from all sorts of rocks. He identified these as almandines—a common species formed in the crustal rocks of the kind he'd mapped all summer. As for ilmenites, they come from many rocks too, and all tend to look the same. He drew a map of Lac de Gras showing exposures of the various bedrocks, which his wife colored with lovely Japanese watercolors. He stuck the minerals in a drawer. After the GSC published his map, he put out a minor paper on his own about the beach minerals. Folinsbee proceeded on a brilliant career as a university professor and, later, a hunter of freshly fallen meteorites.

Forty-four years later, in November 1991, after academic journals had published a few articles about kimberlitic indicator minerals, grayheaded professor emeritus Robert Folinsbee picked up his morning paper and read of a supposed diamond strike in the Northwest Territories. He had not been there for decades.

Most readers figured the "strike" was just another of many scams designed to run up the stock of some small exploration company. But Folinsbee blinked once and thought back to that great, faraway esker of his youth. And the cove with the blood-red garnets. And an obscure journal article about diamond-indicating garnets he half-remembered seeing recently. And again the garnets—which he realized now he may well have misidentified. He dialed a stockbroker he knew. Chuck Fipke's little company, Dia Met Minerals, was still selling for $3 a share. At a time in life when most people are making out their wills, Folinsbee emptied his bank accounts and put his savings into Dia Met.

S oon after geologic mapping of the tundra began, human travel virtually ended. Waves of European-introduced influenza, smallpox, tuberculosis, scarlet fever, and measles continued to chisel at Indian and Inuit alike. It is said some Yellowknife families fled to the Barren Ground during a flu epidemic in 1928, thinking they could outrun the unseen germs there. Most never came back. Their numbers were already decimated by earlier warfare with the Dogrib. The Yellowknives—probably only 400 to 500 at their height—dwindled. In the 1930s, they disappeared, according to anthropologists. Only the Dogrib survived to briefly dominate the Misery Point narrows. The Inuit survived—with the world's highest incidence of tuberculosis. The Copper subgroup, never numbering over 1,000, were culturally wiped out when missionaries, guns, and alcohol from ever-encroaching trading posts exerted their hold. Many settled at the mouth of the Coppermine River in a new, ramshackle town dubbed Coppermine, where many proceeded to drink themselves to death.

Through the 1940s some diehard aboriginals still traveled into the tundra in little motorized canoes to trap furs for the Europeans, but newly available government health clinics, plywood houses, and oil heat started providing an easier life. Compulsory schooling and welfare payments arrived, and in 1951, old-age pensions. To collect them, you had to stay in town. People began losing track of complex inland water routes and the secret spots where their ancestors had mined copper, lead, or stone; these treasures, these places, had been replaced. In the mid-1950s fur prices crashed. Finally, through some inscrutable natural cycle, the caribou population nosedived, perhaps as much as 80 percent. Years later the deer came back, but not the people. In 1958 some of the last nomadic Inuit hunters starved to death on the Barrens. The rest fled or were rounded up by authorities and deported to settlements on the edges. Some natives took up work in new nickel, gold, or uranium mines. Younger ones learned to speak English, but the Dene persisted in calling Yellowknife by its separate Dene name: Somba K'e, or "Money Place."

Prentice Downes, a white schoolteacher who befriended the Indians in the 1940s, described one of their last migrations back into the

tree line: "There was something strange, dark and splendidly bar-
baric about the whole thing. Children rushed about for sticks, and
the camp fires were stoked up. The sky in back of the camp was a
dull, sulphurous yellow with black clouds. The women, in their red
silk bandannas, shrieked and clubbed at the dogs. The men were wild
and shaggy, their long coarse black hair matted about their heads. . . .
Here they were men and we were the timid strangers."

By 1959 the Barren Lands were empty and silent, no longer trodden by the foot of the wandering Indian, or the Inuit, or even the
hardest-bitten prospector.

PART II

CHAPTER 5

Golconda

As the Barrens emptied in 1959, twenty-year-old Stewart Blusson was starting a brilliant, though eccentric, career with the Geological Survey of Canada in the cordillera to the west. Far south, in the backwater town of Kelowna, British Columbia, twelve-year-old Chuck Fipke was just learning to drink, drive, and party, often all at the same time.

I could tell you now how Fipke and Blusson met on a remote mountain lake during the first of their many shared near-death experiences, and how by chance they later came to hunt diamonds for an unlikable Texas billionaire. I could also tell you how they worked their way across the United States and back up into Canada with the help of an ex–De Beers geologist—and a scientific secret even De Beers did not have. There is more after that, much more. And it will all make sense after I tell you about the ancestral diamond prospectors.

Jacques Cartier and others were too ignorant to be real diamond hunters. Well past Cartier's time, many Europeans still believed in the serpent-infested Valley of the Diamond—which existed only as an invention of its promoters. For centuries the world's real diamond capital was Golconda, a fortified city near the banks of the Krishna River in the deep interior of southern India. Here the high ground was dominated by a turreted stone fort, where merchants routinely told nosy outsiders such tall tales; it kept them off the track of where the goods actually came from.

The first known European to see the truth was Jean-Baptiste Tavernier, a Parisian-born gem dealer who lived a century after Cartier. Tavernier made six voyages to India from 1631 to 1669 to buy diamonds for the French King Louis XIV and other luminaries and was so rich, impressive, and persuasive that merchants decided to show him around the mines. These lay within six eastern Indian rivers. Some sites had been worked for alluvial diamonds 2,000 years or more; others were more recent. All the discoveries appear to have been accidental, perhaps made by prospectors looking for gold. Tavernier was told that in 1560 one poor farmer stumbled upon a 25-carat stone while planting millet along the Krishna's bank—the start of a huge mine still then in operation.

The stones had been washed from long-eroded pipes, then carried along the watercourses and reburied in various ways. Along the Krishna, most lay in loose sand; after each rainy season as many as 60,000 diggers descended. Farther north, on the Mahanadi, gems were found only on the left bank, in a less accessible stratum of tough reddish clay near the confluences of rivers coming from the north. On other rivers, workers hacked tunnels with iron bars through bedrock to narrow diamondiferous layers of sandstone and conglomerate as far as 115 feet down. People held that the diamonds grew where they were found, like seeds. Quartz, often found among them, was believed to be immature diamond, or *kacha* (unripe); diamonds were *pakka*—ripe. Small ones were said to combine over time to make bigger ones. Thus for the Golcondans prospecting involved finding not only the right place, but also the right time.

The mines were not freelance enterprises, however. Digging was strictly controlled by nobles and merchants; possession of diamonds by ordinary people was severely punished. So after seeing the sources, Tavernier traveled north to inspect the jewels of the Mogul emperor Aurangzeb and to do what he came for: buy. He purchased dozens of stones over 20 carats—far larger than any yet sent to

Europe—among them one of 112 carats from the Krishna. This went to Louis XIV, and from it was cut the great 44.52-carat blue Hope diamond, later worn around the neck of Marie Antoinette before it met the guillotine in the French Revolution. Such spectacular stones had the powerful effect of redoubling the already great gem hunger of the European nobles. Demand skyrocketed. The Indians, aided by a few lucky finds around this time, were glad to oblige the new demand. The longtime steady production of maybe 10,000 carats a year went to about 75,000 carats annually.

In fact the mines were suddenly too successful. The flow grew so much that they suddenly did what all mines do: They ran out. By 1680, just two decades after the visits of Tavernier, India's millennia-old diamond industry was collapsing. In 1688 the stone castle of Golconda was pillaged and wrecked by Aurangzeb and was no more.

Just around this time, Portuguese adventurers penetrated the precipitous, jungly mountains of Brazil and began panning abundant gold from streams. North of Rio de Janeiro, down in rock crevices, they sometimes found irregular shiny stones along with the gold. Most ordinary men still had never seen a diamond rough or cut, so they used the large ones for card-game chips; smaller ones were tossed away. In 1727 Friar Bernardo da Fonseca Lobo, a well-educated former resident of Portuguese India, saw some of the chips and spirited them to Europe. At first many couldn't believe the New World contained genuine diamonds after all—least of all merchants, who wanted to keep prices high. They insisted these were inferior Indian specimens smuggled through Brazil.

When it became clear they were wrong, the Portuguese called the diamond region Diamantina, and rich nobles were given royal franchises to work the rivers. Newly imported African slaves did the work, digging ten to eighty feet to diamondiferous gravels coating bedrock. If a slave came across an "octavo," a stone of 17.5 carats or more, overseers set a flower garland on his head, paraded him around, gave him a new set of clothes, and declared him free. Significant but smaller finds got just the clothes. Others needed not apply: They were driven in the sun with whips and branded with irons if caught stealing.

Once this Brazilian search got going, all the slave drivers and soldiers in the world could not control it. Freelance prospectors instantly proliferated in the wild back country—many of them escaped slaves—to randomly dig every remote riverbed. The tough, unschooled freelancers were called *garimpeiros*, or, roughly, "snipers." To them, legal status did not matter; they moved from place to place, disappearing when diamonds and gold ran out or when police arrived. Within decades they boosted production to once-unimaginable heights and largely displaced the big owners; Brazil annually produced four times more than India at its height.

The *garimpeiros* survived because they were good inventors, adapting old techniques for their specialized purpose. For centuries prospectors in the mountains of Bohemia had used wooden settling troughs to concentrate heavy gold flakes, copper lumps, and deep-red garnets (a popular Bohemian export) by running water through to float off light sands. The *garimpeiros* dove underwater from canoes to pull up gravel and substituted for troughs a more portable, Chinese hat–shaped pan, known as a *batea* or *siruca*, to swirl gravel around with constant additions of water. They also discovered that nature sometimes did the work for them in stream-bottom potholes or crevices carved from bedrock; one in the Ribeirão do Inferno was said to have held 8,000 diamond carats. The reason was simple: Anything heavy sweeping downstream fell in, then tended to stay put. The *garimpeiros* also searched dried-up watercourses in high plateaus and mountain passes. In 1824 one plateau north of Diamantina yielded stones in the reddish soil. Chickens scratched them out, children picked them up, adults found them clinging to pulled-up roots of grasses and garden plants.

Diamonds rarely revealed themselves by mere visual searches, so over generations the *garimpeiros* developed elaborate systems of signs. These were mainly other heavy minerals (or sometimes rocks) found in the places where diamonds tended to settle. They included garnets and gold and countless others more peculiar to Brazil—the first putative diamond indicators.

English naturalist Richard Burton canoed through rural Brazil in the mid-1800s to find people swearing by a vast panoply of these so-called *cattivos* (captives) or *escravos do diamante*—slaves of the diamond. Many were beautiful minor gems themselves. There was *pinga d'agua* (drop of water), which Burton identified as water-rounded, semitransparent white topaz, pea to pigeon's-egg size; *pedras de leite* (milk stones), water-washed diaphanous agates marked with concentric undulations; *ovos de pomba* (doves' eggs), quartz balls of various colors; *siricória*, long prisms of chrysolite; *fava*, shaped like the broad bean, of quartz or blood-red jasper; *feijao*, a haricot-shaped dark-green to black tourmaline seemingly glazed by great heat, like stones in English tin mines; *casco de telha*, a reddish clay flecked with mica and talc; *pedra de Santa Anna*, cubes of magnetic iron that made compasses go crazy; *palha de arroz* (rice straw), a slice of yellow slate resembling a cucumber seed; *osso de cavallo* (horse's bone), a long, osseous fragment of pure sandstone.

Garnets came in many hues from light pink to almost black, but *garimpeiros* called them all *xique-xique*, perhaps for the shick-shick rattle they made on the pan. However, *xique-xique* was only a minor *cattivo*; among the few locales where it appeared with diamond was the malarial mouth of the Jequetinhonha River, where the gems lay uncharacteristically under two feet of white clay and dead plants. Another minor indicator was *cattivo preto*, or black slave, apparently a variety of ilmenite. It was readily identifiable, for the mineral possesses a deep, metallic bluish-black luster and, when smeared, leaves a black-violet streak. It fragments sharply, so big crystals can have a vicious sharp edge. Some *garimpeiros* embedded slivers of it in wooden slabs and used them to grate vegetables.

Regarding all indicator minerals, Burton repeated the wisdom of the *garimpeiros*: "The[y] show that the diamond may be there, not that it is there." What they did not know was that many *cattivos* led nowhere because most were heavy minerals from common nonkimberlitic rocks; they ended up in streambeds in the same places as diamonds only because they obeyed the same laws of physics. To add confusion, every locale had different common rocks, so every locale had different *cattivos*, few or none with an intrinsic link to diamond. No one could have known that kimberlite had its own suite of heavy minerals, not necessarily identifiable by eye, including very specific kinds of garnets and ilmenites. If there were any of these, they were probably sand-size grains hopelessly mixed with the flood of other debris.

As the *garimpeiros* reached their height in the early 1800s, modern geologic theory took hold. Minerals were classified more carefully, depending on crystal structure and chemistry. Biblical ideas of geology were replaced by the overarching theory that rocks and minerals were formed over great periods of time within seabeds and mountains and the deep earth, then eroded by wind and water, then built up again. By the mid-nineteenth century an even more incredible idea was added: Another major erosive force was ice—glaciers—periodically moving across the surface. In the middle of this revolution, educated people grasped the key concept: Diamonds, like other loose matter, did not originate in rivers. Somewhere was a "mother rock." But where?

Louis Agassiz, the towering Swiss-American naturalist who first proposed the shocking idea that huge ice sheets had altered the landscape of North America and northern Europe, visited Brazil in 1865. He wrote that he was "prepared to find that the whole diamond-bearing formation is glacial drift. I do not mean the rock in which the diamonds occur in their primary position, but the secondary agglomerations of loose materials from which they are washed." But Agassiz admitted: The materials' history "cannot be traced."

Oddly, common quartz first got the most attention as a possible "mother rock." Quartz was in fact so common, it was the one thing almost always found somewhere near diamond. Based partly on this, Max Bauer, German author of the definitive late nineteenth-century book *Precious Stones*, agreed with the Indians: Diamonds grew in, or from, quartz. "The fact that the diamond itself has never been found in a quartz-vein perhaps may be explained by the extreme rarity of its occurrence as compared with other minerals," he wrote. The other popular candidate, also based on its abundance in streambeds, was a sandstone called itacolumite—when carved by nature in the right shape, the *cattivo* horse's bone. Laminated with layers of mica and talc, it is extremely flexible, at least for a rock; a narrow piece a hand long can be bent a quarter-inch. Diamantina's rivers cut through highlands made of it, and diamonds were sometimes found encased in itacolumite—to modern eyes, proof only that eroded-out gems were reburied so long ago that sandstone had enough time to lithify around them. Back then, the presence of diamonds in sandstone seemed an argument for their origination there.

Such wrong ideas were fueled not only by observation but also by experiment. At the same time more diamonds came to Europe, modern science budded. Among the first to investigate them was Boetius de Boodt, who in 1609 published the Latin treatise *Gemmarum et Lapidum Historia*. First thing the audacious de Boodt did was take a diamond—presumably small—and smash it. That wrecked the myth of Pliny's *"adamas,"* the unbreakable substance. He also proved that diamonds did not neutralize magnets or do other things they were supposed to do. Soon after, Sir Isaac Newton, founder of modern physics and math, did a great many experiments on diamonds. Among other things, he measured their unequaled ability to scatter light. Based on deductions from that, in 1675 he advanced the heresy that diamonds were basically the same as wood—carbon—and thus could be burned.

In 1694 C. Targioni of Florence's Accademia del Cimento convened a solemn ceremony. In front of dignitaries he and a partner set out a fair-size diamond, took a great magnifying glass, and focused the sun on it. Presently it split, emitted a stream of red sparks and smoke, and shrank into a black residue. Diamond obeyed the same laws as humans: ashes to ashes. In 1772 Antoine Lavoisier, the main founder of modern chemistry, examined the gases emitted while diamonds burned and detected carbon dioxide—same as burning charcoal. Five years later English chemist Smithson Tennant took the final logical step, burning identical weights of charcoal and diamond and then measuring the volume of carbon dioxide each gave off: equal. Newton was right: Diamond was basically the same as wood, just arranged differently. Further experiments showed that when heated in a vacuum to prevent combustion, the crystals rearranged themselves into a soft, black stuff—graphite.

Thus armed, scientists blundered not closer to the truth, but farther away. They decided that since diamonds resembled plants, diamonds must form from plants. French naturalist Georges Buffon, noting that they had so far been found only in the tropics, speculated that the sun baked decaying vegetal matter close to the surface. Others theorized plants were buried farther down, to be transformed by the same pressures and heats that metamorphosed rocks—a bit closer to the truth. The intermediate product could be the proverbial hunk of coal, or perhaps petroleum; porous itacolumite formations in fact turned out to be major hosts of oil reserves. Since diamonds resembled droplets, others thought they might be compressed amber. In the 1860s paleontologist Johann Göppert peered into diamonds through the newly invented microscope and saw fine green needles, fibers, and cells. He believed these to be remnants of buried algae, the raw material. He even named two diamond-forming algae species: *Protococcus adamantinus* and *Palmogloeites adamantinus*. Actually, the flaws were inclusions of chrome diopside.

Lab scientists also tried deducing diamonds' origin by reviving alchemists' attempts to make them. It was now agreed that heat and pressure applied to carbon must be key, but no one knew how much, or whether other substances might be required as catalysts. French scientist M. Despretz tried for decades. Once he borrowed every battery in Paris to cook a batch of carbon. He got only soot. In the 1850s, thinking to imitate geologic time, he passed electricity through carbon for over a month. This did not work either. No one succeeded at making synthetic diamonds until the 1950s—and even this would not provide any direct clue of how to find the far more valuable natural ones.

Then there was a revelation: Diamonds were discovered in the United States of America. Also kimberlite, though no one knew it at the time.

The new American states were eager to develop mineral resources, so several began their own geologic surveys. In 1836 New York sent out four geologists. That year, one, Lardner Vanuxem, walked 1,000 miles through the then-wilds of the western part of the state, where he mapped endless expanses of ancient ocean sediments—shales, siltstones, and other largely worthless formations whose main resource seemed to be one small lead mine. Around what is now the college town of Ithaca, he clambered through one of the many deep, shady gorges that lace the land there and noticed small vertical spouts cutting up through otherwise horizontal layers. He wrote: "In the ravines east of Ludlowville . . . in the fissures of the slate, are two narrow veins of semi-crystalline rock of a blackish brown color, becoming olive by alteration. It appears to be a mixture chiefly of serpentine and limestone. There are also two similar veins, near the foot of the second falls, in the same ravine. Both . . . traverse the creek at nearly right angles." Vanuxem had just written the world's first description of a kimberlitic rock. It was instantly entombed in the state archives.

Around this time, the first U.S. diamonds surfaced, far south. Early settlers had panned for gold, and by the 1830s the Appalachian foothills of Alabama, Georgia, the Carolinas, and Virginia had produced $20 million worth. U.S. mints were striking gold coins in New Orleans; Charlotte, North Carolina; and Dahlonega, Georgia. It was again gold seekers who found diamonds.

The most compelling—though unprovable—story comes from antebellum Georgia. According to documents in the Georgia state archives, a certain Dr. Loyd had a job overseeing thirty Negro slaves at a placer on Glade Creek, thirteen miles northeast of Gainesville. When gold was cleaned up from the sluices each day, lustrous pebbles sometimes would be in it. They averaged maybe 4 carats each, according to later descriptions. Loyd had no idea what they were, but he filled a half-pint mustard jar with them, which his wife kept in a cupboard so the kids could use them as marbles. Loyd found a much larger stone early one morning while digging in place of a sick hand. It was the size of a guinea egg—perhaps 100 carats. He laid it on the creekbank next to a gum tree, but when he went back later, it was gone.

After four years the mine closed, Loyd's wife died, and his daughter married, taking the cupboard. A dozen years later Loyd was living in Atlanta when someone showed him a rough diamond, recently identified from a Georgia river. Loyd rushed to his daughter, but her own kids had apparently lost the jar. He raced back to the Glade mine—now a cornfield. The gum tree was long gone. He spent months digging where he thought it had been, but never found anything again. Since no stones survived, no one knows if they were diamonds or quartz. However, it was reliably reported in 1952 that banker Rafe Banks of Gainesville, whose grandfather owned the Glade, possessed a local 1-carat diamond from the site. Banks had it cut by Tiffany's and wore it in a ring.

Among other early reports, in 1830 a man was said to have found three gems in Coco Creek, near the headwaters of the Tellico River in eastern Tennessee. Another supposedly surfaced in Indiana the same year. Documents from the Vaucluse gold mine in eastern Virginia say a worker found a diamond there in 1836, near workings that now lie abandoned at the end of State Road 667 in Orange County. All were unverified.

The man usually credited with the first authenticated stone is Dr. M. F. Stephenson, an experienced placer miner and assayer at the Dahlonega mint. Some attribute to Stephenson the old expression, "Thar's gold in them thar hills." He was said to be pointing to Georgia's Findley Ridge, where much gold had been claimed, urging boys not to go to the 1849 California gold rush. In 1843, Stephenson traveled 100 miles to Burke County in western North Carolina and panned a ford in Brindletown Creek. There he picked up a small, clear octahedral pebble—a diamond, valued at $100. He also found a modest stone that same year in Winn's, or Williams', Ferry, at the mouth of a branch into Muddy Creek, a half-mile before it hits the Chatahoochee near Gainesville. Another stone was said to have been found somewhere near the Brindletown ford by the widely traveled English geologist George W. Featherstonhaugh, but the exact details of that are unclear.

The finds aroused little notice at the time; the finders probably kept them quiet. Then, in 1845 another diamond was found west of Brindletown by panner J. D. Twitty. This one ended up with General T. L. Clingman, a wealthy Asheville resident interested in promoting North Carolina mining. He wisely gave it to C. U. Shepard, an eminent Yale University mineralogist, who described it in the prestigious *American Journal of Science*: "a beautiful diamond of fine water, transparent, possessing only a faintly pale yellowish tinge of color; and it is nearly without flaw." It was 1.33 carats. Shepard wrote: "It is to be hoped that the proprietors of gold washings throughout the district, will immediately set on foot a systematic search for this precious gem, which, in the ordinary operations of gold mining, might be overlooked to almost any extent. Henceforth there can scarcely remain a doubt, but that the diamond is to form a part of the available mineral wealth of the country."

Southerners, alerted to the possibilities, now eyed odd stones more closely, and dozens more alluvial diamonds surfaced. Most lay in the Piedmont region, the plateau of debris washed from Appalachia toward the sea. Four counties around Brindletown yielded them, as did scattered parts of Virginia, South Carolina, and Alabama. In Georgia, dozens emerged from gold sluices in Lumpkin, White, and Hall counties. Three gold washers near Charlotte, North Carolina, found the largest known: a lovely, though blackened stone the size of a chinquapin, a small southern nut. They immediately destroyed it by pounding it with a hammer to see if it would break. Plinyesque folklore still ruled.

Only one large gem survived. In 1855 Benjamin Moore, a laborer, was grading a street at the southwest corner of Ninth and Perry streets in Manchester, Virginia, now part of the city of Richmond. There, in dirt left by flooding of the St. James River, was a $23^3/_4$-carat stone, octahedral, a bit greenish, slightly rounded, flawed by one or more big black specks in the middle. Richmond jeweler John Tyler appraised it at $4,000, and Moore sold it to Captain Samuel Dewey, a Philadelphia mineralogist. Like an Indian rajah, Dewey named his acquisition "Oninoor," or "Sun of Light," showed it to newspaper reporters, and had it displayed at a New York jewelry store. Glass casts of it were deposited at the U.S. Philadelphia Mint and other institutions. Dewey then spent $1,500 in Boston having it cut to $11^{11}/_{16}$ carats.

Misfortune followed. Dewey ran short and had to pawn the diamond. The new owner promptly borrowed $6,000 against it from a prizefighter, John Morrissey—and went broke himself. Morrissey took the stone. In 1858 John Morrissey became heavyweight boxing champion of the United States—in those days it was done with bare fists—and had it set in a ring. Experts revalued it: Due to its off color and significant flaws, it was worth not $4,000 or $6,000, but $300. It is unknown who got the ring after Morrissey; since it probably looks like any other low-priced piece of diamond jewelry, it is likely that the current owner does not even know it contains a once-famous diamond.

Some geologists still doubted the diamonds were truly North American. One theory held that they arrived in the stomachs of birds migrating from Brazil. More patriotic (or pragmatic) investigators traced upstream into the Appalachian highlands and spotted belts of the putative "mother rock," itacolumite. Around the sandstone, mineralogist Frederick Genth of the University of Pennsylvania also saw metamorphic schists, gneisses, and beds of graphite—a possible carbon source. This intriguing combination sparked scientific interest but was suddenly overwhelmed by a far greater event: the California gold rush.

When placer gold was found at Sutter's Mill in 1848, everyone ignored Dr. Stephenson's plea to stay home. In spring 1849, 50,000 set out for the West. For pans, they used any shallow vessel they could find, often the one they cooked their bacon in. Like the *garimpeiros*, the forty-niners quickly learned to detect a potential "pay streak" by watching for heavy minerals—in California, most commonly black magnetite, red hematites, and grains of tiny pink garnets.

In some auriferous streaks were also modest diamonds. The first known was a yellowish pea-size specimen found in July or August 1848 by a nameless panner near Placerville, east of Sacramento. In following years about 500 others surfaced in fifteen counties, many in locales that now exist only in history: Jackass Gulch, Hangtown Creek, Fiddletown, Loafer Hill, Volcano, Fairplay, Spanish Ravine. Many were flawless, but without exception small, and further degraded in value by a characteristic yellow cast. In the northern Trinity River bordering Oregon, they were literally microscopic. It is possible that bigger ones existed, but the goldfields were quickly mechanized with powerful hydraulic hoses that tore apart riverbanks for miles, and heavy stamping mills that pulverized gravel so gold flakes could be extracted. It didn't hurt the gold, but miners occasionally spotted tiny, new-looking diamond splinters in the refuse.

As a result, California stones rarely drew more than $10 or $50 from buyers in San Francisco—hardly worth the effort. Many finders instead wore them in rings made by local jewelers or gave them away as keepsakes; customarily they were preserved rough, as found. The majority came from placers around the towns of Cherokee and Oroville, north of Sacramento, where the biggest operation was the Spring Valley Hydraulic Gold Company. An early superintendent of the mine, a Mr. Harris, gave his wife a rough stone in a ring, and when California's first state geologist, Henry G. Hanks, saw it, he opined it was far more beautiful a gift than any expensive cut stone. Clearly, California diamonds were not for making money; they were personal.

For decades longtime Placerville justice of the peace William Pitt Carpender kept a meticulous handwritten record of friends' and neighbors' stones. Excerpts:

"John Lyford, at Smiths Flat, 1865, one white stone, 3 grains, stolen by a woman and set rough in a ring."

"Nathaniel O. Ames, mine on Webber Hill (no date), one stone of 1.3 carats, set in a ring and worn by Shelly Inch, postmaster of Placerville for many years. This stone now in possession of L. P. Inch, city manager of a fire insurance company at 444 California St., San Francisco."

"Fred Bendfeldt, Sr., at Smiths Flat, 1878, one stone of $^7/_{16}$ carat, canary color, sold to W. F. Alma of Boston. Fred Bendfeldt arrived at Smiths Flat, from Hamburg, Germany, via the Horn, in 1853, and operated a deep-gravel mine at the Flat for thirty-odd years. He 'crossed the line' on the Flat Feb. 4, 1916, at the age of 88 years."

There was, however, a connection between the little California diamonds and the rise of industrial-scale diamond mining. Mr. Harris's successor in Cherokee was a clear-eyed, extravagantly mustachioed engineer, Gardner F. Williams. Williams was part one-mule prospector, part modern scientist. When Williams was ten, his father had brought him to the rush from Michigan. He went on to study mining at the new University of California, then Germany's Freiberg Mining Academy, the world's top school, where he helped Alfred Nobel refine techniques of blowing up rocks with Nobel's new invention, dynamite. Williams traveled the deserts of northern Mexico in search of salt, prospected silver in Nevada, and took part in a gun battle with the Apache near the future Tombstone, Arizona, in which the main casualty was a white man's horse. Upon taking over the Cherokee mine, Williams found a number of diamonds and generously gave a particularly fine one to Henry Hanks for the State Museum. Hanks labeled it specimen No. 4033. Williams then moved once more—to South Africa. There by chance he took on the managership of a new outfit then in deep trouble: De Beers Consolidated Mines.

It was gold, not diamonds, that made the Americans rich. Within twenty years Cherokee alone produced a glassful of middling-value diamonds—and 600,000 ounces of gold. Gold strikes swept backward through Colorado, Arizona, and North Dakota, then up the cordillera toward the Klondike. The exploration spurred discoveries of iron, copper, coal, zinc, manganese, nickel, mercury, and petroleum; the continent seemed blessed with everything but gems. The American Civil War of 1861–1865 kick-started exploitation: Without endless cannons, rifles, armored gunboats, ammunition, railroads, and other needed appurtenances of destruction, the killing and maiming of 1.1 million people in so short a time would never have been possible. Minerals made the Americans so rich, they no longer needed to find their own diamonds. Now they could afford to buy them elsewhere. The Canadians caught up later. The United States is still the world's largest producer of raw minerals, with Canada not far behind.

Men's pinky rings became *de rigueur* symbols of success. In the 1850s U.S. diamond imports went from $100,000 annually to $1 million—and Customs suspected five times more were smuggled. 1860s San Francisco was a major market, with prices 25 percent over Paris. Richard Burton, lately of Brazil, wrote: "In [America] . . . gems are eagerly sought by those who have made money. In the Atlantic cities every one that can afford them, even hotel waiters and nigger minstrels, wear diamonds in rings and shirt fronts." In 1874 *The Times of London* diagnosed Americans as having "a weakness for diamonds. Not only do the ladies blaze with gems, but the men wear stones [such] as no man would think of adorning himself with in this country." The first tier of seats at the New York Metropolitan Opera was dubbed the Diamond Circle, for the Astors and Rockefellers who owned so many. Jewelers, led by Charles Tiffany of New York, started buying the crown jewels of beheaded French monarchs and other European nobles and setting them in baubles with American motifs such as acorns. The Hope diamond went to *Washington Post* owner Edward McLean, then to the Smithsonian. Then something happened to make the Americans even crazier. Diamonds were discovered in the southern African desert—more than anyone had ever imagined to exist on earth.

In the nineteenth century the arid interior plateau of southern Africa was thinly inhabited by native Korannas, mixed-race

Griquas, and Boer ranchers—Dutch-Germans fleeing competing English colonists on the rainier, more hospitable coasts.

The first documented stone surfaced in late 1866 or early 1867 when Erasmus Jacobs, a sixteen-year-old Boer, was clearing a clogged drain near the region's central Orange River. The Orange's broad alluvial terraces are lined with great expanses of colorful water-worn pebbles from the chasms of the ancient rock it cuts—garnets, carnelians, jaspers, agates, chalcedonies. One day Erasmus picked up a particularly glittery one, which his younger siblings used as a marble, until a neighbor offered to buy it. Weighing 23.25 carats, it later drew the huge sum of 500 pounds from the governor of the English Cape Colony.

At first this was just like Brazil: Merchants did not want to hear bad news of more finds. By their request, University of London mineralogist James Gregory visited to investigate, then wrote an article in *The Geological Magazine*, the leading journal of the day. Someone, said Gregory, had imported the diamonds as part of "an imposture— a Bubble scheme. . . . It [is] impossible, with the knowledge we at present possess of the diamond-bearing rocks, that any [gems] could have been really discovered there. In these so-called diamond-districts there are no traces of what are usually termed metamorphic rocks, such as mica-slate [itacolumite] . . . nor any of the minerals usually, nay always, found in [Brazilian] diamond districts, such as Zircons, Anatase, Rutile, Brookite, Cassiterite, Titaniferous Iron [and] Gold." When a settler showed Gregory more stones, he snapped: "Brazilian." In 1869 a Griqua shepherd spotted a superb white 83.5-carat beauty. This made worldwide headlines, sparked a rush, and made a fool of Gregory.

The Koranna nomads had apparently always known of the stones; they believed snakes with diamonds on their heads had bathed in the rivers and dropped them. They began picking up the little crystals to sell and soon grew systematic, waiting for rains to wash the surface, then joining hands in long lines to search. In November 1869 the first organized Europeans arrived. Within months the rocky side ridges of the Orange and its main tributary, the Vaal, hosted 10,000 Frenchmen, Malays, Bohemians, Australians, Chinese, New Zealanders, European Jews, and, perhaps most numerous, Englishmen. By consensus, each man worked a thirty-by-thirty-foot claim at a time. The outnumbered blacks were pushed aside and made to work for wages.

Naturally, some of the most numerous, enthusiastic arrivals were Americans, many deserters from the California goldfields, and Canadians. Five ships fitted out in Boston especially to transport diamond diggers, including Irishmen who had already crossed the Atlantic once but failed to strike it rich. At least one Native American from an unknown tribe sailed. His name is not recorded, nor his fate. Well experienced in placer mining, the Americans quickly made their mark. A Louisianan, Jerome Babe, invented a machine called the Yankee Baby that sifted material by size through a series of sieves and doubled working speeds. Other Americans soon engineered even bigger contraptions to increase the pace.

However, the diggers remained just that: They dug randomly and moved on unless they hit something fast. In addition to various bogus indicator minerals, some tried witching sticks and fortune-tellers to locate spots. Many could not identify diamonds anyway; discoveries of quartz bits caused many a finder's neighbors to pull up stakes and pile on furiously, until an experienced hand came along and informed everyone that a mistake had been made. Such embarrassments were called "Gregories." Everyone had a drink, then dispersed to the next rush. A magazine reported that "avaricious spirits refuse to go to bed, but work steadily till daylight . . . diggers gravel the floors of their tents with the refuse washings of the mines, and amuse themselves by looking for diamonds while lying in bed." Churches also had refuse floors, so prayer was mixed with prospecting. One week the local paper, *The Diamond Field*, reported "Master Kidd picked up a diamond in church . . . on Sunday last."

The prospectors declared a "Diggers Republic." President was top-hatted Englishman Stafford Parker, owner of a makeshift dancehall, ex–water diviner, ex–U.S. Navy sailor, and ex–California forty-niner. It did not last; the English saw a good thing and sent in troops to annex the territory, naming the main town Kimberley for John Wodehouse, Earl of Kimberley and Secretary of State for the Colonies. When the Union Jack went up, an irate American hauled it down, but the soldiers did not bother arresting him. They saw they could control this disorganized new colony a lot better than the old one.

It was the diggers who, through sheer numbers and industry, stumbled upon the truth. River deposits thinned fast, leaving latecomers mining the teeth and jaws of hyenas. Some now drifted deep into the vast, empty triangle of land between the Vaal and Orange. In September 1870 a Boer rancher let one English prospector dig at a ranch called Dutoitspan. He turned up plentiful stones fifteen miles from any river. Hundreds joined up. Some moved a quarter-mile away to the next ranch, Bultfontein, and a month later a second non-riverine deposit turned up there. Hordes alit like locusts on these so-called "dry-diggings." At first everyone assumed them alluvium of some sort, if not from an extinct river, then from a flood or glacier outflow. And in fact, the good spots usually were circular yellow-mud depressions forming seasonal ponds (in Afrikaans, *pans*), suggesting the same. After two feet the diamondiferous mud ended in tough limestone—presumably bedrock—so diggers moved on to a new thirty-by-thirty plot when they hit it.

The next ranch over was owned by a couple of taciturn, bearded Boer brothers, D. A. and Johannes De Beer. The De Beers evinced interest only in cows but let a neighbor named Cornelis dig near a *pan*. He found diamonds there too, and made the mistake of innocently showing them to a newly arrived Englishman, who rode back to his river camp to tell his friends—too hastily. By now some 30,000 largely claimless newcomers were about, watching every move. As the men returned to the De Beers' land, they looked behind them and saw pillars of dust rising from followers' oxcarts, as if they were Moses and the Hebrews pursued by Pharaoh. Multitudes swarmed in. The De Beer brothers thought the world had gone mad; unable to evict anyone, they fell back on collecting what royalties they could. Finally, on the evening of July 20, 1871, a party led by English digger Fleetwood Rawstorne discovered yet another deposit on a small hill, also on the De Beers' land, and staked a plot by moonlight. By sunset next day the hill was stripped of vegetation, and thousands of diggers were into the dusty red soil, on their way to removing the hill by hand.

Not far down was a dusting of chalk resembling the limestone under the *pans*, but more easily penetrated. Under this was a layer of a peculiar yellowish material that might be called a breccia—a soft, seemingly decomposed stuff resembling *pan* mud, but mixed with bigger chunks of all sorts. It was easily knocked apart by a pick, and when sieved it was full of diamonds. Many men became millionaires literally overnight. The breccia was dubbed "yellow ground." At the older dry diggings, people decided to see what was below the limestone "bedrock." It was yellow ground. Everyone piled back on.

The farmers all figured things would soon run out as usual, and so, hoping to get out from under their now-ruined land at the peak, all sold out to venture-capital syndicates. The syndicates then sold claims back in thirty-by-thirty parcels to the diggers for vast profits. The De Beer brothers got £6,000 from Dunell & Company of Cape Town for a property they had bought for fifty, built a three-room mud-and-thatch house at a quieter place, and kept raising livestock. One brother later said he could not think of anything he could buy with more money except maybe a new wagon and some ox yokes.

Dutoitspan, Bultfontein, the two De Beer strikes: This now made four dry diggings, all inexplicably in the same three-mile radius, all of the same odd yellow ground—which some people still assumed some sort of alluvium. Unknown to them, beneath their feet yawned the deep epicenter of the diamond world. It would soon be controlled by a few far-seeing men borrowing the name of two simple Boer farmers. A hundred years later, an enthusiastic young Canadian geologist would get a guided tour of the mostly emptied pits and collect leftover minerals from them in a jar—Chuck Fipke. At the time, he would not know how handy those minerals could be.

CHAPTER 6

The Great Hoax

Within the United States reaction to the African strike was swift and sure: false-alarm rushes, get-rich-quick schemes, and, as the crowning achievement, one of the most sweeping and elaborate frauds of all time.

In mineral-hungry America mining scams were already such an exalted tradition that all mines were presumed fake until proven otherwise. Most ruses involved "salting"—livening up some small, worthless prospect for quick resale with a bit of planted gold, silver, or copper. In 1860s Nevada, Mark Twain noted, "Every man . . . had his new mine to boast of, and his 'specimens' ready. . . . [H]e would fish a piece of rock out of his pocket, and after looking mysteriously around as if he feared he might be waylaid and robbed if caught with such wealth . . . dab the rock against his tongue, clap an eye-glass to it, and exclaim: 'Look at that!'" The owners of Nevada's North Ophir mine salted the place using melted silver half-dollars, but got caught when someone spotted an undigested rim reading "-ted States of." Salters obviated signs of suspicious digging in gravels by shooting them up with gold dust–loaded shotguns—the true "gun that won the West," say some. Canadians loved this game, too, especially if rich Americans could be victimized. Around Yellowknife old-timers still tell of one early resident who customarily flicked ashes from gold dust–laced cigarettes while walking buyers around.

Fraud instantly spread to Africa. One early digger nicknamed Champagne Charlie salted claims with pieces of broken glass bottles and sold them to greenhorns. When this cheap trick stopped working, crooks used a few real stones. At a desert place called Lanyon Kopje, crowds followed a mere rumor of a strike. There were no diamonds, but by some odd coincidence there were many vendors of food, liquor, and gambling, already set up for business and waiting. These crafty merchant-engineered events had a name: "canteen rushes." A contractor was hired by the city of Kimberley to quarry gravel by the new cemetery, and hit diamonds—supposedly. Crowds sifted everything but the dead until the truth emerged: They had just dug and sorted the contractor's gravel for free.

None of these schemes rivaled the Great Diamond Hoax of 1872. Playing off the Africa frenzy, it had everything: true-life Wild West panoramas, grand historical characters, heart-stopping audacity, and plenty of real diamonds.

Diamond stories about the western deserts circulated as soon as explorers got there. Kit Carson, the 1840s hunter and guide, said they lay all around New Mexico. Jim Bridger, the early trapper and storyteller, claimed that on the gravel plains of Colorado ancient fruit had "peetrified" into diamonds big as walnuts; in Yellowstone country, one so big was embedded in a high rock, its glint was visible fifty miles off. People regarded these as the tall tales they were. Then came South Africa, and people were ready to believe anything.

The Hoaxers included a couple of semiliterate prospector cousins from backwoods Hardin County, Kentucky, Philip Arnold and John Slack, along with a rich onetime Kentucky neighbor, Asbury Harpending. All were southerners to the core. Harpending had once unsuccessfully conspired to turn Nicaragua into a slaveholding dictatorship headed by Tennesseean William Walker; then to declare California a country, allied with the South. During the Civil War, Confederate President Jefferson Davis sent Harpending to hijack

Union gold ships out of San Francisco. Afterward, Harpending became a San Francisco mining banker.

In spring 1871, the height of the African rush, Arnold and Slack were riding their horses through still-wild Navajo territory in Arizona. With them was James Cooper, employee of a San Francisco firm that sold miners' drill bits tipped with flawed industrial-grade diamonds—a commodity then blooming into huge demand with the simultaneous rise of U.S. mining and the flood of African stones. Cooper got the idea: Why not a fake gem mine? Psychologically the timing was perfect, and they were in the perfect place to gather salt. The desert floor of Navajoland is pierced with spectacular volcanic needles and spires, around some of which the Indians were said to know anthills filled with crystals: bright-greens that could be mistaken for emeralds; blood-reds resembling rubies; shiny black chips like half-burnt diamonds—actually, chrome diopsides, pyrope garnets, and ilmenites. It is not clear if the men really believed in the stories of anthills, but they acquired some of the minerals, probably through trade, and continued to San Francisco.

Arnold pored over the latest books on diamonds and studied specimens around Cooper's office—in short, he became one of the few Americans to really study rough diamonds. The conspirators then lifted a few from the company stockpile, mixed them with the Navajo minerals, and took them to a jeweler. He misidentified the garnets as rubies but got the diamonds right. They let rumors spread on their own, then went to eager local bankers for "advice." They let on that an unidentified Indian had shown them anthills in the desert that were loaded with these gems. The Indian had mysteriously drowned; now they alone had the secret. Could the bankers help develop a mine? As proof, Arnold and Slack offered to go fetch a couple million dollars' worth more.

Harpending was in London when he started getting frantic cables about the anthills from fellow bankers; he visited Nathan Rothschild, the vastly powerful British-Jewish banker. Rothschild mused: "America is a very large country. It has furnished the world with many surprises already." Harpending offered to head home to "investigate." There, bankers had already provided guns, horses, and provisions for Arnold and Slack's three-month "expedition." Secretly, Arnold and

Slack also had money—possibly from Harpending. They took the two-year-old transcontinental rail back to the desert—then continued on to Halifax, Nova Scotia, and sailed to London, where in July 1871 they visited Leopold Keller jewelers to buy $19,440 worth of cheap, heavily flawed South African stones. The dealers tried not to smirk. These bumpkins did not inspect or negotiate; they just put diamonds in piles and asked how much—typical nouveau-riche Americans out to dazzle their fellows, they assumed. The two then returned to San Francisco looking suitably weather-beaten and half-starved, bearing horror stories of attacks by the Apache—and a buckskin sack. Harpending, as bearer of the Rothschild connection, took over. Before San Francisco's illuminati he spread a bedsheet on his billiard table, cut the sack's fastenings, and spilled out a cataract of gems.

Harpending then led a delegation with 10 percent of the stones—undoubtedly the best 10 percent—for Charles Tiffany to appraise. Tiffany, ensconced in his fashionable Manhattan shop, was then America's—indeed the world's—reigning merchant prince of luxury. Present also was a stunning collection of prospective investors: Horace ("Go West, young man!") Greeley, editor of the *New York Tribune*; General George McClellan, former supreme commander of the victorious Union forces; and Union General Ben Butler, who had mercilessly hung a civilian protestor in wartime New Orleans for wearing a torn scrap of the Stars and Stripes and had plundered the city so thoroughly that southerners called him "Spoons."

Tiffany sorted the stones into little piles with a knowing air, holding some to the light, then valued them at $150,000—ten times their worth. The truth was, Tiffany did not know what the hell he was doing, and the plotters knew it. Tiffany's field was *cut* gems; rough ones, still rare here, were easy to overvalue. Harpending handily multiplied by ten—$1.5 million for the buckskin bag—and armed with this unimpeachable appraisal set up the Golconda Mining Company. Mine consultant Henry Janin, the Tiffany of his field, was hired to inspect the "mine." Janin was considered the ultimate fraud insurance; if he said a mine was real, it was. Of course he had never seen a diamond mine in his life.

Arnold sailed again, this time for both London and Paris, for $30,000 worth of rough diamonds, this time using the bankers' money, plus ten pounds of assorted low-grade rubies, emeralds, and other gems. For good measure, on the way back he stopped in Navajo-land for fifty pounds of spinels, sapphires, and more pyrope garnets. These were planted in a secret place.

In June 1872 the plotters took Janin and a few investors there. They rode the transcontinental rail to a remote stop in the Wyoming desert, then picked up horses and rode for four days. At times every-one but Arnold and Slack was made to wear blindfolds. It was wild, barren country, washed with shifting colors, punctuated by the occa-sional movement of a jackrabbit or snake. Finally they reached a high, arid mesa in the extreme northwest corner of Colorado, off the north-east flank of a tall piney mountain now called Diamond Peak. Sum-mer days there are broiling; winters, and nights year-round, brutally cold. Capping the 3,000-acre mesa was a great outcrop of coarse red-dish, iron-rich sandstone—itacolumite, the presumed mother rock.

When the capitalists arrived at a scrub-covered basin near the outcrop, they jumped from their horses and began scratching the sand with shovels, picks, and a Bowie knife. In three minutes Harp-ending gave a great shout. Diamonds were on or near the surface, all over. Henry Janin was cautious at first, but the Hoaxers had planned well. The geologic setting was good, and this quantity of diamonds was unlikely to be planted. Furthermore, some stones were nearly microscopic—and thus by Janin's logic also unlikely to be planted. By suppertime he had swallowed the whole thing. Within two days the searchers logged 1,638 diamond carats, four pounds of rubies, and assorted other finds. Janin called the mine "wonderfully rich . . . safe and attractive" and urged that "one great corporation" move in immediately. He predicted this one great corporation would "cer-tainly control the gem market of the world."

The location was kept secret to ward off claim-jumpers, but McClellan, Rothschild, Tiffany, and about twenty other eminences blindly invested a total of $2 million. They knew only that the place was on federal land—a possible sticking point—so they bribed "Spoons" Butler—now U.S. Senator "Spoons" Butler—with $40,000 in shares to pass legislation ensuring they could stake it for minerals. It was more than fraud; it was a legacy. This giveaway of public prop-erty, part of the General Mining Law of 1872, became a cornerstone of U.S. mining regulation. To the present time it has resulted in the basically free extraction of more than $240 billion in minerals from federal land by large companies.

Arnold then led a secret party of surveyors to lay out the 3,000-acre future "Brilliant City," complete with neighborhoods: Discovery Claim, Ruby Gulch, Diamond Flat, Sapphire Hollow. On this trip there were of course more diamonds. Alfred Rubery, a rich Englishman, found one of the fabled anthills. At first he couldn't believe his eyes. It was a sure-enough anthill—natural to this area (as were the rumored ones in Navajoland), a foot or two high, made of coarse rock and mineral grains. But this one was sprinkled with many red and white stones of uniform size. Looking closer, he realized they were rubies and diamonds. Practically screaming with ecstasy, he shoveled the whole thing into a sack, ants and all.

News spread fast. Entrepreneurs formed some twenty-five companies, most intending to piggyback onto the claims. These included the San Francisco–based New Golconda Diamond Mining Company, selling shares at $2 each and falsely claiming to know the location; the Original Diamond Discovery and Mining Company; First Choice Diamond Mining Company; Pacific Diamond and Ruby Company; Arizona Diamond Company (actually based in Nevada); and Denver Diamond Company. Advertisements appeared in papers for "well armed" men, willing to pay $250 each to ride out on diamond-hunting expeditions. "Off to the Diamond Fields!" read one.

Many tried following Arnold, but he shook them by dividing his men and covering horses' hooves with canvas to deaden tracks. He planted rumors the mine was in Arizona—or Utah, or Wyoming, or New Mexico. Harpending even sent out a decoy horseback "expedition" from California for competitors to uselessly tail 600 twisting miles through several states. Curiously, some returned from these exploits with large "diamonds" of their own, which helped sell stock in their own "mines." During these days of copycat scams, various fake western diamond ranges, mountains, and valleys received names they still bear.

As the fever spread, the *Denver Post* announced that Mexicans in the states of Chihuahua and Sonora were massing to emigrate to the mines. The *New York Sun* got a dispatch from "the frosty north." The Canadians, "not to be outdone," had discovered a "real diamond field" at Hot Creek, British Columbia; stones were on their way to New York. Colorado Governor William Gilpin even gave a faux-scholarly speech to promote his state's new "Diamond Regions," of which he knew absolutely nothing. Among other extemporized expostulations, he said success "was assured" by signs of all sorts of peculiar "carbonaceous formations" and "loose pieces of brown hematite." Even Dr. M. F. Stephenson, the original Georgia diamond finder, got back into the act. Now an old man, he instantly published a 244-page book, *Geology and Mineralogy of Georgia, with a Particular Description of Her Rich Diamond District*. Georgia had produced only a handful of stones, but he performed complex statistical analyses to show that by the "immutability of human events" it would produce quantities surpassing "all other nations"—if only people invested in techniques promoted by his own diamond company.

Amid the brouhaha, *The Times of London* published a letter from diamond brokers Pittar, Leverson & Co. It told of a couple of crass Americans who had come in the year before with a huge bankroll and bought up piles of cheap, industrial-grade stones. Now we know where those stones went, said the letter. Few heeded it.

Late in 1872, Arnold wanted to dig one last batch of salt, but by now spies and followers were everywhere. Fearing he could trust no man—at least no white man—he went home to Kentucky and recruited two dozen uneducated ex-slaves to dig and fight Indians. The African Americans took the train to Laramie, Wyoming, where each received a Winchester rifle and Colt revolver. They spent days and nights hunkered down in rented houses, drinking, eating, and smoking, waiting for a chance to slip away unnoticed. It never came. After a month Arnold disbanded his army, handed $150 to each man, and sent them home without retrieving the rest of the salt.

There is no telling where all this might have gone if not for a thirty-year-old geologist, Clarence King, who happened to be friends with Henry Janin. The United States had yet to organize a federal geologic survey, but the government had King leading a temporary survey fifty miles either side of the new transcontinental rail—

the so-called Fortieth Parallel Survey. They were now just wrapping up, and King was horrified; if a mine was on their surveyed land—and he had heard it was somewhere off the railroad—his reputation would be ruined for missing it. He set out to investigate.

In early October, Survey geologist Samuel Emmons and a comrade ran into Arnold's staff on the train. The friendly Survey men chatted them up and, from separate conversations, put together seemingly harmless, unconnected data points. They heard that the round trip to the "mine" from an unidentified rail stop took less than three weeks and that it was at the foot of a piney mountain, with no high mountains north or east. Nights were so cold that water froze in June. Later, King added other facts, including the one that certain rivers had been unfordable in June, when Arnold's men had set out. Based on conventional theories, he guessed there must also be outcropping sandstone. The Survey men had maps of their long desert travels burned in their minds, and collective rumination settled on only one place with all the right geology, geography, topography, and climate. In late October they told friends they were going fossil hunting and saddled up.

Four days later they were on top of the now freezing, deserted diamond mesa looking at Arnold's claim posts and many storm-worn footprints, among which King recognized the slender boot of Henry Janin. In the center was a high, bare, iron-stained table of sandstone where all tracks converged. Throwing down their bridle reins, they jumped onto hands and knees in the fierce wind. Emmons found a small ruby almost as he hit the ground. By darkness they had a hundred rubies and four small diamonds. Trembling with desire and excitement, barely able to hold the tiny stones between their frozen digits, they too believed.

At first light King got up to resume the search, but soon realized something was wrong. Diamonds kept turning up in perfect mathematical proportion to rubies. They lay only in bare soils—never in undisturbed ones covered by live plants. They were never buried more than a few inches; in fact, some sat under the open sky on inclined ledges, where winds of a single season should have knocked them off. In steep Ruby Gulch, gems were at the head—but not the foot, where erosion should have carried them. And ten feet beyond the claimposts, you couldn't find a thing. Over several days they tore open the mother sandstone itself, smashed it, ground it, and peered at it through magnifiers. Nothing. They turned to the mythic anthills, sprinkled with bits of garnet and diamondlike quartz. The hard crust of each had a volcanolike vertical hole on top where the ants went in and out, but some also had a horizontal hole. The geologists chipped through into the interiors and found the extra holes always went straight in, as if someone had made them with a long stick. At the terminus of each was a precious stone, awaiting discovery.

Just then a stout man in city clothes galloped up with eight horsemen at his back. This apparition was a New York City diamond dealer, J. F. Berry. Berry had also been trying to find the site, and had shadowed them. King told him it was a swindle, but this only made Berry happier; he could make a killing on the stock market before it came out. The gleeful diamond dealer wheeled and galloped to tell the world where the "mine" was.

Clarence King felt terror. Unless he beat this fool back, thousands of greenhorns would soon fill this wintry plain. Lacking heavy clothing, provisions, or arms, many would die of exposure, starvation, or Indian attacks. The U.S. stock market would go crazy, then, once the truth came out, crash.

The geologists packed up and drove their horses day and night. Instead of trying to catch up with Berry, they navigated by stars and sun in a straight-line shortcut over impossible plateaus and canyons. They beat him to the rail, flagged down a freight to California, and, on the night of November 10, 1872, barged in on Asbury Harpending and Henry Janin.

The Great Diamond Hoax, as newspapers dubbed it, collapsed like a building dynamited. Faced with King's highly detailed report, the Golconda Mining Company was liquidated. Preachers, politicians, and editors acclaimed King a hero and great man of science. When the United States Geological Survey (USGS) was organized a few years later (four decades behind the Geological Survey of Canada), King was made its first director.

No one ever owned up. James Cooper testified against Arnold and Slack, but he had been pushed out of the plot early and went free.

Harpending had been seen walking with Arnold in London during one gem-buying spree, but he claimed he was there on other business. No one could prove otherwise of this rich and powerful man. Following the collapse, Harpending's fellow banker (and possible co-conspirator) William Ralston was found floating face down in San Francisco Bay, a possible suicide. Of the anthills, Harpending said indignantly: "They weren't anthills at all. They were fakes, the work of a sinful man, not the moral insect."

Phil Arnold blew town fast for Hardin County bearing $550,000. With the Civil War still fresh, here the deeper meaning of the Hoax emerged. Instead of being arrested, Arnold was hailed. "If two men have committed fraud, they have only been successful at out-Yankeeing the Yankees, or out-Heroding old Herod himself," said the *Louisville Courier-Journal*. Hardin was home to the most famous, deadliest southern ex-guerrillas; no northern police or capitalists touched Arnold here. Harpending later wrote a book intimating that the likes of "Spoons" Butler got just what they deserved.

In 1878 Arnold tried opening his own bank in the county seat of Elizabethtown, but a brawl with rivals on the street saw him shot-gunned to death. His handsome brick antebellum-style house still stands on East Poplar Street, and local legend says there are diamonds buried somewhere down in the cellar.

John Slack simply dropped out of sight. After he died, in 1896, his whereabouts emerged. He had quietly moved to White Oaks, New Mexico, to become a coffinmaker. At the time he occupied a piece of his own merchandise, his estate was $1,611.14.

The American mine had panicked Kimberley gem buyers, who feared it would flood the market. Some temporarily closed shop. When the Hoax collapsed, they reopened and the pandemonium multiplied.

Sixty thousand humans swarmed like ants over the dry diggings, where the mysterious "yellow ground" occupied four charmed circles of ten, fourteen, twenty-four, and thirty-two acres. Thirty-by-thirty claims became so valuable that diggers sold them in sixteenths, and everyone dug straight as plumb bobs, resulting in a devilish honeycomb, all cells in various stages of descent. The De Beers' onetime hill, with 1,600 separate workings, became known as the Big Hole—the greatest hand excavation in history, works of ancient Egypt and Babylonia not excepted. Miners dragged ore up through vertical labyrinths of ladders and pulleys as winds swirled lung-clogging fine tailings. Amid the heat, flies, and sewage, fistfights broke out in the claustrophobic pits. Narrow roadways between claims carried carts, but when drops penetrated eighty or a hundred feet, they often collapsed and sent people and animals to their death. Landslides lumbered from the sides, burying diggers alive. One man took over the claim of a neighbor who had mysteriously disappeared, and found him some time later, in a caved-in tunnel. With roads gone, claimholders strung wires to perimeter wheeltowers to carry buckets of ore, tools, and people, forming a monstrous cobweb. When they got deep enough to hit groundwater, everyone bailed to keep from drowning.

Worst of all, at sixty to a hundred feet, the yellow ground was giving way. Below was a greasy, slate-blue stuff—barren bedrock, everyone again assumed, "blue ground." Even if it did hold diamonds, mining looked fatal. Claim prices plunged, and by 1876 entrepreneurs were buying plots wholesale. These buyers included the son of an Anglican minister, Cecil Rhodes, who started out selling ice cream. When the pits flooded, he acquired the Cape's single power pump and made a fortune renting it out. His chief competitor was Barney Barnato, son of a London Jewish secondhand clothing dealer, who at first sold cigars and did standup comedy for the diggers. Barnato had heard a theory from a visiting geologist: Perhaps "the diamonds come from below through what was once a sort of tube in the earth." If this were true, they would never run out. The entrepreneurs gobbled claims and headed for a showdown, though they were not sure how they would keep going.

It was in 1884 that the restless California forty-niner Gardner Williams showed up. Williams was not interested in diamonds—he had come to look for South Africa's abundant, more dependable, gold—but he discussed with Rhodes how the Big Hole might be deepened without killing everyone. The smart, pragmatic Williams impressed Rhodes so much that Rhodes asked him to manage the operation. The salary offer was enormous, and in the end the American threw in his lot with the Englishman. Rhodes then carried out a lightning series of buyouts, maneuvers, and dirty tricks to gain control of the two De Beer mines. On March 13, 1888, De Beers Consolidated Mines was born, with only four main shareholders: Rhodes, Barnato, and two others, with Williams as manager. De Beers then systematically bought out the other two Kimberley mines. About fifty miles off were two more dry diggings, by this time recognized as similar; De Beers bought those too. Old-time diggers burned Rhodes in effigy for wanting everything, but it was no use.

In 1890 prospector Henry Ward observed some yellow ground a meercat had thrown out of a burrow near Kimberley, and staked out another dry digging. He did not want to sell, but De Beers searched for deficiencies in his title and, after a complex, bitter court battle, took over in 1891. Diggers Percival Tracey and Thomas Cullinan found yet another spot by tracing minerals up a streambed 300 miles north of Kimberley. At first De Beers thought they had salted it, but by 1902 the two had a gigantic pit and were offering real competition. De Beers governor Alfred Beit visited and was so shocked to see its scale that he suffered a paralytic stroke right there. His partners did not worry much, though. Cullinan and Tracey carelessly sold too many diamonds, depressing prices. De Beers, with its deep pockets, sat back for nine years, absorbed the declines, and watched them flounder. In 1911, De Beers bought them out.

The pattern was set. De Beers did not need to find diamond mines; it had only to wait for the little guys to do it, then move in. "Whenever you hear that a new mine has been discovered ... if De Beers are not there, they are very near," Rhodes boasted. A few smaller independent producers would survive, but De Beers would seduce or force almost all into a complex De Beers–controlled international arrangement that bought the stones and wholesaled them at set prices—the cartel.

With understanding of the mines' geology growing, Williams dynamited the Big Hole's dangerous old workings and built massive terraces and multilevel tunnels. He ordered new custom steam equipment, which made mining relatively safe and extremely profitable. Blue ground—which it seemed was just unweathered yellow ground—was spread a foot deep over a three-mile-long "farm," turned by tractors like a Kansas wheatfield and passed through gigantic washing plants, which disgorged diamonds not by carats, but by tons. Through subsidiaries and related companies, De Beers soon branched into gold, copper, tin, manganese, coal, fertilizer, explosives, vineyards, fruit orchards, breweries, banks, railroads, and automaking. When Cecil Rhodes died, his estate endowed the famous Rhodes scholarships. The partners and their descendants made a tight circle, ruled by powerful family loyalties and common memories of old digger days. The "one great corporation" foreseen by Henry Janin had come into being. But not on a windswept mesa in Colorado.

German geologist Ernst Cohen was the first to declare the deposits cylindrical magma columns erupted from below—Barnato's "tube in the earth." That was 1872, with the diggings at only 130 feet. The following year Australian geologist E. J. Dunn of the Cape Colony Geological Survey coined the term "pipes." The ore itself was variously called "adamasite" (from *adamas*) or "orangite" (for the Orange River). Henry Carvill Lewis, a geologist at Haverford College in Pennsylvania, came up with the name that stuck. Lewis, fascinated with diamonds, had traveled the States to investigate domestic finds, but never made it to South Africa. In 1888 someone shipped him African ore, which he described as "porphyritic volcanic peridotite of basaltic structure." He called it "kimberlite." He died soon after of typhoid fever, but the name survived.

Where kimberlite came from and how diamonds got into it became an instant raging debate, with the deepening mines providing fuel. At Lewis's death the Big Hole was 600 feet deep. By 1898 it had reached 1,800 feet; by 1905, 2,100, and the orebody was narrowing to a funnel shape. Alongside lay a perfect cross section of rock layers: 90 feet of basalt; beds of black, coal-rich shale 200 or 300 feet thick; 10 feet of pebbly conglomerate; 400 feet of hard olivine diabase; 700 feet of quartzite; then dark, heavy slate, continuing to unknown depth. Clearly they all had been pierced from below.

Kimberlite was a mind-boggling breccia of fragments seemingly from everywhere and nowhere, glued with a grainy blue matrix—the "blue ground," which actually also ran to green and black. In addition to diamonds, the matrix held xenoliths, or intact rocks ripped from elsewhere in the eruption. Some matched the rock on the visible sidewalls; others seemed to come from layers not yet exposed. The matrix also contained eighty-some types of free-floating mineral crystals, most of unknown origin: tourmaline, zircon, rutile, sapphire, epidote, olivine, enstatite, chrome diopside, smargadite, biotite mica, perovskite, magnetite, chromite. Particularly abundant were shiny metallic ilmenites and garnets of many colors—hyacinth red, pinkish red, yellowish red, golden yellow, reddish brown, bright orange, pale bluish violet, purple, cinnamon brown, dark brown, or brick red, apparently from all sorts of rocks. Very dark red garnets, tinged almost a bluish black like fresh venous blood, were especially numerous; some were as big as human heads. Some looked identical to the long-popular Bohemian garnets, so De Beers marketed them separately as "Cape rubies."

Garnet identification was very confusing. Jewelers alternatively labeled Bohemian, or Cape, garnets as "pyropes," from the Greek *pyropus*, or "fiery-eyed." The best did shine almost like rubies and in ancient times had been held equal, under the name *carbunculi*. Rubies and garnets could be hard to distinguish visually, so the still-developing new fields of chemistry and mineralogy now added more tests. Each gem—in fact, each garnet type—theoretically had its own chemical composition. However, mineralogists found this not strictly true. Compositions often varied—none more than garnet. All appeared built on silicon and oxygen, but they might also have aluminum, iron, manganese, calcium, or other elements stuck on. Pyropes seemed distinguished by a high content of the metal magnesium, but tests were inexact, laborious. By look, pyropes could be confused with other minor gems like zircon and spinel, or other garnet types, known at various times as spessartite, hessonite, essonite, rhodolite, almandite, grossularite, andradite, uvarovite, schorlomite, succinite, demantoid, vermeille, leucite, melanite, colophonite, Syrian, Chian, Trozenian, Corinthian, and Ceylonese.

It was known that *garimpeiros* and others had long used garnets and other minerals as indicators, but even now, with the "mother rock" staring everyone in the face, there was no telling which were the true indicators—those minerals formed with the diamond, as opposed to those that had simply tumbled in from elsewhere during the chaos of the eruption. And kimberlites obviously contained much extraneous matter: petrified trees, plants and fish, scorched ostrich-egg shells, and, in one, a whole fossil anthill, its business day cut short by the lava. The rarest constituent of all was diamond: in the Big Hole, 1 part in 2 million; in minor pipes, 1 in 40 million.

In 1892 a De Beers diamond sorter found a diamond partly grown into an adjoining pyrope. Some took this to mean that the two had crystallized together in the kimberlite; thus dark-red pyropes might be good indicators. Then, in 1897, at a minor pipe forty miles from Kimberley a miner found an ellipsoidal football-size xenolith. Rounded as if waterworn, it was made almost wholly of shimmering, sea-green chrome diopside crystals and garnets—but chiefly orange garnets. Sticking from various surfaces were ten diamonds. American geologist T. G. Bonney, an old friend of Henry Lewis, identified this unearthly-looking thing as an "igneous" rock called eclogite and said it was from a deep, unknown region. He concluded kimberlite "is not the birthplace either of [diamond] or of the garnets [or diop-sides] . . . it incorporates. . . . Eclogite [is] the parent rock." Kimber-lite magma must have snagged some on the way up, he said. If true, orange garnets and green diopsides were indicators. He could not prove this seductive theory; more eclogite turned up diamondless; and some diamond-rich pipes held red garnets, but no orange ones.

Scientists agreed now only that diamonds came from great depths. Gardner Williams retired to San Francisco in 1905 to write his multi-volume *The Diamond Mines of South Africa*—long the world's main reference on diamonds. It demolished every theory. "I have been fre-quently asked, 'What is your theory?'" wrote the old man. "I have none. . . . All that can be said is that in some unknown manner car-bon, which existed down deep in the infernal regions of the earth, was changed from its black and uninviting appearance to the most beautiful gem which ever saw the light of day."

Williams had a son, Alpheus, who grew up around the mines and inherited the managership of De Beers—a post he held for thirty-some years. Alpheus ended his own career by writing an update of his father's book, *The Genesis of the Diamond*, with far more infor-mation—but he was equally as puzzled. The Big Hole had long petered out into a deep, narrow fissure filled with water and was closed for good, but still Alpheus Williams could not figure out how diamonds were made.

He was perhaps missing the point. To the countless small-time prospectors buzzing around his periphery, the central, holy mystery was not how diamonds were made. It was how to find more.

It did not take long for the Americans to find their own kimberlite—the real thing this time. Before typhoid took Henry Lewis, he mentioned a startling fact: He knew of some apparently identical rock close to home—in backwoods Kentucky and in Syracuse, New York.

In Syracuse a local professor saw it in 1837, when a subsoil excavation was dug on Green Street. Everywhere else was limestone; in this hole jutted a dike of something else. The hole was long covered up now, but Lewis had a sample: Dark green, full of xenoliths and all the right accessory minerals, it was the undeniable twin of kimberlite. Then, in 1885, geologist A. R. Crandall was investigating a shale formation in Elliott County, Kentucky, and found several small igneous knobs spread along a mile near rural Isom's (sometimes called Ison's) Creek. He sent some to Lewis. There it was again.

This caused huge excitement. Shortly Syracuse dug new sewer lines, reexposing the rock in a series of fingerlike dikes. Local amateurs mapped a "south crater" 800 feet by 1,200 feet along Green and James, and a "north crater," running about 1,200 feet around Griffith and Highland. Many jumped into the sewer trenches to carry off chunks. People dug up their own yards. City businessmen proposed investing $50,000 for a mine, but soon a problem loomed: No diamonds could be found, and any further digging had houses and streets in the way. The sewers got covered back up to do what sewers do.

Hope never quite fled. Around 1900 one Fred Patterson was screening sand in a quarry near Brighton Avenue when he found a diamond, which he sold in Springfield, Massachusetts, for $1,700. At least that was what he said; the next year Patterson found a smaller crystal and drew a local offer of only $50. A Syracuse geologist believed it was quartz. Several years later Syracuse University mineralogy student Frank Brainerd was digging along Green Street and found what the local paper called "one simon pure diamond," extremely tiny. For a second opinion, his academic adviser sent it to the office of a more expert colleague in New York—where it was promptly mislaid. It was never seen again. In sum, no one knew if diamonds had been found in Syracuse or not.

Reaction in Kentucky was more dramatic. When word of the rock there leaked in late 1886, farmers who owned the knobs had to fend off hundreds of invaders who combed surrounding areas for precious stones. Soon Kentucky newspapers were screaming about "Kentucky diamonds," and the search extended to every corner of the state. "It was pretty soon believed that all one had to do to find diamond was to cross the line into Kentucky," wrote John Branner, a geologist in nearby Arkansas. Enthusiasm grew so much that farmers there abandoned their land to hunt Kentucky gems. Branner toured Arkansas for months trying to convince people to stay home. "The amount of time, labor and money squandered on this wild scramble nobody knows," he wrote.

In summer 1887 a colleague of Crandall's, Joseph Diller, went with a mineralogist to inspect the Isom's Creek pipes. They were decaying rockpiles no more than 30 feet high and 130 feet across. James Maggard, owner of the biggest, had dug into debris next to one and discovered something strange: Indians had preceded him. In the refuse was a trove of stone arrowheads, chisels, and fireplace ashes. No one knew why. The scientists systematically dug the narrow, wooded gullies around, deputizing locals to comb the high ground for gems. They found yellow ground on top, blue below, and determined the rock was 8 percent pyrope garnet crystals, many fine enough for jewelry, plus diopsides, ilmenites, and tiny unidentified pale-yellow spheres of uniform size. The spheres looked like diamond, but were not; they dissolved in hydrochloric acid. The farmers leased the pipes to a succession of entrepreneurs, including one Austin Q. Millar, a coal-mine geologist from Ohio. Over dirt roads Millar and his teenage son hauled in three-ton African-style washing vats, dug an eighty-foot shaft, washed 50,850 cubic yards of material, and used so much water that they ran Isom's Creek dry and had to stop.

Naturally geologists ransacked old literature for other odd igneous intrusions and soon came across the 1837 report of New York's Lardner Vanuxem. Geologists headed for tiny Ludlowville, fifty miles southwest of Syracuse, and spotted Vanuxem's veins, plus fourteen others he had missed. They were crumbly things, a few inches to a few feet wide, cutting straight up through 400-million-year-old shales and siltstones. In nearby Ithaca there were four in Cascadilla Gorge, on either side of the Central Street bridge bordering Cornell University. A company formed to mine one dike but found no diamonds.

In California, it was the reverse: They had diamonds, but no kimberlite. Prospectors traced up creeks around Oroville and Cherokee, Gardner Williams's old haunts. The gold was gone, weathered wooden mine buildings turning ghostly. Hydraulic hoses had ripped once-lovely creeksides for miles, leaving only wide, lifeless tumbles of boulders. In one of these wastelands, prospector M. J. Cooney found what he thought were kimberlite pebbles, traced to hosed-off outcrop on the west bank of the Feather River. He formed the United States Diamond Company and sank shafts. The new USGS reported the rock did look like kimberlite—but was actually common serpentinized amphibolite schist. Cooney refused to listen; he continued digging for nine years, reaching 300 feet. A few small diamonds did turn up—all from nearby placer workings run by diehard forty-niners. After Cooney folded, one old miner, Alexander Wilson, recalled a tunnel he had dug forty years before into a local butte, the Sugar Loaf. He said it contained a buried chimney of bluish rock, and the foolishness started all over again.

At this point, the United States finally had an honest, competent diamond expert: mineralogist George Frederick Kunz. Kunz stated the obvious: "It has not been proved that [where] kimberlite . . . exists it must carry diamonds." By 1905, worldwide searches had revealed 250 kimberlitelike formations, including a dozen in the United States. Most lay in southern Africa, where most searching took place; others lay as far away as Kakanui, New Zealand. Only African ones yielded riches, and only a handful of them. Coal miners southeast of Pittsburgh, Pennsylvania, hit one in the middle of a tunnel 300 feet underground. But coal intersected by kimberlite did not become diamond. It was baked and ruined for twenty feet around.

Through a sixty-year scientific and commercial career, George Frederick Kunz assumed the role of muse of North American gems, though a somewhat ill-fated one. He believed the land was full of them, and he went everywhere looking. As top expert, it was he to whom the Syracuse "diamond" had been sent, he who accompanied Joseph Diller to Isom's Creek, he who inspected the California "kimberlite." Despite disappointments, he believed that someday the big strike would come. When he was old, he prophesied it would happen not in his own land, but in unexplored northern Canada. He had his reasons.

Kunz grew up across the Hudson River from New York City, in Hoboken, New Jersey. He remembered Hoboken as precursor to Coney Island—a cheap resort where hoop-skirted women crowded dance floors with faces ablaze, little boys vended poisonously brilliant penny ices, and, beyond, "in the green and still beautiful distance" baseball players hollered to each other. Here young Kunz walked alone and oblivious, eyes glued to the ground. At ten, he had visited Phineas T. Barnum's American Museum on Broadway. Amid the Siamese twins and bottled embryos was a case of exotic mineral crystals. Hanging over them, he felt suffocated by some unknown pleasure. The next day he began his own mineral collection, and with exquisite timing. New York above 42nd Street was then one large construction site, with rail cuts and foundations everywhere. Manhattan's hard, stable rocks are filled with minor gemstones of all kinds. During this brief, fabulous window, Kunz filled his pockets. By the time he was nineteen, he had two tons of material, trade partners all over America, and his own huge specialized library.

Kunz was a self-made genius of underappreciated domestic geology. He was smitten not merely with minor gems, but more particularly with American ones. No one had given them much thought, but in 1876 at the age of twenty, Kunz visited Charles Tiffany and sold him a fine tourmaline from Maine. Tiffany saw the business potential here, but recognized Kunz as the real find—a youngster with a God-given talent for appraising any stone at a glance, up to and including diamond. Still stinging from his role in the Great Diamond Hoax—it was just four years previous—he hired Kunz. Within three years he was vice president and chief mineralogist. Kunz remarked that the Hoax might never have happened had he been around. He was so good that the USGS gave him a simultaneous appointment as "special agent" in charge of gems, responsible for annual reports on all U.S. activities. He and his wife, Sophia Hanforth, had three daughters, the eldest named—naturally—Ruby.

Kunz wrote that Europeans "sailed all the way to the New World to open up a new India, [rich] in precious stones"—an allusion to Cartier—yet the continent remained "poorest of all." There was a cornucopia of minor gems, his first love: opals, turquoises, whole fossil-wood forests. Montana had sapphires, North Carolina a small, genuine emerald mine. In upstate New York, quartz crystals were playfully sold as "Lake George diamonds." During the 1890s on Tennessee and Arkansas rivers, Kunz witnessed Klondike-like rushes when it was realized native shellfish carried pearls—a discovery that dramatically advanced the extinction of many American freshwater species and wiped out the pearl booms as fast as they came. Kunz even found the Navajo anthills that supplied the Hoaxers' salted pyropes; garnets lay in them by bushels, mined by ants from parts unknown. They were eventually traced to the nearby buttes—diamondless kimberlitic rocks. Kunz patriotically opined that clear, blood-color Navajo garnets "worn in the evening" far outstripped any Cape ruby. But he had to admit: The Africans had the diamonds. Why had America been cheated of its favorite jewel?

Kunz assembled histories of all the random finds to date, tracked down specimens, bought them for study, and visited sites of kimberlite real or suspected. He investigated each report with the same care, taking extra care to prevent any more scams. He was on the trail.

In 1886, twelve-year-old Willie Christie found his $4^1/3$-carat diamond at a spring in Dysartville, North Carolina. Kunz received the stone by mail, certified it as real, and was soon on the scene at the Alfred Bright farm in Dysartville. He took an affidavit from Willie. Then he took an affidavit from a local jeweler who had examined the stone. Then he took affidavits from a mayor and judge to attest to the characters of the other affidavit-givers. He compared the stone and found it looked like other North Carolina stones. Satisfied it was not planted, Kunz bought it for $150 and preserved it rough for scientific study, as he always did. (It is not recorded whether young Willie received the cash himself.) Kunz then removed all the sediment from the bottom of the spring, sifted it, and dug at hollows in the adjacent hillside for traces of kimberlite or garnet, which he suspected had some unspecified connection. He found nothing. All he could say was that the diamond had been transported from soil or rocks upstream, and he urged the locals to carry on.

Unfortunately, they often did so too enthusiastically. Near Dysartville and other diamond locales, Kunz found many "diamonds" hopefully displayed in homes—transparent zircons, smoky quartzes, and bits of glass worn smooth by running water or the gizzards of chickens, in which they were sometimes found after the chickens were slaughtered. Kunz dreaded this part every time. In one town he met a spinster who for thirty years had kept a "diamond" for her retirement. Another family had one saved to send their son to college. James M. Smith of Gibsonville, North Carolina, found what a local "expert" said was an emerald with diamonds inside. Refusing $1,000 for it, he made an expensive train trip to New York, where Kunz told him it was quartz and byssolite, worth $5. Every time he watched people's faces fall, he went home feeling crushed and guilty, like a doctor who has just informed a patient a fatal cancer is growing inside him.

He felt less sorry for the endless stream of people he encountered eager to capitalize on diamond mania. In one collection near Dysartville he spotted real diamonds and learned the owner hoped to sell his land for a high price. These did not look like North Carolina diamonds; they had the characteristics of South African industrials. "It is to be hoped that the few legitimate finds which have actually occurred at this locality will not lead to any deceptions, which would greatly retard any natural development," he wrote in his 1886 USGS report. He courteously omitted the offender's name.

To promote honest prospecting, Kunz approved of a scheme conceived by J. Adlai D. Stephenson of Statesville, North Carolina: deployment of children. Just like the permanently ten-year-old Kunz, kids were always picking up things off the ground. Stephenson was carrying out "the happy idea of interesting the children of his vicinity in the search for minerals. A trifling reward [is] sufficient to awaken a keen interest, so that healthful exercise certainly, and often valuable specimens, are the result. [Many finds] we owe to the[ir] industry and sharp sight." To promote amateur identification, Kunz seconded the idea of Boston jeweler Dwight Whiting: $5 or $10 rings mounted with uncut industrials, aimed at familiarizing wearers with the look of rough stones and providing a ready surface against which suspected gems could be scratched for hardness. "Several thousand searchers thus prepared would soon ascertain whether diamonds really exist" in any locale, he said. Kunz tried getting farmers around Isom's Creek to sport them. In California, some ditch diggers did take to wearing them, just in case.

As the nation's population grew and more ground was turned over, Kunz was busy. He authenticated a diamond found by a Chinese man near Deer Lodge, Montana, and a beautiful one of 4.25 carats, found by a little girl in a pile of garden dirt by her house in Shelby, Alabama. In 1899 he recorded Tennessee's first proven stone—perfectly white and flawless, found along the south bank of the Clinch River. It lay next to an ancient Indian mound. However, he denounced a 3.5-ounce "gem" found by one J. S. Keyser while digging coal near Ponca, Nebraska, and suggested that a diamond found by a hunter while drinking from a Missouri brook had somehow arrived from Brazil. He tracked wild rumors of diamonds at Mount Edgecombe, Alaska, to a geologist who had merely mentioned rocks resembling kimberlite. Kunz helped show that Arizona's Canyon Diablo meteorite held diamond dust, presumably created on impact, by using bits of the meteorite to polish a diamond in front of flabbergasted onlookers at the 1893 World's Columbian Exhibition. In his spare time he investigated freshly fallen meteorites and made the world's first major collection of them, though he found no more that contained diamonds.

Nothing offered a consistent pattern until a new diamond region suddenly appeared: the northern Great Lakes states. By bits and pieces, it emerged that people had been finding diamonds for decades there. And if you plotted the finds, they traced the lower edges of the glaciers.

Kunz at first believed the Great Lakes diamonds to be another fraud, and he may have been half right.

In 1876 tenant farmer Charles Wood was deepening a well in Eagle, Wisconsin, southwest of Milwaukee, when seventy feet down in glacial debris he found a wine-yellow stone the size of his thumbnail. His wife, Clarissa, kept it on a clock shelf for seven years. When her husband died she took it to Milwaukee jeweler Colonel Samuel Boynton, who told her it was a topaz and paid her a dollar. It was a 15.37-carat diamond. Within months Colonel Boynton bought the farm and told people he intended to raise chickens, but when he started digging with partners, including Milwaukee's mayor, the truth leaked out. People talked of renaming Eagle "Diamond City," land prices quadrupled, and Clarissa Wood sued. Things soured fast, though. Underground water gushed in, halting operations. The diggers claimed to have found more stones, but Kunz pronounced one he saw "African" and said the well was in "ordinary glacial drift." The whole thing, he said, reeked of "very peculiar circumstances." Possibly Boynton had salted the place with more stones, but Kunz was missing a crucial puzzle piece: Clarissa Wood's stone, which Boynton kept after retiring from the scene in considerable disrepute.

Ten years went by. In 1893 geologist William Herbert Hobbs of the University of Wisconsin in Madison read in local papers of a new find. Five-year-old Stanley Devine was playing in a rocky field in Oregon township, fifty miles west of Eagle, when he found a 3.83-carat stone at the base of a cornstalk. Amid Stanley's tears and howls of protest, his father confiscated it in exchange for a penny. Hobbs authenticated the stone, then started poking around and discovered other finds that had not made the papers. Fifty miles northeast of Eagle, near Saukville, Hobbs met a German farmer who had picked up a large stone in 1881 while plowing with his horses, but kept it unidentified. He discovered a huge one, 21.25 carats, found by a farmer in 1886 in nearby Theresa. Also, from 1887 to 1889, gold panners 200 miles west had found tiny diamonds in the bed of Plum Creek, Pierce County. Hobbs went back to Colonel Boynton and documented the diamond from the Eagle well. Then, the state geologist in neighboring Indiana decided to investigate old lore. He confirmed that as far back as 1863 gold panners in central Brown and Morgan counties had found small diamonds in glacial drift. They had kept quiet, apparently heeding the words of the previous state geologist, Maurice Thompson: "Every person who claims to have discovered in Indiana . . . gemstones should be treated with the utmost caution. He is dangerous if he is not ignorant, and if he is not crazy he soon may be."

Hobbs publicized his research, and this brought out new finds. In Dowagiac, Michigan, a building contractor found one in a gravel pit. Outside Cincinnati, Ohio, two young sisters found one in a glacial moraine. In 1899 Hobbs wrote a long paper, demonstrating that all the diamonds lay in a great arc of glacial debris 600 miles long and 200 miles wide—the outer edge of the last ice sheet that presumably had come from Canada. He theorized they were scraped from pipes, carried down, and dropped when the ice retreated. That any had been found at all in the vast piles of debris signaled to him a huge lode somewhere along the path.

Kunz admitted he had been wrong. In Indiana he bought up gold panners' specimens. He paid $50 for the diamond found by Stanley Devine in the cornfield. Then he went back to Colonel Boynton, who still had Clarissa Wood's stone. It was of only fair quality, but of great historical value. Kunz forked over $850.

Egged on by Kunz, Professor Hobbs led the charge to trace the source. He sent notices to every newspaper for hundreds of miles around to arouse interest. Like Kunz, he suggested "children particularly be urged to use their keen eyes" in search of more stones, and enlisted a corps of geologists to identify them. Foremost, he said, was to study "directions of glacial movement . . . to discover the law of the [diamonds'] distribution and to glean . . . the ancestral home."

Across the northern border, the GSC was in fact just now publishing rudimentary maps of glacial movements. Their main man in the field was none other than Joseph Tyrrell, the explorer who in 1893 had found the starting point of the Keewatin, or North Wind, glacier in the faraway Barren Lands. Based upon such work, the GSC rapidly determined that part of the northern ice had descended from around Hudson Bay, then split into two massive lobes that ended up on opposite sides of the state of Michigan. Hobbs had established that diamonds lay at the termini of both lobes. Assuming only one diamond source, that meant it must lie at or above the apex where the lobes had calved off like spokes on a fan. Slightly to the right of the Keewatin, on the east side of Hudson Bay, lay another big ice spreading center, the Labradorean. This was where Hobbs believed the pipes lay.

Others were not so sure. There was no proof the diamonds had only one source—and so no reason the pipes could not lie further south. Also, it was acknowledged that previous glaciations had probably taken place, each wrecking traces of its predecessor. Any one of them could have transported diamonds from anywhere. Theories soon abounded. Some thought glacial diamonds could be tracked via boulders of red jasper conglomerate spread through the northern states. These seemed to come from north of Lake Huron, along Ontario's Ottawa River. Others were interested in chunks of native copper, thought to be transported from the shores of Lake Superior. Yet others believed the diamonds had not traveled at all, but had bled up from kimberlites buried under glacial till on the American side. After all, no one had yet found a diamond in Canada.

Dr. Robert Bell of the GSC liked the idea that diamonds might be found on his side, but he found it unlikely glaciers could carry anything from the tundra to a Wisconsin cornfield. Besides, he said, Tyrrell and others were showing "the whole of [the north] consists of monotonous primitive granitoid gneiss which is peculiarly barren of minerals of any kind." Bell favored more convenient southerly sites. The GSC already had a brief 1888 report mentioning kimberlitelike rocks in Ontario's Rainy Lake region, just above Minnesota, and Archibald Blue, director of the Ontario Bureau of Mines, mentioned volcanic dikes intruding carbon-rich shales. "I think we ought to look for diamonds [in Ontario] and expect to find them," said Blue.

It so happened that the Canadians were surveying for their own transcontinental railway at this time. It ran north of the Great Lakes, conveniently across the paths of all the glaciers—a fantastic opportunity to intersect evidence. In 1905 Dr. H. M. Ami of the GSC went to railroad camps and instructed over a hundred surveying and construction parties how to spot and identify diamonds. Everyone kept their eyes open, but that year a bombshell went off. Canada's hopes seemed dead.

Newspapers announced that in Plum Creek, Wisconsin, where diamonds had been panned decades back, an experimental shaft had hit a huge diamondiferous kimberlite thirty feet under water—the apparent source. The Minneapolis-based American Diamond

Mining Company was organized and bought up mineral rights to 2,500 acres. The newspapers said they now only needed to raise money for caissons. It was projected that $100 invested would easily grow to $100,000. The owners had intended to sell shares for 25¢, but at the last minute, decided to let people of moderate means become rich quickly. And so, for a limited time only, it was 10¢ a share—12$\frac{1}{2}$, on the monthly payment plan.

The reports were thinly camouflaged advertisements, complete with clip-out coupons. ("Of what use to you is your $100 in a bank? If you die what will it buy your widow or children? We are careful, sincere, cautious men, and have worked hard for what money we have. It has come to be a saying that one can't lose in diamonds, and after what we learn of the tremendous fortunes made in South Africa, we believe that statement to be literally true.") Among endorsing experts: "Mr. Geo. F. Kunz, gem expert of Tiffany & Co."

Plum Creek was, of course, just another American criminal enterprise made from whole cloth. There was no kimberlite, no diamonds, no connection with Kunz.

Kunz was tired. He calculated that in one generation South Africa had produced twelve *tons* of diamonds—twice those from all previous human history. From 1867 to 1905, Americans imported an astounding $333,111,094 in precious stones, including the majority of those African diamonds. Total domestic diamond finds: $10,000. And prices were doubling, despite skyrocketing supply; the laws of free enterprise did not seem to be working. De Beers Consolidated Mines, holding 97 percent of the market, had a stranglehold on its biggest customer.

By now Kunz was an international star. He went about calling himself Doctor, having several honorary degrees to choose from. He had hundreds of scientific papers to his credit and a newly discovered California gem mineral—kunzite—named after him. He had bought great gems in Russia, Thailand, Mexico, and India and entertained visiting royalty in New York. In his apartment at Riverside Drive and 86th Street he contemplated a turquoise-encrusted human skull and collected prominent people to discuss his quest: banker J. P. Morgan, who helped finance some of Kunz's major American gem collections;

inventor Thomas Edison, who collected them himself; writer Oscar Wilde, who cooed over samples ("But, my dear Kunz, these are exquisite, charming!"). "Wealth may still lie at our back door, and any day a new vein of gems may be opened up by the native who searches earnestly and intelligently," Kunz insisted. Only President Theodore Roosevelt stopped him. At a New York club one day, Kunz started in about American gems, and Roosevelt sat for a moment looking puzzled. Then he roared, "You're talking to the wrong man, Kunz. Ivory is my life, Doctor, ivory!"

In early July 1905, as Kunz was trying to figure out what to do next, a lean, absentminded-seeming professor loaded with test tubes and books came through the worn-out silver town of Tonopah, Nevada. Going by the name E. W. Hews, he mumbled vague, impressive formulas about diamonds and local geology. On July 22 the *Tonopah Sun* announced Hews had staked a $5-million diamond claim three and one-half miles from town. Hews's diamonds, clear blue and radiant, went on display in the window of the Miner's Drug Store. Within three days Hews laid out a new boomtown: Kimberley, Nevada. Townspeople were wary at first, but soon many bought interests for lack of anything else to do, built sagebrush huts, and armed themselves against claim-jumpers. A tent saloon went up. Just then John Hassell, a mining engineer back from a stint in South Africa, was passing through, and Hassell and a South African friend obliged requests for an educated opinion. The two men heated one of Hews's diamonds slightly, dipped it in water with tweezers—and watched it dissolve. Hews had dissolved, too, with the cash. Kimberley, Nevada, had lasted exactly one week.

It was hard to believe anyone still fell for it. But more telling was the reaction afterward. Soon a conspiracy theory floated: The fraud had been perpetrated not by Professor Hews but by Hassel and his friend. Locals developed the idea that De Beers had sent them as counteragents to discredit what was in fact a real diamond field. The theory went that when everybody pulled up stakes, De Beers would slip in and take the ground. Some people truly believed this even though soon nothing was left of Kimberley but broken whiskey bottles and torn scraps of canvas. It was a measure of how deeply the Americans now suspected the suppliers of their own obsession.

CHAPTER 7

Crater of Diamonds

Then a real diamond pipe surfaced.

The finder was none other than John Branner, the Arkansas geologist who had tried to keep farmers from ruining themselves on the barren Kentucky kimberlites. For his good deed, Branner was appointed Arkansas state geologist, but then, probably to his horror, he discovered a predecessor's 1842 report: an unidentified igneous intrusion amid miles of otherwise sedimentary rock. Near the muddy, isolated town of Murfreesboro, he found it: three wooded knobs of rubbly rock covering about twelve acres. In 1889 he identified them as kimberlite—but pointedly added there was no evidence of diamonds. He spent hours combing the muddy gullies for them, but later admitted he had not been very systematic; the last thing Arkansas needed was another rush. Most people were tired by now anyway, and the rocks were soon forgotten.

Around Murfreesboro lived a gangly, harelipped sharecropper's son with a speech impediment named John Wesley Huddleston. Given to far-off looks and surliness to strangers, he spent most of his time in the woods hunting game, medicinal plants, and nonexistent gold. Even his deeply rural neighbors thought he was a little touched. Seventeen years later, in early 1906, the illiterate Huddleston gave a mule as down payment on a farm containing part of the knobs and started panning for gold. On August 1, 1906, he came up instead with two large crystals from the black soil down near Poor Man's Branch. The grindstone he used to sharpen his pig-butchering knives did not hurt them. A Little Rock jeweler forwarded them to Kunz, who confirmed they were 3- and 4.5-carat gem-quality "deemints," as Huddleston called them.

Arkansas businessmen optioned Huddleston's farm while Kunz and a Smithsonian Institution petrologist, Henry S. Washington, rushed down, taking the Iron Mountain Railroad as far as it would go, riding a logging train through hilly forests, and finally hitching rides on farmers' wagons the last miles on unpaved roads. Test borings were already underway, showing the knobs extended underneath Huddleston's fields and those of the Mauneys, next door. At least 140 diamonds had been taken from the top forty feet of soil and crumbled rock; more convincing, Kunz and Washington witnessed diamonds taken from the intact rock itself. They inspected thoroughly, and the evidence was incontrovertible: A source had at last been found. Of the diamonds, Kunz wrote: "They are absolutely perfect and are equal to the finest stones [of Africa] or that were ever found in India. . . . [E]conomical extraction [should be entirely] successful." His only reservation: White Arkansas laborers probably would never submit to the body-cavity searches and purging with emetics routinely inflicted on blacks working African mines. That left Murfreesboro vulnerable to theft.

Kunz and Washington reported for the syndicate and helped draw up a mining plan. A corporation was formed, the Arkansas Diamond Company, to take up the option. Along with the Arkansans, it appears that the in-laws of Kunz's eldest daughter, Ruby, came in as ground-floor stockholders, and Kunz himself probably invested. He was so excited that he quit his twenty-five-year tenure at the USGS, presumably to avoid conflict of interest.

Huddleston sold out for $36,000—all $10 bills, he insisted—and moved to town, where he kept the money in a safe. He couldn't work the combination, so he left the door open and just told people not to touch it. He acquired various pieces of real estate, the county's first automobile (a Ford Model T), and a new blonde wife, whom locals referred to as a "carnival girl" he'd met in Arkadelphia. Of an evening the new Mrs. Huddleston drove the car repeatedly around the town square while "Diamond John" waved delightedly to all. One day he disembarked to buy a cigar, and the new Mrs. Huddleston drove west and just kept going. Miraculously she left the cash, but Huddleston ran through it quickly without help.

The main orebody was mapped at seventy-three acres, and soon other farmers discovered apparent outliers: separate plugs and dikes two or three miles away, disguised by heavy oak and pine forest. Within months a half-dozen small companies and family operations were stripping trees and sinking holes.

Ten thousand diamond seekers descended on Murfreesboro that first year. Next door to the Arkansas Diamond Company, the Mauneys built a large washing plant of homemade design, fenced in their portion of the pipe, and charged diggers 50¢ a day to keep any diamonds they found. During popular nighttime "diamond hunts," picnickers paid a whole dollar to roam woods and fields bearing candles. Tent cities sprouted in neighbors' fields, so the Mauneys founded the misspelled town of Kimberly, Arkansas, and sold lots at ridiculous prices. It soon sported stores, restaurants, and a three-story hotel. Old man Walter Mauney advertised by having the dentist embed a diamond in one of his eye teeth and smiling a lot. Walter Jr. replaced his coveralls with fancy riding breeches, a jacket, a string tie, and a massive-brimmed snow-white hat, which he wore all the time, including when he went to milk the cows.

The Mauneys and most outlying farmers seem to have produced at least a few diamonds, but they were suspicious of outsiders and tax-men, and so refused any accounting. The Arkansas Diamond Company, on the other hand, had a serious mine. At the outset the company produced at least 1,400 stones, worth $12,000. The problem was, no one knew how to run things; the expertise was all in South Africa. Work was confined mostly to the regolith, the top layers of soil and loosened rocks, from which hired laborers pulled tree stumps, augured holes, and dug trenches with mule-drawn scrapers.

The sorting plant was so crude and security so poor that many stones were probably left in the tailings or stolen. True to Kunz's fear, rough diamonds soon floated for sale all over Murfreesboro. In 1908 the company hired a man with the proper credentials: John Fuller, an American engineer who had run De Beers's Dutoitspan pipe in the real Kimberley. Fuller took over forcefully but warned: "It is impossible to make assays of a diamond mine. Bore holes [and trenches] only give an indication of the nature of [kimberlite] and tell nothing of its richness. The only way . . . is actually to mine." The stockholders put up $250,000 for a more sophisticated washing plant, and Fuller set to larger-scale excavating.

From here on, nothing went right. The plant never seemed to work correctly. Tens of thousands of dollars were wasted. Thousands more diamonds appeared, but for all the money, they were scarce and small—less than half a carat on average. Fine, big ones occasionally surfaced, but many seemed burnt, industrial-grade. Instead of selling good ones, the company stockpiled. Capital started running out, and no one wanted to put up more. By 1912, the mine was forced to shut. Only a watchman and his assistants were left to sift for occasional stones. The Arkansas Diamond Company had collapsed.

Aside from incompetence and thievery, many geologists saw another reason: The pipe had at first seemed richer than it was. It appeared the top had weathered, releasing diamonds to the regolith. Then erosion washed away most light elements, leaving diamonds and other heavies distilled near the surface. Below, the source rock was quite poor. However, some townspeople and state officials could not accept this. They suspected what ex-residents of Kimberley, Nevada, suspected: The pipe was sabotaged by De Beers. This time, if you assembled the evidence just right, it did seem suggestive.

For one thing, John Fuller came straight from De Beers. If he was an expert, why was the plant a disaster—unless he made sure it was? For another, in summer 1910, with the mine still running, the company's president, Little Rock banker Samuel Reyburn, had gone to New York to raise more money. Here he was approached by two men from London mining financiers Lewis and Marks—a close ally of De Beers. They offered him a $50,000-a-year job with their firm and said they would give $500,000 for 51 percent of the mine. All he had to do was walk away—the implication being they would shut it once they had it. The sale never happened, and Reyburn always insisted there was no secret deal. But after this, the company rebuffed approaches by others to join forces, and watchmen were cautioned to say nothing of the diamonds they kept finding. When a USGS geologist came to study the mine, they wouldn't let him in. Several years later Reyburn took a lucrative job with another financial firm that had by now closely allied itself with De Beers: J. P. Morgan, started by Kunz's onetime patron.

In 1919 the stockholders of the Arkansas Diamond Company managed to reorganize and hired a new engineer, Stanley Zimmerman, to reopen. But before Zimmerman did anything he traveled to South Africa, where he met—only by chance, said Zimmerman—Sir Ernest Oppenheimer, De Beers's soon-to-be chairman. When Zimmerman came home, he suddenly condemned the mine as hopeless. This brought a second collapse. Most of the stockpile of 3,000-some gem-quality stones was spirited to Tom Cochran, a J. P. Morgan vice president who now somehow held a mortgage on the assets. One was the largest diamond ever found in North America—a faint rose gem of 40.23 carats, later dubbed the Uncle Sam. By some sleight of hand, Cochran "auctioned" the stockpile to himself for $19,000. New York jeweler Schenck & Van Haelen later acquired the bulk; the Uncle Sam alone, cut into three pieces, was valued at $75,000. The tortuous conspiracy-theory thread led nowhere, though: Nothing was ever tied to De Beers. Besides, most neighboring outfits went bankrupt on their own.

There was one short-lived exception: Austin Q. Millar and Howard Millar, the father-son team who once fruitlessly mined Isom's Creek, Kentucky. This time the Millars refused to give up. Starting in 1912 they rented or bought various properties around Murfreesboro, including the Mauney place, and sank their own capital and sweat into an elaborate mile-long ore railroad and state-of-the-art washing plant. Over seven years, the hardworking Millars produced thousands of modest stones. But someone did not like the Millars.

Unknown arsonists tried burning them out several times, until finally their insurance was canceled. Howard Millar kept ducking mysterious blasts of buckshot from the woods, and once narrowly missed getting blown away while sitting in the outhouse. One week the flow of diamonds stopped so abruptly, Howard suspected the sorters. He strapped a .45 under his hunting coat, strode into the plant, and fired everyone on the spot. He did the sorting himself and the flow resumed, but misfortunes kept coming. One landowner took them to court to get his property back—thirty-six times. The finishing blow came on the night of January 13, 1919, when an unknown woman lured the night watchman off the job. Around 10 P.M. the Millars' washing plant, crusher, pumphouse, manager's house, and railroad went up in flames. In the morning they surveyed their last batch of crushed ore, baked like a giant brick in its vat, and knew they would never rebuild.

It could have been just a rural feud of locals against outsiders, but some people insisted: De Beers had carried its plot to the bitter end. Howard Millar said that many years later, a local man in the final throes of tuberculosis called him to his deathbed and made a confession. He had been paid $50 and a new pistol to set the fire by the man behind the plot. He whispered the name, but no one could ever get Howard to repeat it.

During the 1920s the Arkansas Diamond Company once again reacquired control from creditors and put up $125,000 for a small pick-and-shovel operation. They recovered $30,000 in diamonds, and by 1932 were shut again. Kimberly, Arkansas, dwindled to a ghost village of fifty-six aging inhabitants. After the buildings burned, fell down, or were moved into Murfreesboro, a cow pasture reasserted itself along winding State Road 301, where it once had been.

It seemed like the end for Kunz. Early, when hopes were high, he had personally acquired some of the most beautiful Murfreesboro diamonds—exceptionally pure, pellucid fragments, like tiny stars. For Tiffany's he bought a great many weird, beautifully shaped irregulars, ranging from pure white to lemon yellow, gray, cinnamon brown, and mahogany. The firm cut them and drew extraordinary prices from patriotic buyers wanting to own American diamonds. Kunz collected and preserved many rough, for posterity. Then, as operations halted in 1912, almost simultaneously Kunz's beloved wife of thirty-three years, Sophia, died. Two of his three daughters followed; his youngest, Elizabeth, was struck and killed by a runaway horse. Only Ruby remained.

New discovery sites were tapering too, and precisely in 1913, they ceased. In 1911 and 1912 lone prospectors picked up two small stones in Texas—that state's first and last. The last from Wisconsin were reported on January 19, 1913, in the *Milwaukee Sentinel*. A hermit named Peter Zagloba had passed away near the hamlet of Collins, and neighbors cleaning out his things found a handful of rough gems inside his coffeepot; where he found them, no one knew. The USGS kept issuing yearly gem reports for another decade, but there was nothing to report. In 1923 the reports were suspended.

It is perhaps no mystery what happened. Fields once new to the plow had been turned for decades; most easy finds had been made. Farmers no longer followed on foot behind plowhorses, laborers no longer dug with shovels; a man on a tractor did not see much. One-mule prospectors disappeared, too, for placer gold was gone. The nineteenth century had been an era of great rushes and instant wealth, but those days were over. The land was being paved. Only children played as before, picking up unidentified bright objects and, most of the time, losing them again.

Kunz remained a widower for twelve years. Nearing seventy, he met Miss Opal Giberson, a thirty-year-old dilettante from a rich old New York family. It is hard to believe her name did not inspire in him unrealistic notions of romance with the young woman—he had written a whole tract on opals—but in 1923 she consented to marry him. Kunz, revived, looked around again. While he was occupied in Arkansas, ideas on northern diamonds had quietly gathered steam. Some now speculated that even those as far south as southern Appalachia could be attributed to glacial outflows originating from the north.

In 1910 American prospector Cab Tabernor was getting a haircut in Detroit when the barber mentioned two German diamond prospectors who had traveled to the boreal forests near Hudson Bay, to near where Hobbs predicted the Great Lakes' diamond source lay. For some odd reason, the barber had a map. Tabernor offered to split 50-50 if he could have it. After checking in at Tiffany's—Kunz still had a standing order to buy all North American gems—Tabernor headed with two companions by canoe up the maze of rivers emptying into Hudson Bay. They made it to remote northwestern

Quebec—and there to their astonishment ran into two others. Perhaps the barber's map had been a sort of apocryphal chain letter. In any case, the five men traveled together up the Nottaway River and found a tumbledown cabin full of quartz and feldspar chips, next to a tree blazed with a diamond shape—obviously the spot. For the rest of the summer they pounded at the hard rocks with picks and crowbars as the blackflies and mosquitoes that rule the north woods ate them. Then the leaves turned. Ice formed on the Nottaway. They fled south—too late. Winter came so fast their canoes were frozen in. Exhausted, stranded by blizzards, and down to a few tea leaves, two went ahead on foot to find help. They were saved by a brand-new rail line that had been cut across the woods. The others were rescued by an Indian couple out checking their trapline.

Tabernor never went back, but the following summer others headed up. Facing the same hardships and lack of success, most gave up quickly; it never developed into a rush. Then, around 1920 a worker was helping digging a rail cut between Ottawa and Toronto when he spotted what everyone was looking for: the first Canadian diamond. Found near the city of Peterborough, about thirty miles above the border with New York state, it was hard to miss—33 carats, the size of a robin's egg. A Canadian jeweler bought it and took it to Kunz. Kunz thought the diamond, broken and poorly colored, was worth little as a gem but priceless as evidence. It proved that a source, or sources, lay at least that far north.

Kunz announced that diamonds were "given to the United States by Canada in the Glacial Period." "I have no doubt that a diamond mine or mines of great value are to be found in Canada," he said. He added: "[Great mines] are never found in the vicinity of towns and cities. As if with a sure foreknowledge of where man would build, Nature has always, infallibly and with great secrecy, hidden [them] far from these sites." To publicize the idea, he went to Massachusetts and lifted a fragment of Plymouth Rock, the glacial erratic upon which the Pilgrims were believed to have landed in 1620, and took it to a scientific meeting in Ottawa. To much merriment, GSC men soberly described it as "biotite granite with altered plagioclase feldspar." They claimed ownership, saying it had been moved by ice from Labrador 500,000 years before the Pilgrims landed.

Kunz's extensive collections of domestic diamonds were the supreme demonstration of North American geologic unity. While still alive, he arranged to have them donated to New York's American Museum of Natural History, where he had for years served as honorary gem curator. The gems were given a place of honor in one of the great halls. He took the British Lord Gray to see them at a reception one evening, and Gray only half-jested: "Why not return these to Canada?" Gray pointed out that British troops had stolen a portrait of Benjamin Franklin during the 1776 revolution, and that he had recently been gracious enough to return the painting to the original owner's American descendants. "Why not return those diamonds?" he repeated. Kunz demurred.

In the end, Kunz did not see things go right. One day in 1929 the *New York Times* reported Opal had startled friends by learning to fly—the "patriotic duty" of all "modern" women, she said, in case of another world war. Sixteen days after she got her license, a biplane went down in fog over Morris Plains, New Jersey. Opal and an unidentified young man rushed away from the wreck—unharmed. Two weeks later Mrs. Thomas Edison christened another plane at a 500-person hangar party. Opal called it the *Betsy Ross*, in honor of the Revolutionary War flag maker. Kunz, who now devoted more and more time to memorializing dead mineralogists and criticizing the way young ladies dressed these days, was not present; things were not going well between them. Four months later, they had their marriage annulled. A few months later Opal's engine conked out on takeoff from Bethlehem, Pennsylvania. She plowed through a fence and turned upside-down. Unhurt again, she directed onlookers to pry her out. Thirteen months later, in May 1931, she crashed at Washington's Hoover Airport while reportedly on her way to organize the Betsy Ross Women's Emergency Reserve Air Corps. This time the nose went into the ground and gasoline sprayed her face and arms. She was only slightly burned. She got up, walked away, and went to her hotel. Opal's number was not up yet.

In June 1932 George Frederick Kunz was admitted to Manhattan's Post-Graduate Hospital, suffering general weakness. On June 29 he died of a cerebral hemorrhage at seventy-six, without ever seeing the big strike. He was buried in upper Manhattan's Trinity Church cemetery, next to his first wife. It was a modest grave of grass and dirt, marked by a fist-size stone reading KUNZ, set among the otherwise opulent above-ground mausoleums of the rich.

His American diamonds remained in the Museum of Natural History until closing time on October 29, 1964. That night a Florida beach bum and sometime cat burglar named Jack Roland Murphy, a.k.a. Murph the Surf, broke in with two colleagues and took much of the museum's diamond collection, including huge foreign stones. Murphy was arrested and confessed the following year, and with great fanfare the police recovered the most glamorous stones, including the famous Star of India and other royal baubles from afar. But Willie Christie's stone from the North Carolina spring; the diamond five-year-old Stanley Devine wept for in a Wisconsin cornfield; the gem from the deep, dark well in Eagle; the pellucid fragments from Arkansas—these and many other homegrown miracles were gone, and they have never, ever been seen again.

If De Beers paid America much mind, it was not evident. The minutes of the directors' meeting for September 1907 briefly noted Arkansas, a full year after the discovery, then dropped the subject. De Beers had bigger fish to fry, mainly keeping track of the thousands of small-time prospectors now combing the vast and hostile landscapes of southern Africa.

Weathered old-time diggers and their descendants had now formed a permanent gypsy subculture, traveling in lonely twos or threes, looking for the next big strike from the sand dunes of the Kalahari desert to the west, or Skeleton, coast, named for its violent seas and the silent, barren shore confronting mariners who survived frequent shipwrecks. In 1906 Sherlock Holmes creator Sir Arthur Conan Doyle helped sponsor an expedition to some tiny coastal guano islands, using a secret map from a Cape Town sea captain. They were driven off by waves, poor results, and, in one case, massed birds. Then there was the unlucky Fred C. Cornell, an English ex–hotel owner with a bit of geology training. Cornell's exploits were famous: He searched everywhere, confronting quicksand, spotted cobras, sand leopards, hostile German soldiers. In the Kalahari, where his misadventures peaked, a sudden downpour came, and then, far as the eye could see, scorpions and snakes emerged from their holes to flee the water. Afterward he lost half his teeth chewing desert plants to keep from dying of thirst. Cornell wrote a beautiful autobiography, *The Glamour of Prospecting*, and never found a thing.

De Beers's Alpheus Williams kept careful track of such men. Many, he noted, adopted garnets, particularly purple ones, as their favored indicators: Common in kimberlite, they were weather-resistant, easily panned, and distinguishable even by the amateur eye. But garnet trails usually led nowhere; forked divining rods were about as successful. In fact, sometimes the two were *combined*. The prospector held garnet or another hoped-for indicator in hand with the rod. The "aura" of these minerals was said to be somehow attracted to the aura of identical minerals in the ground, producing a flow of energy through the rod and a subtle tug—possibly from miles off. Such prospectors were "led away by lack of scientific knowl-edge," observed Williams dryly.

Unfortunately, he did not know much more than they did—nor did Ernest Oppenheimer, the German-British diamond buyer who by 1929 had become chairman of De Beers and owner of many related companies. Perhaps Oppenheimer's smartest move, helping him into this position, had been acquiring beachfront property on the Skeleton Coast as an independent geologist, Dr. Hans Merensky, was searching there. Merensky was far more systematic than others, finally zeroing in on a horizon of sediment laced with fossilized shells of the oyster *Ostraea prismatica*. Among them lay incredible concentrations of allu-vial diamonds, origin unknown.

No one was sure what erosive forces created the mysterious Oyster Line, as it came to be called. But Sir Ernest—he was knighted by the king of England soon after—recognized the value of having his own prospectors and scientists to figure out such phenomena. Upon taking office he decreed, "A new era of scientific diamond prospecting has been inaugurated." Otherwise, he said, "irrational exploitation" might take place—that is, small prospectors might eventually uncover deposits De Beers could not control. They must, he said, prevent "surprise discoveries." Oppenheimer assigned most of De Beers's prospecting to a division of his own giant gold company, Anglo-American. Following this, Alpheus Williams wrote his massive, multivolume *The Genesis of the Diamond*.

Chemists had by now analyzed countless whole-rock samples from pipes both barren and diamondiferous, hoping to find some compound useful as a diamond tracer. Compositions were fantastically complex, with no pattern. The task was complicated by the wild assortment of xenoliths and minerals, many of which chemists subjected to separate analyses. These included the green diopsides found in T. G. Bonney's xenolith of eclogite decades before. But even if chrome diopside was associated with diamond—not proven—it was useless. It quickly disintegrated in weather and thus was rarely found outside a pipe. As for harder, highly varied garnets, chemists broke down all kinds but found no secret formula.

Williams looked to diamonds themselves. He and his father had collected many containing inclusions, which almost certainly had crystallized with the diamonds. Most were unsightly blackish masses, viscous drips like lava-lamp blobs. Many looked like ilmenite. Others resembled quartz, mica, rutile, gold, copper. Some were almost certainly chrome diopside and pyrope garnet. Williams desperately wanted to analyze them to know for sure, but could not. Chemists required a chunk at least the size of a sugar cube for grinding, boiling, mixing, separating, and all the other procedures of "wet" chemistry; most inclusions were flyspeck-size. Peering at them was like pounding at a thick bulletproof enclosure to ask someone inside a vital question; but the person always had his back turned and heard nothing.

Sir Ernest's new "scientific" prospectors limped out with limited information, but at least he got what he considered the best: Canadians. Canada had about the world's longest tradition of exploration geology, and Canadian mining was now coming into its own; with an excellent educational system, many highly trained Ph.D.s were developing resources fast and fanning out all over the world to work. To head his team, Oppenheimer picked Dr. J. Austin Bancroft, a jovial geology professor from McGill University, the top Canadian school. Bancroft, or "Banky," as he was called, imported a handful of his countrymen, including a fellow Quebec native, John Thoburn Williamson. It was at first a low-key operation, limited to within a few hundred miles of Kimberley. For a long time they found nothing. Also, Williamson turned out to be a most unfortunate hire.

I n late 1938 Sir Ernest's son and successor, Harry Oppenheimer, traveled to the United States and hired the advertising firm of N. W. Ayer to increase sales even more. It was the start of the eternal, unmatchably successful "A Diamond Is Forever" ad campaign— and serious hostilities between the United States and De Beers.

With the outbreak of World War II, U.S. President Franklin Roosevelt told the Oppenheimers he needed a 6.5-million-carat stockpile of industrial diamonds, vital for making gunsights, gear grinders, radio wire, airplane engines—in short, all industrial goods requiring fine abrasive machine tools. Sir Ernest, willing to feed the Americans plenty of gem-quality stones at high, fixed prices, refused; selling that many industrials at once would undercut the market, he said. In a letter, he termed Roosevelt's request "farcical." Roosevelt, furious, asked the British to intervene, but Sir Ernest got away by sending a fraction of the order—inferior goods at that, at the usual monopoly prices.

The U.S. government retaliated. The FBI and the Office of Strategic Services (forerunner of the Central Intelligence Agency) summoned U.S. diamond dealers for interviews, stole items from the mail, and sent spies to Africa. They found a letter suggesting the Oppenheimers be "ruthless" in stamping out competition. There were documents suggesting De Beers had sued Nelson Rockefeller and other Americans to keep them from prospecting Venezuela for diamonds. Most shocking, agents traced industrial diamonds going to the Nazis through Tangier and Cairo to a cartel mine in the Belgian Congo, where they were coming out in Red Cross parcels. Moldering conspiracy theories about Arkansas were revived. The Department of Justice went over the evidence and came out with an astounding memo: "An inference could be drawn . . . that the [Arkansas] property was sabotaged and then closed at the insistence of Sir Ernest Oppenheimer."

The diggings were long abandoned. The Arkansas Diamond Company had sold its machinery for almost nothing and demolished the buildings. Only a custodian, one Lee Waggoner, remained to farm the fields around the pipe, in return for keeping any diamonds, which still sometimes turned up in the dirt after heavy rains. In 1941 Charles Wilkinson, an Indiana manufacturer of U.S. Army tank parts, bought control. He hoped to produce industrial diamonds, and to ensure support, a delegation went to the White House: Arkansas Governor Homer Adkins, Senator Hattie Caraway, and the head men of Schenck & Van Haelen, still owners of many surviving Arkansas gemstones. Early one morning in 1942 they found President Roosevelt lying in bed with a bad flu. As he sat up, the governor opened a black ebony case absolutely stuffed with Arkansas diamonds. Roosevelt was taken considerably more than his cousin Theodore ("Ivory is my life, Doctor, ivory!"). He ordered the War Production Board to allocate a priority for $506,000 worth of scarce machinery to reopen the pipe.

The USGS suggested the pipe be retested before large-scale mining, and the U.S. Bureau of Mines (BOM) was assigned the task. But the BOM dragged its feet, taking nearly two years to locate equipment and drill out 435 tons of ore—all inexplicably from shallow depths of fifty feet or less. The washing was done in an open shed in winter, when temperatures made it unlikely the process would work, and poor security invited theft. Total yield by early 1944: a miserable thirty-two industrials averaging a quarter-carat each. The Arkansans and some federal officials cried foul, but the mine stayed closed. By this time the war was almost over anyway. Some saw only one explanation: The cartel had gotten to someone again.

With the close of World War II, angry Department of Justice attorneys filed a massive lawsuit under the Sherman Antitrust Act, which prohibits monopolies on U.S. soil. It named De Beers Consolidated Mines, eight related companies, and the largest U.S. shareholders, including Solomon Guggenheim, benefactor of New York's Guggenheim Museum. They were charged with conspiring to maintain "monopolistic and exorbitant prices on diamonds within the United States." In addition, said the complaint, "De Beers directly or through subsidiaries also has acquired many less important pipe mines with the sole intent and purpose of closing them down so as to keep their production off the market." The lawyers never got to argue whether that meant Arkansas. De Beers successfully countered that

U.S. courts lacked jurisdiction. Besides, the Oppenheimers and their minions could never be forced to the States to stand trial. The suit never proceeded, but the flow of diamonds and diamond ads kept up, conducted through intermediaries to avoid trouble. Justice lawyers maintained that if they ever caught De Beers on U.S. territory on direct business, they would reopen the suit. This included prospectors.

More dangerous foes now laid siege to De Beers from all sides—not lawyers but geologists, better or luckier than the cartel's. The Canadian John Williamson landed the first blow. English-speaking son of a Quebec lumber dealer, he bore a spooky resemblance to then–movie star Clark Gable, especially after he grew a look-alike moustache. Ill-tempered and arrogant, after getting hired by the cartel, he spent a year quarreling with colleagues, insisting they were looking in the wrong place. Feeling rejected, he quit and went north to brood and prospect alone in the British protectorate of Tanganyika, now Tanzania.

In an old truck Williamson traveled thousands of miles of highland savanna with the occasional rock or baobab tree sticking out. He followed miles-long streaks of red soil that native women made pots from, on the theory they signaled buried dikes—weaknesses kimberlites might have emerged through. He ran out of money to pay his African help, and they all deserted but for two men, Issa and Ibrahim. Williamson, usually dressed in filthy shorts, was interrupted for weeks at a time by recurrent malaria and dysentery worsened by heavy smoking and drinking. He borrowed money anywhere, including from married European ladies he favored sleeping with while their husbands were away. When his hotel bill in the town of Mwanza fell behind, the manager interrupted him at the bar and told him to leave. Williamson rose with quiet rage and told the man he would return one day, buy the place, and kick him into the street. He had one savior: his brother Percy. Percy lived a far more ordinary life back in Canada, and did not have much contact with John. Nevertheless, he showed enough faith to send money.

Around this time the role of ilmenite as a possible indicator was reinforced. Back in 1895, old-time diggers had found a small pipe pierced by a long-caved-in tunnel. Inside were the skeletons of several large men along with primitive tools and mysteriously inscribed stones. Later, people realized they were not ancient diamond miners, nor even men. They were women, mining ilmenite—*sekhama* in the language of the native Basuto people. Ground to powder and mixed with grease, the grains make a luminescent deep purplish-black cosmetic used by Basuto girls in puberty rites. The *sekhama* mines were sacred, their secret locations known only to older women, but diamond prospectors caught on. By Williamson's time a Basutoland government geologist tracked down a number of *sekhama* mines and found that 60 percent were kimberlites—not necessarily diamondiferous, but at least pipes.

One story goes that in 1940, Williamson was resting under the shade of a baobab when he lifted his hand and saw a diamond. Another says his truck sank in mud and churned up two gems. Secretive and suspicious to the end, he encouraged these silly stories. Issa and Ibrahim had sifted some ilmenite grains from a prospect trench along one of his red-soil streaks and gotten lucky: They were right on top of a diamond pipe. Williamson claimed the world's largest mine—361 acres. Next biggest was De Beers's Premier Pipe near Kimberley—seventy-nine acres. He opened for business with 1,000 laborers and a marmalade jar to keep the diamonds in in a mineside mud hut. He bought the Mwanza hotel, strolled in, and threw the manager into the street. Then he bought steam shovels. By 1946 he had a ten-mile-square principality with its own flag (a red pennant with a white diamond), a 200-man army, an airfield, and Tanganyika's best-stocked bar, where he wooed the wives of imported engineers and played assorted cruel practical jokes, including scattering faux "diamonds" for green newcomers to find. Percy, still in Canada, was made a silent partner.

Sir Ernest offered huge sums for the mine, but Williamson refused to even listen; he viewed the Oppenheimers as backstabbing thugs. Nor would he market his stones through them, like the few other independent producers did. De Beers decided it had better search Tanganyika, too, and built a house for its prospectors. But one of Williamson's drinking buddies was the British territorial secretary, Sir Edward Twining of Twining's tea. Twining made sure that house stayed empty. Williamson then imported his own team of Canadians and sent them to find more mines. Finally, in 1952, Sir Ernest dispatched Harry to sue for peace. When Harry landed in his fancy private de Havilland Dove plane, Williamson made sure his own *two* identical Doves were drawn up on the airfield. For days Williamson tried scrambling young Harry's brains by calling sudden parlays in the wee hours, then refusing to speak. Whenever Harry took a walk to escape, a flock of Williamson's trained storks stalked behind, causing Harry to fear they were spying. Finally he gave Williamson what he wanted: a guaranteed market share and high prices. The Canadian nodded assent, and Harry flew home, beaten and exhausted.

No sooner was Williamson neutralized, at least for the moment, than the Russians hit. No one had even known the Russians were looking.

From at least 1735, Russian children and gold panners had been finding isolated stones from the Finnish frontier to the Vilyuy River in Yakutian Siberia, 2,000 miles east. But no one paid much attention until World War II, when the Soviets needed industrial stones just as desperately as the Americans and had to work placers for whatever they could get. After the war, Josef Stalin ordered that the source be found so that this would never happen again. Everyone knew he was serious: Stalin had already had N. M. Fedorovskii, the country's chief mineralogist and only real kimberlite expert, shot for supposedly resisting the regime. He also executed a prominent Yakut shaman named Kriukov, whose main crime appeared to be owning a single placer diamond.

The Soviets could muster all the might of the state, and so dramatically advanced the science of diamond prospecting. The new leader was the brilliant Vladimir Stepanovich Sobolev, who had long pushed to prospect Yakutia, the frozen region of steppe and taiga where the dead shaman had lived. Neither Sobolev nor anyone else—except the dead Fedorovskii—had seen a foreign diamond mine, for the xenophobic Soviet Union did not permit it; but through reading and rock samples he recognized this section of Siberia was made of the same sort of stable, ancient rocks as in South Africa. He was also aware that as early as 1914, geologists had suggested that diamond pipes might erupt only through such regions, for whatever reasons. There was now a scientific name for them—cratons.

Sobolev started a military-scale operation in 1947 with hundreds of small field teams. Labs were rigged on remote river barges, underemployed waitresses and ballet dancers shipped up to work. Leaning heavily on Alpheus Williams's *Genesis of the Diamond*, Sobolev renewed the study of indicators, or satellites—in Russian, *sputniki*—which he detailed in his own 126-page prospector's handbook, *Geology of Diamond Deposits in South Africa, Australia, Borneo Island and North America*. Teams checked out all sorts of possible diamond fellow travelers. If any river contained grains of a suspected one, they traced upstream until they hit a tributary where they seemed to originate. Then they traced up to a second-order tributary. And so on. It was a old procedure on a vast new scale.

Conditions were deadly. Sampling frozen riverbeds meant hacking through deep ice, then permafrost—and hoping pockets of river water did not burst through. More than one person froze, drowned, or did both simultaneously. During long, dark winters work halted. Starvation and typhus took over isolated camps. The chiefs were so anxious to test one possible pipe that they ordered a party to parachute onto the spot in December, with air temperature at 55 degrees below zero Centigrade.

In 1953 a female geologist with the U.S.S.R. Geological Survey found an area near the Daldyn River that contained blood-red garnets—quite different from more common pink ones found all over the Siberian craton, and thus a favorite with the geologists. A colleague, Natalia Sarsadskikh, narrowed the area. The following year Sarsadskikh was off on maternity leave, but she passed the data to a third woman, Larisa Popugaeva. In August 1954 it was Popugaeva who followed the trail to the first diamondiferous kimberlite. The following year other teams uncovered fifteen more pipes. By 1958 they had 120. Most were poor or barren, but four were rich enough to make the Soviets a force in the diamond market, including their flagship mine, the Mir (Peace) Pipe. They had proved that widescale, systematic indicator searches worked even in the Arctic—if you could bear the expense, human suffering, and loss of life that came with finding scores of uneconomic pipes for each good one.

To improve the odds, they excavated their treasures like archaeological digs. They calculated how far indicators could travel—garnets 125 miles from known pipes, ilmenites further. In the cold, dry Arctic, even chrome diopsides sometimes made it thirty miles. They quantified a key concept: The bigger the grain, the closer you are. Modern spectroscopes could now analyze larger grains, so they also revived the search for a secret formula by which diamondiferous pipes' minerals might be distinguished from those of barren pipes. While working on that, they also showed that pipes often had magnetic fields different from surrounding rock. (American geologist Noel Stearn first tested the idea in 1930, at Murfreesboro.) Canadian geologists had already adapted war-vintage submarine-hunting magnetometers to look for magnetic metal deposits; now Soviet pilots started flying low-altitude grids, seeking small circular or elliptical magnetic anomalies in bedrock that might signify pipes. Botanists pored over air photos when it was noticed that the taiga's thin trees often grew thicker over pipes, perhaps because nutrients were better.

Diamond prospectors were now heroes of the Soviet Union. Larisa Popugaeva, already holder of several medals for her bravery during the war, was given several more. Prospectors were depicted in a string of semifictional paperbacks, some illustrated with pictures of sexy female geologists, including *Almaahy Kordöchüler* ("The Diamond Hunters") and *Taina Yakutskikh Almazov* ("Secret of the Yakut Diamonds"). There was at least one major motion picture, starring the beautiful tragic actress Tatiana Samoilova. By the end of it, almost everyone dies in the snow looking for the diamonds.

D e Beers made another deal. Officially the Russians were hostile to apartheid South Africa and decadent Western luxuries—but they were hungry for hard currency. They agreed to sell De Beers gems at high prices, camouflaged by murky holding corporations and secret transfers. The two powers also instituted a foreign-exchange program. One prominent Soviet scientist quietly resided at a De Beers mine to study. In return, the Soviets supplied torrents of new technical diamond-prospecting papers. De Beers got to know Vladimir Sobolev's son and successor, Dr. Nikolai Vladimirovich Sobolev. Just as brilliant and driven, he was said to have particular interest in the chemistry of garnets.

Meanwhile, John Williamson went downhill. He now rarely left his mine and obsessively counted each day's diamonds himself. He hungered for another mine to disgrace De Beers further, and whenever his prospectors disappeared into the bush, he tortured himself about their whereabouts. Still racked with periodic malaria, he drank and smoked ever more. Eventually he stopped leaving his house. One day he was examining the day's production when a former lover, now his secretary, upset him with a small remark. He scattered the stones, burst into tears, and locked himself in his bedroom. While he wept on the other side of the door, the ex-lover and the mine foreman quietly got down on their hands and knees to retrieve the diamonds. Soon after Williamson was diagnosed with throat cancer, and he stopped leaving his bedroom. On January 8, 1958, he died in terrible pain at the age of fifty.

Williamson's main field man was Mousseau Tremblay, a dashing, mustachioed French-speaking Quebecois. Williamson had taken a great shine to Tremblay, who had gone to his alma mater, McGill University. Tremblay lived at the mine with his wife and five young kids, but was out when the boss died. He returned while the body still lay in bed, and it fell to him to clean out the effects. Beside the corpse was a heavily chewed pencil and a 5¢ schoolchild's notebook—Williamson's diary. Tremblay opened it. "Where is Tremblay?" read the last page. "Hasn't he found anything?" Tremblay put down the notebook and looked under the bed. Underneath was a quart-size marmalade jar, its inner lid still sticky with jelly. Under the lid was one of the world's greatest collections of rare colored diamonds—ones Williamson had apparently been unwilling to share. The awed Tremblay catalogued them and had them locked in a bank vault to await further disposition.

Sir Ernest died at almost the same time, but differently: peacefully, full in his years, leaving behind a strong organization and family.

Percy Williamson inherited control of the mine. He flew in from Canada and briefly tried running it, but it proved unmanageable. Production fell, theft skyrocketed. Without John, the place blundered on like a madhouse, and Percy had neither experience nor strength to change it.

In 1958 Harry Oppenheimer flew to the mine again, met with Percy, and instantly bought him out. Percy split the money, rumored between $50 million and $200 million, with his two sisters. Then he retired in obscurity to Kelowna, British Columbia—boyhood home of a certain Chuck Fipke.

Tremblay could have quit. But then one day shortly after, Harry walked in alone and unannounced and asked Tremblay to show him around. Harry was not what anyone expected—short, friendly, dressed in ordinary clothes. He had a slight stammer, warm brown eyes that drew you in, and was respectful to one and all, including the lowliest black laborers. Tremblay liked him, and said yes when Harry asked him to stay on.

Williamson and the Russians did more than find diamonds; they propelled De Beers into truly massive, modern prospecting. De Beers stock had been hammered down almost 30 percent by the discoveries. Now De Beers expanded its prospecting team into an army. Starting in 1955, battalions began probing beyond South Africa into the Kalahari, the Ivory Coast, French Guinea, and Zambia. In Tanganyika the syndicate gave Tremblay fifty-five geologists, fifty-five Land Rovers, a fleet of five-ton trucks, and, to dig holes for the all-white geologists, 1,800 Africans.

Tremblay's boss was an American, Dr. Arnold Elzey Waters Jr. A blue-blooded Baltimore native and grandson of a Confederate general, Waters knew all about George Kunz and North American diamonds; that was why he had left around the time of Kunz's death for better pastures. Now, twenty-five years later, he was the Oppenheimers' "consulting geologist"—top man. Upright, skinny, and bursting with energy, Waters had a terrible stutter. Most of the time he just kept quiet and worked, often taking the unusual step of digging holes himself. He was cruel to subordinates who showed any incompetence, but he loved Tremblay. Waters rarely criticized him or even offered advice; he just turned him loose and paid the bills. Tremblay, whose father had died young, soon came to idolize Waters as his replacement dad.

De Beers did aerial photography and magnetic surveys of vast expanses, starting with known cratonic rock and working outward, so all possibilities were covered. On largely open savannas, teams measured grids of perpendicular lines, cutting bushes or trees in the way, marking off every kilometer with a bicycle-wheel odometer, and scooping a shovelful of surface soil at each kilometer intersection. The samples were panned for any heavy-mineral grains larger than a pinhead. Each bush camp covered 450 square miles a month, and many camps now moved across the land. At exploration headquarters in Johannesburg, a high-security walled compound, phalanxes of microscopists went through samples with fine tweezers, picking out indicators and counting them—so-called "pickers." Another unit assembled the equivalent of contour maps, showing grain numbers, sizes, and types. With still no dependable way to tell "good" minerals from bad, chemists looked ever harder for the solution. De Beers did much of this work in house, contracting bits to university scientists. Within the compound it built a multistory library to house diamond data and publications from around the world.

With so much information now at stake, the management heightened security. Some big European corporations had now also taken to diamond prospecting, so to keep out spies, De Beers screened not only scientists, but scientists' wives. Any man—and they were all men—whose wife talked too much, slept around, drank, or indulged any other vice that might invite indiscretion or blackmail was rejected. To minimize leaks, units worked as guerrilla cells. When Tremblay sent minerals for analysis, it was understood he would not have the bad taste to ask about the results. Sometimes Waters sent Tremblay to check a spot, and Tremblay later found by accident that some colleague had checked the spot with a different method. Tremblay thought much of the secrecy was laughable, but he figured Arnold was only doing what he thought best for the company.

Tremblay made little progress toward another mine, and in 1960 he and his wife, Claire, decided to return to Canada so their kids could grow up at home. Gently, he told Waters he had accepted a job offer from the GSC. He expected the old man to protest, but Waters made a curiously weak pitch for him to stay. Just before leaving, Tremblay received a glowing letter from Harry. It read, in part: "I am so sorry you are leaving. . . . I am sure we will work together again some day."

Shortly he was at GSC headquarters on Booth Street in Ottawa, down the hall from a space occupied seasonally by a field man, Stewart Blusson. Blusson was mostly just a rumor around here; he showed up only during cold-weather gaps in his adventures in the high Yukon, and while in town buried himself in the library to scan aerial photos of rocks and write up reports. Tremblay never met him.

After six months on the job, Tremblay was getting ready to go to lunch one day when a familiar, skinny man walked in unannounced.

"Arnold. Arnold. It's you. How wonderful to see you."

They shook hands warmly.

"What are you doing here?" asked Tremblay.

Waters got right down to it. "I've come to see you. Would you come back and direct our exploration program here?"

Tremblay was stunned. He did not know De Beers had an exploration program here.

"What would be my title?" he asked.

"I guess you're . . . hmmm. Chief, chief, chief geologist . . . diamonds . . . North America."

Beyond the stutter, Tremblay thought the hesitation meant Waters was offering a job spontaneously, making it up as he went along. Years later he suspected Waters had allowed him to go home so he would be in position to be rehired without arousing much notice.

On the surface Tremblay's new territory looked pathetic. The only known diamond pipe, in Murfreesboro, had by now passed through a string of failed operators, including a Texas oil wildcatter and a wealthy widow whose main adviser was her personal astrologer. Howard Millar, the one-time miner who had survived attempted assassination in the outhouse, had clung to his own part of it for over forty years, and in 1949 reopened it as the semieducational "Diamond Preserve of America." Visitors received a lecture, then could prospect if they wished. It did poorly. Then came the age of automobile tourists, and Millar invented a name that looked good on postcards and billboards: "Crater of Diamonds." He dumped the lecture, built a snack bar, and just turned people loose. Fields were plowed regularly with a tractor to bring up fresh material; everyone got a special little bag to put gems in; finders were inducted as cardcarrying members of Millar's "Finders Keepers Club." The only rule: no power equipment. Clergymen prospected free. The place roared.

Some retirees showed up with metal detectors (which do not detect diamonds), many others with pans or just their eyes. They managed to find thousands of misshapen, low-value crystals, and very occasionally a larger, more valuable stone. But it all seemed by chance. In 1963 a local man set his fourteen-month-old daughter on a blanket, and soon found an 11.92-carat stone—in the toddler's mouth. Millar most remembered a young ex-soldier badly scarred, crippled, and missing an eye from the Korean War. He spent days dragging himself through the dirt on his side while his pretty wife ignored him and silently read in the car. At the end he screamed incoherently, and Millar ran to him in panic. The man was only calling to his wife; he had wanted to prove to her that he had spotted a diamond, six inches from his face.

Millar wisely turned good finds into press events. He appeared on television's *I've Got a Secret*, in which celebrity contestants vied to guess some odd fact about a person. (Millar's secret: "I own the only diamond mine in the United States.") He made *Johnny Carson* twice. Now old, he came to love the soft chip of shovels, the cries of delight when somebody hit, the childlike smile on an adult face when he told someone they had just found a genuine diamond and handed them their Finders Keepers card. But he also felt sad. To him, the mine was an unfinished masterpiece, a still-veiled mystery that no one had yet plumbed. It was just another roadside attraction.

For Tremblay, it was a starting point.

CHAPTER 8

De Beers in America

De Beers was plotting a vast worldwide expansion, with North America high on the list. Other regions had placer diamonds, but as yet no sources: Australia, Venezuela, and Brazil, to name a few. But Canada was of particular interest. GSC mappers had now shown much of the nation was underlain by old rocks from Ontario on up. Kimberlites had surfaced farther north, as had assorted new rumors of diamonds, though some were wacky. The United States was of somewhat less interest. More barren kimberlitic rocks had been uncovered in Virginia, Tennessee, and Illinois, which locals unsuccessfully dug from time to time. Then, in the 1950s, a few more new diamonds surfaced near the Great Lakes. The North American gem craze had been moribund for a generation, but much early work had been done on glacial diamonds. De Beers thought it might be able to walk in now with modern methods and bring it all to stunning completion.

This was Arnold Waters's brainchild, and the perfect excuse for him to go home. He was old enough to "retire"—but spry enough to get around. He already had a nice lakeside place picked out in scenic eastern Vermont, conveniently near the international border, where he could keep an eye on both sides. He moved there but did not spend much time fishing or boating.

Canada still claimed exactly one known stone, from the Ontario rail cut a generation before. But in summer 1946 a mining company had accidentally found kimberlite in northern Ontario, 400 miles north of Montreal at the end of the road grid. Drillers looking for zinc and silver struck what a provincial government geologist described as a "dykelet" of kimberlite. A few years later another team 200 miles east in Quebec hit a series of them, each a few feet wide. Several small Canadian syndicates staked around them and renewed the search.

In 1948 the *British Sunday Express* mentioned a "group of international geologists" preparing to test volcanic outcrops apparently even farther up, in an unspecified "rugged, remote area inhabited by only a handful of Indians and Eskimos, where few white men have set foot." Over in northern Saskatchewan, Canada's rough equivalent of the upper Midwest, a prospector named Johnny J. Johnson claimed in 1948 to have found five gems in a "wall of blue clay" within 100 miles of Flin Flon, an end-of-the-road mining town. His strike petered out, though, when provincial officials asked to see stones in intact rock. In 1958 another prospector, Einar Opdahl, did find a 1-carat diamond in gravel in the westerly province of Alberta, again near the northern roadhead, but he kept it hushed. Shortly after that, the first real northern diamond rush occurred. It was led by Max Pellack, who had taken a prospecting course while a resident at Saskatchewan's Prince Albert Penitentiary. After getting out, he claimed he had found two small diamonds in a river just outside town. At least 500 claims were staked, but no one else struck.

Things briefly pushed farther north when the RCAF declassified World War II aerial photos of the northern tundra—not the Barren Lands, which lay west of Hudson Bay, but regions east, in Labrador and Quebec. Toronto prospector Fred Chubb pored over them and spotted a startlingly circular two-mile wide lake with a raised rim— possibly the world's largest kimberlite. It lay 300 miles beyond the tree line, on the northern tip of the Ungava Peninsula. Chubb, without backing to reach it, showed the picture to V. Ben Meen, director of the Royal Ontario Museum. Meen was intrigued. He borrowed a huge amphibious plane from the Toronto *Globe and Mail* in return for letting a reporter come along. In July 1950 Meen and Chubb got to this brooding, boulder-littered place, and Meen realized it was not the world's largest kimberlite; it was the world's largest meteorite crater. The discovery made headlines and led to a wave of new research on craters around the world. There were no diamonds near this one—only long-abandoned campsites of Inuit.

Soon geologists spotted more big craters, some real, some imagined. H. P. Schwarcz, a geologist at McMaster University in Ontario, thought he saw one in the semicircular Nastapoka island arc, which lies within lower Hudson Bay, paralleling the shoreline. He concluded that Hudson Bay itself might be a vast meteorite crater. Then he took another leap and hypothesized that the meteorite had dug a hole so vast, it had shattered the earth 110 miles down. The impact, he said, had jetted up great masses of deep rocks containing earthly diamonds. These stones, he speculated, were the ones carried by glaciers. Alternatively, the impact might have hurled the diamonds directly through the air to Ontario, Wisconsin, and points south.

Some scientists guffawed openly. Others allowed the meteorite crater, but not the excavated diamonds. Then there were those who collected all data, theories, and speculations, no matter how preposterous, and silently filed them for future reference in their library.

The cartel already had partly or wholly owned operations in Canada dating to 1949: Hard Metals Canada, which dealt in diamond-tipped industrial drill bits; Canadian Rock Company, a tunneling contractor; and Hudson Bay Mining and Smelting, a lead and zinc outfit. Their connections to De Beers were low-profile but legal; unlike the Americans, the Canadians were not at war with the cartel. Tremblay's outfit used the companies as a cover until it got its own corporate identity, called Diapros—as far as Tremblay could tell, some executive's stab at combining "diamond" and "prospecting" without having it sound like either. Later it was renamed Monopros, apparently to signify the search for one thing, and one thing only.

Tremblay refreshed himself on domestic diamond literature at the McGill library. He decided to first reinvestigate any spot where a diamond had been found; then to resample known kimberlites, even if barren; then to cover all of Canada with a huge African-style grid sweep for indicator minerals. The first two required visiting the States.

This was not exactly illegal; it was just, well, touchy. In 1957 FBI chief J. Edgar Hoover had opened another big antitrust investigation into the cartel. Johannesburg lawyers advised that De Beers employees nabbed within U.S. borders could be made to testify about company secrets if they were there on "ongoing business." But there was a loophole. It appeared that under certain legal precedents, "ongoing business" meant more than three people or so, working openly. Three or less coming and going quietly once in a while with no fixed abode passed legal muster. Tremblay was told that whenever he wanted to cross over, he should ask permission via telex to Johannesburg. A head count would be made. Then he would get a one-word reply: "yes" or "no."

This was how the Tremblay family got to see the U.S.A. in a beat-up Volkswagen van during the 1960s. During fall or spring school breaks, Tremblay piled Claire and the five kids in with lawn chairs, a picnic basket, and other tourist paraphernalia to obscure shovels, pans, and canvas sample bags. At the border, he told U.S. Customs in his thick Quebecois accent that they were on vacation. He figured the kids were perfect decoys, a kind of human shield against suspicion. It being the 1960s, the VW van added a nice hippie tinge. At worst, Customs would mistakenly search for drugs.

They first visited Eagle and other Wisconsin hamlets where diamonds had been found. The towns were still tiny, unchanged. They headed to Dowagiac, Michigan, and spent one Easter week in Indiana. The scrupulously Catholic Tremblays went to Easter mass in Bloomington, then headed out again.

Most spots were unmarked but easily findable; Kunz, Hobbs, and others who had originally investigated had left excellent notes and topographic maps. Whenever Tremblay got to a historic place, he rummaged in the cornfield or whatever it was on the off chance he might find a diamond, then cruised until he found a creek draining downhill from the spot. Here he dug out about sixty pounds of gravel, panned away coarse rocks and light stuff, and bagged the rest. When passersby started asking what they were doing, he picked up a useful American slang word: "rockhound."

"We're rockhounds," he replied.

By each trip's end, the van's springs practically dragged on the road.

Sometimes the older kids helped, but more often they all lined up behind him, oldest to youngest in order, and watched sullenly. Then someone would begin complaining.

"Let's go do something fun!" cried his daughter.

Then the others would chime in. "Let's do something *fun!*" Tremblay, usually kneeling in several inches of cold running water and panting from exertion, turned around and tried to convince them that this *was* fun. He gave them his biggest smile to prove it.

In Tennessee they visited a barren kimberlite discovered in the 1920s, now under shallow water behind a new Tennessee Valley Authority hydroelectric dam. To see it, you had to wait for low water, rent a rowboat, land on a roadless shore, and wade onto the exposed muck. They visited still-rural Isom's Creek, Kentucky. Upon scaling one kimberlite hill, Tremblay heard chickens clucking in a shack down through the trees. He began daydreaming about moonshiners. Suddenly a man appeared right next to them from nowhere.

"I haven't seen you around here lately," said the man in what to Tremblay was a nearly incomprehensible southern accent. Tremblay caught on instantly; trespassers must come all the time, looking for garnets or diamonds.

"We're rockhounds," he said, and smiled. The man shrugged and moved off through the woods.

They ended up in Arkansas, where he visited the state geologist and decided to drop the act. He stated who he was and his mission. To his surprise, instead of being hostile, the official was so delighted anyone would show scientific interest in Arkansas geology that he loaded Tremblay with copies of old papers and even bags of ore he'd collected himself. Afterward the family drove to Murfreesboro and bought tickets to Crater of Diamonds. At last, the kids were excited; they knew the odds of finding a diamond here were not impossible. Tremblay panned expertly for hours, while the rest of the family competed to see who could find the first one. To Tremblay's disgust, he and his massed force of child diggers struck out.

They returned home. He could not very well comb Canada with a cast of thousands, as in Africa; it would attract too much attention, and Canadian wages were thirty times higher. He contented himself with hiring several geologists and beefing up the crew in summers with ten or twelve college students. He had the leisure to do it this way; no one else seemed hot on any trail, and most of his few competitors were small-timers with no expertise. He noted with satisfaction that one firm below Hudson Bay was spending huge amounts drilling directly for kimberlite instead of looking for kimberlitic minerals—a sign they did not know what they were doing. Another "firm" was just an ex–taxi driver from Toronto who liked to go camping.

Tremblay and his new main man, a fellow Quebecois named Joe Brunet, worked methodically east to west, looking for mineral trains coming from the north. The first few summers they drove the Trans-Canada Highway, digging samples of sediment from roadside streambeds every five or ten miles. With a few side trips, they covered a ribbon fifty miles wide this way, then moved north a notch to cover another swath. The western prairies' arrow-straight back roads and open land made this easy. Each year they took 5,000 samples across southern Quebec, Ontario, Manitoba, Saskatchewan, and Alberta and into the lower British Columbia Rockies. Other years they covered the Maritime provinces and circled Prince Albert Penitentiary. Tremblay heat-sealed each sample into a special black plastic tube and freighted it back to Johannesburg for the pickers.

This time he was allowed to see the results. Some contained indicator minerals. In 1963, he latched on to a perceptible train and traced it to a pipe 250 miles north of Toronto—the continent's first kimberlite found by design, not accident. It was barren; but it was a start. In fact, finding indicators turned out to be no problem. It soon became apparent they were everywhere. Sometimes there were just a few grains in a fifty-mile-square grid, sometimes a tantalizing patch where they lay thick. But except for that one early success, numbers swelled and diminished without pointing anywhere.

It did not take long to exhaust the road grid, so Tremblay hired a couple of cars from an east-west freight rail traversing the high woods, and the train put them off at sidings to work. In places with minerals, the following year he hired tracked vehicles called Bombardiers to move around the forest. When the rails ran out, Tremblay broke out canoes. Many rivers flowed north, so they drove to where a road crossed one and put in there or had themselves hauled in by float plane. Within a few years they canoed a whole network leading to lower Hudson Bay—the Little Abitibi, the Moose, the Mattagami. Eskers lay thick in places, covered with scrubby trees, and Tremblay dug them; the GSC, he knew, was studying techniques of tracking ores—though mainly metals—through glacial debris. He and Brunet were often tangled in underbrush, soaked by rapids, and eaten by blackflies, but Tremblay did not think it was too bad. They had radios, plane resupplies of food, even wine for dinner. At each river mouth, the plane whisked them home.

Eventually they had a line across the map at 52 degrees latitude, about level with James Bay, the lower reach of Hudson Bay. Everything below had been sampled on a reconnaissance scale. The indicators continued.

Despite their hard work, small-timers kept galling them with stupid good luck—nothing serious, but annoying. In northern Ontario and Quebec, De Beers optioned over 100 square miles of ground around known kimberlite dikelets, but the only person to find anything nearby was Reno Jarvi, a prospector from South Porcupine, Ontario, who dug a quarter-carat industrial diamond from an esker. Then in 1966, four McGill geology students were on a field trip to Ile Bizard, a river island near Montreal. It contained many weekend homes and a golf course with a peculiar hump-shaped hill, whose side had recently been bulldozed. In the cut the students saw red, green, and black minerals. One had done some work in South Africa and recognized the minerals: garnets, chrome diopsides, ilmenites. The father of another was in the mining business. They staked out mineral rights, which in Canada are allowed to anyone with a $15 prospector's license, even on most private land.

Arnold Waters personally came up from Vermont to deal with the discoverers. The hill was definitely kimberlitic. He paid them well to option the ground, then had his people stake Ile Bizard end to end. Come the weekend, many homeowners returned to find stakes in their backyards. They angrily yanked them out. Joe Brunet started restaking the golf course to make sure everything was legal, but the manager screeched up in a station wagon and blocked the road. Other employees surrounded Brunet with golf carts and screamed at him in French not to move. Brunet screamed back in English. The golf course people called the cops. Brunet flashed his prospector's permit, written by federal law in both French and English, and they had to let him finish. After that the stakes kept disappearing. Often when Brunet returned to the little kimberlite hill, he found that a couple of large-fanged guard dogs had been turned loose by someone and were roaming around.

Arnold Waters hated publicity—he had already given Tremblay a rare dressing-down for filing required government documents reporting their activities, and Brunet got one for daring to wear a nametag at a scientific meeting. Ile Bizard produced tons of publicity. The circus atmosphere was complete when Moe Berman, a Montreal restaurateur, staked out Ile Ronde, the nearby island hosting the Montreal Expo '67 amusement park, and threatened to dig up the rides. Others staked the McGill University Library when renovations uncovered rocks that looked vaguely kimberlitic. Best of all, the whole thing was happening a few miles from where Jacques Cartier had made his big mistake some 430 years previous. However, the golf course had the only real pipe. Tons went to Johannesburg, and this time there were diamonds—ten microscopic bits weighing a total of .537 carat. De Beers clumped off to look elsewhere.

Waters came by once in a while to check on Tremblay. Once Harry Oppenheimer came to Ottawa to give a speech and brought along his son and heir, Nicky, to meet the Tremblay family. They were thrilled, but in the end Tremblay decided this was not for him. The trail did not seem to go anywhere. No one blamed him when he quit to go into private consulting. Waters and the others kept going.

A fateful crossing of paths now occurred in the Kalahari.

In 1955 De Beers had sent Ph.D. geologist Gavin Lamont to search the desert in what is now Botswana, after another company found three diamonds in a dry riverbed. Twelve years later Lamont was still out there, amid sun and drifting sand that buried all rocks hundreds of feet deep, making conventional prospecting more or less impossible. Finally he found an unexpected ally: ants. In one region the insects had built countless pointy anthills, higher than a man and

carved by wind like gothic spires. Lamont realized the ants were mining the faraway water table for moist particles so they could live.
Near the remote cattle-trading outpost of Orapa, he discovered what
they were mining: diamondiferous kimberlite. Around some hills just
under the surface of the sand lay halos of deep-red garnets and even
a few tiny diamonds. Three hundred feet down were the pipes. The
rich, monstrous Orapa Pipe opened for business in 1971—the first
mine De Beers had ever found on its own.

Lamont, however, spilled some secrets. While in the desert he
got friendly with one Chris Jennings. Jennings was a tall, bespecta-
cled South African who worked as a geologist for the Botswana government. He held this position partly because he had been denied
one at De Beers. Dr. Waters had once deigned to grant him a job
interview—then rejected him. Jennings now hunted water instead of
diamonds. However, things came around. He was eventually
assigned to collect the regular prospecting reports Lamont made to
the government. Over the years the two became friends, and they
went on for hours about diamonds. Jennings learned De Beers's
techniques, and sometimes if Jennings was going to a particularly
remote area, Lamont even asked him to bring back samples. Later
Jennings tried taking credit for pointing Lamont to Orapa. After the
discovery, Jennings would lie on his side in the hot sand and peer at
ants emerging from their holes, carrying mineral grains far bigger
than themselves. It was then, in middle age, that he decided to find
his own diamond mine.

In 1969 Jennings joined the African branch of Falconbridge
Explorations Ltd., a Canadian mining firm with a global sweep of
metals properties. Falconbridge was based in Toronto but had just
come under the control of an oil billionaire in Houston, Texas. This
was Howard Keck. Keck was not just any Houston oil billionaire; his
father had been William Keck, the world's greatest oil prospector, a
man whose instincts about the location of petroleum were so uncanny,
some believed him clairvoyant. Starting as a penniless roustabout, he
rose in the 1920s to found the Superior Oil Company. He pioneered
deep offshore drilling, was first to find commercial deposits in the
Gulf of Mexico, and practically ran the oil-rich nation of Venezuela.
Even in his later years, when a drilling rig brought up a slimy core, old

man Keck sniffed and tasted the rock to gauge the prospect. In 1964, Howard overrode his siblings to inherit his father's mantle. Some said it lay too large on him.

Recovering from alcoholism, thin red hair now turning gray, Howard collected racehorses and farms, fired people at will, and continued to jockey for power. Outsiders quipped that the Kecks were not really a family, but a collection of people with the same last name; the dastardly deeds of the family on the TV show *Dallas* could easily have been modeled on them. Howard's brother claimed Howard would sooner run him down with his private jet than offer a ride. Most disappointing, Howard was no good at finding oil. Thus pursued by stockholders and his father's shade, he was now buying up mineral-exploration companies in hope of finding something more glorious. Jennings could not have pitched his idea to a better man.

In 1973, when Keck made a swing through Africa, Jennings latched on and talked up diamonds. In the private jet, he gave diamond slide shows and educated Keck in diamond science. Jennings's cause was furthered by a grudge that only a billionaire could hold: Just as Howard had been buying up stock in Falconbridge's parent a few years before, Harry Oppenheimer had foxily started buying shares, too—enough to keep Howard from taking over. Harry sat on the controlling shares for a while, then let Howard have them—at an obscene profit. Shortly Jennings had himself a bottomless diamond budget.

Another South African then signed on with Keck: Hugo Dummett, a heavy, towering man with a booming voice. Dummett's father had been a well-off mining engineer. Endlessly self-confident and blessed with an apparently genuine photographic memory, Dummett cut a figure somewhere between a U.S. senator and a grizzly bear. He had a De Beers connection, too: He had spent part of the 1950s and 1960s prospecting for them. He hated it. The secrecy deeply disturbed him. Not only did it smack of paranoia, but it denied him the opportunity to learn geophysics, geochemistry, or anything beyond his job, which was mineral sampling. He hated his own part in the apartheid system. In his opinion the Oppenheimers did not even pay very well. He renounced his South African citizenship, became an Australian, and bounced through jobs in New

Guinea, Canada, and the United States. There, in 1977, he met Jack Langton, a Texas geologist hired to run Keck's brand-new Superior Oil Minerals Division, aimed at finding U.S. minerals. Langton recruited Dummett to co-manage.

Just then, Jennings triumphed. By seeking finer sizes of indicator minerals than De Beers did and importing the latest geomagnetic instruments from Canada, he spotted sixty-two new kimberlites under the Kalahari. Tests were begun to see if they were economical. Meanwhile, with Falconbridge and Superior now closely linked, Dummett traveled to the Kalahari to meet Jennings. He was bowled over—and extremely envious. In 1978 Dummett convinced Keck to let him revive the long-lost search for the North American diamond mine.

Jennings and Dummett were united in their dislike of De Beers— but by the same token, they were competitors for the boss's attention. Both ached to make a major find, and neither had much affection for the other. However, for the moment they had to stick together. Jennings said he would help Dummett, and added a finishing touch of genius to the whole anti–De Beers plot. He said he knew a South African geochemist on the verge of a momentous breakthrough: the secret of the pyrope garnets. This was Dr. John Gurney of the University of Cape Town. Gurney was working on a formula that would reliably tell whether garnets or other indicator grains came from a diamondiferous pipe—even predict how rich it was, and thus whether a trail was worth following. Jennings and Dummett agreed to support him. They pooled $100,000 from their respective budgets, sent it to Gurney, and waited to see what he came up with.

There was one person who knew all the plotters: J. Barry Hawthorne, now the Oppenheimers' chief geologist in Kimberley. More or less the successor to Dr. Arnold Elzey Waters Jr., Hawthorne was a towering figure. He had hunted diamonds in sixty-two countries. He commanded all De Beers's deepest secrets. He was also gentlemanly; eager to work with outside scientists, he contributed to standard geology texts and hired professors to do some of the company's scientific research. Hawthorne knew Gurney because he had given him diamond-related grants, and through that a friendship developed. Jennings and Dummett he knew also because they had all been at university around the same time, some twenty years before.

Now, one day, a small plane buzzed slowly at low altitude over one of the Kimberley mines. Hawthorne surmised it must be someone calibrating geomagnetic instruments to hunt kimberlites elsewhere. He found out without much trouble who had chartered the plane: his old acquaintance and ex-colleague Hugo Dummett. He knew who Hugo worked for now, and he was not amused. It also came back to Hawthorne that Gurney now worked for Hugo's outfit as well. He reserved his fury, though, for Jennings. He knew of Jennings's connection to Gavin Lamont and the Kalahari anthills. He could not say if Jennings had actually used any De Beers data, but he was convinced they had set him on the trail. It was not illegal; but he did not think it at all ethical.

Without realizing it, Hawthorne had also met the most dangerous and unlikely plotter: a very odd young man from Kelowna, British Columbia, who was not involved yet, but would be soon.

PART III

CHAPTER 9

Broken Skull River

Way back in his youth near the top of a knife-edge ridge in the barren Yukon peaks, Stew Blusson was about to hammer a fossil from a rock when he heard a strange chewing sound. He looked upslope in time to see the hairy rump of a grizzly coming backward from behind a boulder. Then the head, gobs of snow falling from its mouth. Upon seeing its first human, the bear startled, lost its balance, and slid down head-first on its belly in a gulleyful of snow past Blusson. Then it recovered and came back up. Blusson viewed the back of a roaring blackish-purple throat, and a pawful of daggers poked at his face. The bear missed, keeled over again, and collided sideways with Blusson, knocking the young man on his butt. Blusson had by now gotten the camera out of his pack. He pulled off the lens cap and shot. The bear stared, panted, then fled as suddenly as it had appeared. After some time Blusson noticed a neat, deep inch-long claw excavation in his left bicep—one of his nine lives gone, and a souvenir of another fine day on the job.

Blusson loved this job. Before this, in the remote British Columbia logging settlement of Powell River, he had had to endure violent drunken rages from his father. He adored his bright and opinionated mother, Edith, who hand-sewed his clothes, doted on him, and said Anglican prayers; but she dominated in her way, too. Since they were poor, Mother always sent Stew to the store with a list and exact change. If he got one item of Mother's explicit commands wrong, he would run out of cash and shamefacedly have to put an item back. To escape from home, he and his older brother Ross often camped in the surrounding woods for days. Bright-eyed, skinny little Stew hated violence and, especially, guns. He'd read about David and Goliath. Instead of a shotgun to deal with bears, he practiced sending huge cobbles whistling with a biblical sling. He became deadly.

One day Stew met a bearded hiker, Angus MacDonald. Mac-Donald, who carried a pack so big he could hardly stand up, said he was heading to the bush to look for copper. Weeks later he emerged loaded with rocks stained green and laced with sequinlike things, sure he had hit this time. It turned out he had not, but Stew was intrigued. He envisioned geology as a sort of permanent camping trip. A few years later, at the top of his geology class at the University of British Columbia, he scored a coveted summer job with the GSC. Thus in 1959 he found himself on top of this mountain in the upper continental cordillera, several hundred miles west of the Barren Lands.

Too high for trees, the central Yukon peaks were an inaccessible world of crevassed glaciers, secret hanging valleys with high waterfalls and devil's-horn peaks. Wind songs wailed against cliffs and dense mists floated through. From reading, Stew knew Indians kept out to avoid rumored man-eating giants in the sunless gorges; post-Klondike gold panners had probed them but had not gotten far. One of the few named places was Headless Valley, for the skulless skeletons of two prospectors found once. It was too vast for mountains to have names, only whole ranges—the Tsezotene, Dahadinni, Sombre, Sayunei, Carcajou, the Funeral. There was the dark, icy Valley of the Imponderables; Broken Skull River. The GSC was mapping bedrock here to help pinpoint mineral deposits—but he hoped they didn't find any. It was too beautiful to mine. Sometimes it was so quiet a pencil on a notebook sounded like thunder. He saw the veins in the back of his hand pulse and swore he could hear his blood run.

When Stew got home, Mother pushed him to look for a more lucrative job—and, by the way, to quit working once in a while and take out a girl. Stew put his fist through the wall at her well-meant suggestions; the only thing he wanted was to keep seeing those mountains.

He gained his permanent GSC slot after getting the requisite Ph.D. from Berkeley at age twenty-four. In the cordilleran section, he fast emerged as strongest, smartest, and, often, strangest. In these tail-end days of foot and horseback travel, he was sometimes alone for a month or more. He always smiled through the May-to-September field season, outpacing others on daily twenty-mile scrambles, lugging rock samples and darting like a chipmunk to check side canyons. He worked through sunny subarctic nights because the sun dipped, dropping temperatures to where remnant snow crusted hard enough to walk on. He had a natural eye for the uplifted granites and seabeds of the mountains, grasping huge histories and drawing fine color maps of the bedrock they produced. Having learned from his mechanic father, he repaired any machine, improvised any engine part. He kept his David sling. Only incompetence or waste brought out his rare explosive tantrum. He buried leftover bacon fat and caribou heads, digging them up weeks later if food ran low. Everyone loathed his usual breakfast: leftover oatmeal mixed with canned sardines.

The terrain was so impossible that Blusson naturally adopted a second calling: bush pilot. The GSC was just starting to use a new craft, the helicopter. The Americans had pioneered them for use in combat during the Korean War of 1950–1953, and it was obvious they were perfect for checking out rocks infinitely faster than airplanes could; they could directly land a geologist almost anywhere for minutes, then whiz to the next spot. In 1952 the GSC sent out its first helicopter expedition, in the Barrens, where low relief and lack of trees made them easy to use. Previously, the agency had estimated basic geologic mapping of Canada would be done in about 400 years. In 1952, 57,000 square miles of the eastern Barrens were mapped on a reconnaissance scale in one season, and the estimate was shortened to twenty-five years.

Helicopters were something else in mountains, however; fogs, big cliffs, treacherous sidewinds, and surprise downdrafts made them dangerous. They could lose control far more easily than planes and plummet like a rock. Nevertheless, the cordilleran section brought in a couple—small, still-primitive, underpowered machines piloted by ex–U.S. commandos from Texas. Blusson spent hundreds of hours in them. He soaked in every move. He loved how they set down on otherwise inaccessible summits like dragonflies, hovered within feet of sheer walls, and rode up on powerful updrafts. He tried not to think about the downdrafts. Years later he would gaze at a happy group photo of himself with nine buddies and realize everyone but him was dead—mostly from air crashes.

But by 1969, ten years into the job, things were still perfect. No one had died. He now commanded his own dozen-man party. They had worked their way east from the Yukon into the Sayunei Range, in the Northwest Territories, and still were utter failures at finding mines of any kind. He never followed Mother's wishes to find a nice girl because outside contact was not usually part of his world. Best of all, Blusson had gotten a flying license. The budget-minded GSC would not give him his own aircraft and he could not afford a helicopter, so he started off with his own tiny float plane. He skimmed over peaks to photograph rock formations, then drifted down into whatever cramped high lake he could find and unstrapped a canoe to paddle around in. Stew Blusson: commando geologist.

One day he got a radio call from another party leader working the boreal forest at the mountains' feet. The leader had a young geology student there for the summer. "It's your turn to take him," said the transmission. Translation: "He's getting on my nerves. You handle him for a while."

A few days later Blusson stood on the beach at his lakeside base camp and watched a float plane taxi into five feet of water with the new guy. The door popped open. A leg emerged, and onto the plane's pontoon toddled a clean-shaven, very short, very muscular teenager. He was attired in a bright red field vest, its many pockets jam-packed with notebooks and other objects, several crammed backpacks, and assorted compasses and cameras that swung chaotically from his neck. He smiled hugely and waved with both hands as if Blusson were a beloved family member picking him up at the airport. The teenager then began walking obliviously forward on the pontoon toward Blusson—and the plane's still-swinging propeller. Panicked geologists on the shore screamed at him to stop, stop, stop, before he was dismembered. Not appearing to hear and still waving, the young man missed a step, went off the pontoon, and ended perfectly upright in freezing water up to his chin, destroying all his cameras. Stewart Lynn Blusson had just met Charles Edgar Fipke.

S tew quickly found the overwhelmingly friendly, incessantly inquisitive Chuck Fipke to be one of the most annoying junior assistants he had ever had. Inattention to detail could be deadly here, so he sent Fipke where he thought he could do no harm: sitting in a tent pencil-coloring other people's bedrock maps. Red for granite. Blue for limestone. Brown for shale. Chuck scrawled like a four-year-old with his first coloring book—outside the lines, or too far inside the lines, too dark, too light. He kept up a constant banter, laughing at his own jokes and everyone else's with the slightest provocation. He never stopped. Blusson ranted at him for four days and gave up.

Chuck was greatly impressed with Stew. Here was a man only eight years his senior who flew his own plane, identified faraway rocks at a glance, had a fancy degree, and was tough as a nail. He found out Stew had a lesser-known side, too: During time off, he dabbled in mining stocks. Chuck learned that Stew had once made a good bet—and a quick $1 million. Then the stock switched direction; Stew failed to sell on time, and he went back to ground zero. Chuck found that astounding—the guy had made a million bucks. If you thought about it, the second part—cashing in—almost didn't matter. Without thinking, Chuck wolfed down a bowlful of oatmeal and sardines offered him.

Stew had him exiled via the camp helicopter to the top of a high, bare mountain to help a paleontologist, Bill Fritz, collect trilobites. This did not bother Chuck either: He was a fanatic collector of all sorts of minerals and rocks, and he loved smashing the boulders to discover the ancient creatures inside. Fritz, a world-famous trilobite expert, would not let him keep any of the specimens, though—highly unfair, thought Chuck, since he was doing a lot of the work.

They stayed a few days on the windy peak, and the day their helicopter was supposed to come, it didn't. Nor the next day, nor the one after that. It had been the only way onto this spot; now it was the only way off. To pass the time, Chuck asked Fritz endless questions about trilobites, plied him with sardines and oatmeal, and scattered stuff all over the tent. A blizzard buried them and they had to dig to avoid suffocating. Fritz drew a line down the middle of the tent and told

Chuck to keep on his side. Things looked bad as they got to the last jar of peanut butter, but Chuck just kept laughing.

Stew had been taking notes next to the pilot on the way to pick up the trilobite collectors, when at 2,000 feet something suddenly went terribly wrong with the collective. That is the part that controls the main rotor, which keeps the helicopter in the air. First they began gyrating back and forth as if buffeted by opposing winds; then they drifted from the sky like an autumn maple leaf released from its branch. The pilot tried controlling the stick, but it controlled him. He was screaming. At 100 feet, they were about to come in upside-down. At the last second, the pilot somehow managed to change the pitch of the blades. They slowed, righted, hit a gentle ravine, and stopped. The transmission flew through the firewall behind the two men, roared between their heads and out the bubble. They removed their seatbelts and stepped out. The rear rotor was dinging at a rock like a dying animal, and a wisp of smoke came from the engine. Blusson, who had been silent the whole time, looked down and saw the pencil in his hand, snapped into pieces. Afterward, he kept that pencil as a reminder.

When the helicopter did not return to base camp, Stew's men mustered another, and everyone was rescued. The pilot never flew again, but Stew saved for his own helicopter. He also decided Chuck Fipke was not so bad. Not only had Chuck survived the mountain with a smile, but Stew saw that nothing got him down. His laugh was infectious; it lifted you up. He couldn't imagine Chuck as a depressed person. Field geologists need the patience of a rock themselves sometimes to figure out tiny clues, and they must be prepared for ordeals, so laughter is perhaps the greatest asset. The two men shook hands at the end of the season and promised to keep in touch.

In cold weather, Stew customarily went to GSC headquarters on Booth Street in Ottawa to write reports and pore over aerial photos in the library. He did not actually live anywhere; he slept in short-term rented rooms with his few possessions pulled from storage. At the office, he met an intern who mentioned he'd had an interesting summer job in eastern Canada: looking for diamonds. Stew had never heard of this. It was true, said the intern; they were looking between Hudson Bay and the Great Lakes. His own job was to sift bags of glacial debris for certain rare minerals thought to indicate sources. Stew did not recognize the name of the company when the intern told him, but the young man said that didn't matter: It was actually just a front for the De Beers cartel. Then the intern realized he had said too much, and shut up.

Chuck Fipke's trajectory off the trilobitic mountaintop took him around the globe before he remet Stew.

Chuck grew up in Kelowna, a small, pleasant town of vineyards on the edge of a big lake in southern British Columbia. It was about a day's drive west of the Rocky cordillera, which continued on up to the Northwest Territories. Kelowna was full of active, retired, and would-be prospectors—the rural BC economy is built on mining—but Chuck's own hard-drinking father was not one. Among other freelance businesses, Edgar Fipke had invented for himself the odd profession of chartering small planes and photographing people's farms from the air. Hand-colored by his wife, the photos made them a modest living. They and the four kids, Chuck the oldest, inhabited an old farmhouse next to a neglected apple orchard. They had a few hayfields, an irrigation tank for water, an outhouse for the necessaries, and a lot of junk cars for company sitting under the apple trees. When not working, Chuck's father disappeared, drank, and came back to smack around his sons in rages. Edgar Fipke wanted his sons to be tough. Like the Blusson boys, Chuck and his younger brother Wayne fled for solace to the woods and learned to snare rabbits and build winter shelters.

Hurt child that he was, Chuck took to capturing and caring for injured pigeons, falcons, and hawks. He stacked the cages to the ceiling of his and Wayne's shared bedroom; Wayne was allowed to stay as an afterthought. In his secret dreams Chuck was a famous ornithologist. He got a horse. He excelled at rugby, but not school. Runty and with exceedingly strong forearms and bumpy hands too big for the rest of him, Chuck alternately defended Wayne from their dad—or else beat Wayne himself, with fists, mop handle, whatever. One day Chuck saw a pile of clear broken glass on the ground and spontaneously shoved Wayne in it face-first. After that, he accurately called Wayne "Scarface." Wayne always looked up to his big brother, but he never quite forgave him.

Chuck developed a bad stutter and was held back a grade. He became nearsighted and got glasses. As a teen he began drinking a lot, souping up old cars and driving them way too fast—part of the reason for the auto graveyard under the apple trees.

The elder Mr. Fipke was actually a charming and intelligent man when not tormented by drink, and he began worrying about his sons' future. One day he met a rich gentleman farmer in Kelowna whose place he had photographed. They talked for hours. By pure coincidence, it turned out the farmer was Percy Williamson, brother of the diamond prospector John Williamson. After hearing the tale of the family fortune, Ed Fipke went home and pushed Chuck to go riding with one or both of Williamson's two teen daughters because, he said, they had such an interesting family. Each owned their own Mercedes Benz.

Chuck was not much of a reader, but at the Kelowna public library he checked out *The Diamond Seeker*, Williamson's story. Unfortunately, *The Diamond Seeker*, aimed at young readers, was bowdlerized. It left out most of the good stuff—the adultery, the heavy drinking, the satisfying revenges. Besides, as Chuck knew, the ending was not happy; Williamson went crazy, the De Beers cartel won, and now here was Percy, rotting lakeside in boring Kelowna. Presented this way, diamond prospecting did not appeal. He was not interested in the Williamson girls, either; they did not really seem to be his type.

Chuck was instead attracted to Marlene Pyett, a slim, dark-eyed Catholic girl from a middle-class family. She had a reputation for virtuousness and was one of the best-looking girls in Kelowna. He was ecstatic when she agreed to go out with him. Then he got her pregnant.

It was 1966. Abortion was out of the question. Her parents arranged a wedding, but Chuck literally left her standing at the altar, not showing up.

Some months later Marlene had the baby and named him Mark. The Pyetts started paperwork to put Mark up for adoption. When Chuck heard this, the guilt was too much: He proposed marriage, and showed up this time.

Chuck, a teenage dad struggling just to finish high school, now received more advice from his father: geology. At the moment there was an exploration boom for copper; companies were hiring students during vacations. Chuck listened for once, got through high school, and took up geology at the University of British Columbia. His eyes were opened.

During semesters he studied just enough to get through, but during summers exploration companies shipped him off for fabulous wilderness adventures. One summer he got the GSC job and met Stew Blusson. He found himself being paid a man's wages for the permanent boyish pursuit of treasure-hunting. He threw himself into mineral collecting and prospecting and did it so thoroughly that he became a campus character with a reputation for monomania and absentmindedness. Chuck once forgot all about a major exam because he was in the library studying some esoteric copper-finding technique. On another occasion, into some other private project, he could not find the ratty pickup truck he drove and phoned Marlene with a string of nasty curses to tell her it was stolen. Marlene—steady, organized, loyal, unfazed by Chuck—reminded him she had dropped him off that morning, still had the pickup, and would now pick him up if he was ready.

"OK?" she said.

"Oh. Oh. OK, Marlene. OK," replied Chuck, instantly mild and smiley as a talking doll—his customary manner when corrected. In addition to caring for Mark, Marlene's full-time job became retrieval of the bread-crumb trail of things Chuck lost—money, papers, vehicles, tools, shoes.

Some people thought, fairly or not, that Chuck Fipke was not very smart. However, at graduation in 1970, a professor got him a job in Papua New Guinea looking for copper. Chuck, Marlene, and four-year-old Mark were given plane tickets and arrived with just some extra clothes. These were all stolen from their parked car a few hours after they landed.

Marlene would eventually bear him five children. She followed him anywhere. It was good for Chuck: The bush got him away from what he often considered the drudgery of domestic life. In

New Guinea, Marlene and Mark stayed in town while he disappeared for weeks. American vets fresh from the nearby, then-ongoing Vietnam War flew the copters that dropped him into jungle clearings. From here he waded in reptilious swamps to dig standard sandwich-size bags of sediment, to be searched for traces of copper minerals. He happily went anywhere for a sample. Monsoons, bugs, hostile natives, and leeches did not stop him. His guides included spear-carrying tribesmen who bit the heads off live frogs for lunch and attired themselves with two-foot squash gourds strapped upright to their penises. At villages, he pounded on their drums and blew into their flutes. He traded so avidly for their regalia that one pilot set down to find him naked except for shell necklaces and a pile of spears. The gourd men stood by smiling, wearing Chuck's field vest, khakis, and boots.

The bosses said getting the sample was everything, but early on he recognized that was only half the equation: You had to analyze it. His company's method was to cull the sandwich bags for heavy indicator grains containing copper minerals. He developed a fascination for them, for he saw heavies were versatile; you could use similar grains for zinc, silver, gold, nickel. Doing separations was both art and science. At its most basic, you could hand-pan like a California forty-niner, and maybe use a hand magnet to pull out some bits. The next step up was a variety of mechanical shaking tables, cyclones and rockers, and interesting automated gizmos that sorted magnetic particles. Or you could pour the sample into a container of dense, bad-smelling, carcinogenic liquid with a high specific gravity, like methylene iodide. Common lightweight grains floated, while heavies sank slowly to the bottom for retrieval, like snowflakes in a shake-up winter diorama. There were endless variations on the sorting exercise, all satisfyingly concrete, because telltale particles appeared like magic as you watched. That was the real payoff. He often fooled with these various techniques.

Then he caught an especially virulent strain of malaria. He had to leave the field. When he got worse, the company sent him to an Australian hospital. There, he hovered in a semi-coma for weeks. Drugs would not work. His condition worsened. The doctors apologized to Marlene and sent him to their house to die. He was twenty-four years old.

Marlene sent Mark from the house. She hand-fed and bathed Chuck's insensate body for many days. She said countless Hail Marys.

One day Chuck opened his eyes and said: "Marlene."

The previously clean-shaven Chuck grew a goatee, sprouting from his chin like a bush hanging upside-down off a cliff ledge. The Fipkes bounced around. There was tungsten prospecting in Australia, freelance opal digging on an outback mountain where crowds were at a new strike. He innocently showed up at the gates of the huge main lead-zinc-silver operation of Australia's largest mining company, Broken Hill Proprietary, and asked what they were doing. Instead of kicking out the tatterdemalion, the managers took him in and gave him a huge tour, free lunch, and whatever rocks and minerals he wanted for his growing collection. When Chuck got restless again, they moved on to South Africa. Here he got the idea that he could go to Kimberley and tour the famous diamond mines. Sure enough, he knocked on De Beers's door—and it opened. After a three-day security check to make sure he was not a spy, he was assigned a Dutch geologist to take him around. He first took Chuck to meet a friendly, broad-faced South African with a penetrating gaze: J. Barry Hawthorne, the man who had authorized the visit.

Hawthorne liked people who took initiative. He figured he would interview the young man, and if he liked him, offer a job. Chuck, however, seemed not exactly sure whom he was meeting. When he got to Hawthorne's office he indecorously began joking, talking about his adventures, and in general acting like he and Hawthorne were kids about to trade marbles. Chuck pulled out a big opal from Australia and offered to trade with Hawthorne for a small diamond—preferably one still embedded in the original kimberlite. Hawthorne was too polite to mention that diamonds are rarely seen that way. He was gracious but noncommittal about the trade and said he would see Chuck later for a more serious talk.

Chuck got his tour of the mines—vast open cones following the steep pitches of the pipes. You could throw a stone in and wait seconds before it hit bottom. Distant birds could be seen flying around the middles. Each cone disappeared into underground workings among the kimberlite fissures—all that was left now. De Beers obviously needed to find more mines. Chuck asked his usual endless questions, stopped at every turn to examine things, and cadged free kimberlite chips off the Dutchman. At the processing plant, there were great heaps of blackish gravel filled with ilmenites, low-quality garnets, and other minerals left after ore was concentrated for heavies and relieved of diamonds—the dregs. Chuck asked if he could take some, and the Dutchman said sure; they scrounged a glass jar and filled it. In fact, Chuck got so caught up collecting stuff and asking questions that it all took too long, and by the time they got back to the office, Hawthorne had left for business in Pretoria. Chuck shrugged. Within days he blithely got himself a job with an unconnected South African outfit looking for antimony, and never did get his employment interview with Hawthorne.

As long as they were in Africa, Chuck made numerous trips for back-country adventures in Malawi, Namibia, and Rhodesia (named for De Beers founder Cecil Rhodes). Often he brought the family. Young Mark got to meet Kalahari Bushmen, rain-forest Pygmies, and, in Uganda and Zaire, scary teenagers with automatic weapons who said they were government soldiers and took most of their belongings once again.

A year or so later the Fipkes landed in Brazil. The glass jar of kimberlitic minerals had survived, and it came along. Here Chuck got a job with the Vancouver-based mining company Cominco, this time hunting zinc. It was here, while panning a river for zinc indicator minerals, that he ran into some *garimpeiros*. They were still common in the backwater provinces, though Brazilian diamond production was way down. They showed him their own personal diamond indicator minerals, and he realized that some did not look that different from the ones in the glass jar. A light went on in his head.

He experimented more diligently now with heavy-mineral separations, sometimes mixing the contents of the Kimberley jar with masses of light sands and practicing until he could get every grain back into the jar. He set the poisonous heavy liquids on the balcony of their Rio apartment so Marlene and Mark would not have to breathe fumes. He learned Cominco had an interest in diamonds, too—and was gearing up its own heavy-mineral lab in Vancouver. He convinced his superiors to let him do a stint there and spent months honing his techniques. At the end of this trip, he made a pitch for Cominco to let him prospect diamonds in Brazil—and was laughed at. He lacked credentials for such a specialized pursuit; besides, some at the lab considered him not the brightest penny in the barrel, as one of them put it. This opinion grew when he returned to Rio with a glass container of methylene iodide in his checked luggage, which broke in flight, dripped into his clothes and the surrounding luggage compartment, and seriously upset the airline people. At the end of his two-year contract, Cominco let him go.

After nearly seven years, Marlene had tired of travel and was toting a second baby. Chuck had an idea: They could go home, where he would open a business separating heavy minerals for exploration outfits. In 1977 they headed up by way of second-class buses through South and Central America, with various attendant adventures. Upon arrival in Kelowna, Marlene put half their $12,000 savings down on a beat-up farmhouse. Kelowna was small; the house was not far from the Williamson spread. Chuck took the other half to start his company, C. F. Mineral Research Ltd. He began with only a few thin contracts.

Things had changed. Chuck's father had quit drinking and become a Jehovah's Witness; he was nice all the time now. Marlene's dad forgave all. He helped build a clunky mechanical jig for shaking up mineral samples in the backyard that was hooked up to the garden hose and broke about every other day. Marlene dried wet samples in the kitchen gas oven, usually cradling a small child with her free arm. When it was time to make dinner, she took out the minerals and put in food. Wayne (a.k.a. Scarface) was now a creative-writing student at the University of British Columbia. He and Chuck often went drinking together with Wayne's best friend, another arts student named Dave Mackenzie. Chuck enthralled them with tales of foreign exploits. One night they got into a fight with other bar patrons over possession of the pool table, and Chuck spilled some blood in defense of Wayne and Mackenzie before they were all thrown into the street.

Mark, now eleven, was unaware that any life but prospecting existed; he spoke fluent Portuguese and some Spanish, and knew a fair bit of geology just from watching. He got a paper route with the

Kelowna Courier—the receipts of which Chuck borrowed to pay for lab equipment. Chuck spent ten or twelve hours a day on the phone hustling everyone he had ever met, and many people he had not, with a stuttering, wildly enthusiastic pitch about heavy minerals—old teachers, ex-colleagues, geologists he'd met once in bars, people at big companies.

His pitch: Forget the usual sandwich bags of sediment. Dig tons, and send it in. I will sort it all with my soon-to-be patented process. I will isolate whatever indicator grain you want—even if there is only one among billions. By sheer volume, I will set you on the trail to anything.

The contracts, mostly for copper, were barely enough to get the *Kelowna Courier* off Mark's back for the $500 he owed them. Things looked bad. Among those Chuck desperately tracked down in early 1978 was a fellow student from UBC, Mike Wolfhard. Wolfhard now worked in Tucson, Arizona. His employer was the brand-new Minerals Division of Superior Oil. After the pleasantries, Fipke gave his pitch.

CHAPTER 10

Deliverance

Wolfhard had not heard Chuck's excitable voice for eight years, but he remembered Chuck well. Sitting in Superior's new mineral headquarters, on the second floor over a Tucson bank, Wolfhard knew that few labs did the kind of work Chuck was proposing. One was De Beers's in Johannesburg. Superior certainly had no such heavy-mineral operation. Wolfhard knew it would take someone with a one-track mind and the industry of an insect colony to make it work. He went down the hall to talk to his new colleague Hugo Dummett.

Hugo had charged ahead with single-minded aggression and thoroughness. First, he wanted to assure himself that North American diamonds here were nice enough to search for. George Kunz's gifts to the Museum of Natural History were gone, but Hugo made endless phone calls and discovered that the Smithsonian Institution had a few other American specimens, stashed in a back room. He flew to Washington and persuaded the curator to pull them from their flat drawers: a pale, flat fragment from 1840s North Carolina; eight from nineteenth-century Brown County, California; a dodecahedral crystal from 1911 Foard County, Texas; twenty-four fragments in fantastical shapes ranging up to 17.86 carats, all from Murfrees-boro, Arkansas. He spent a morning photographing them, handling them, savoring them. They were nice.

With the U.S. search basically dead these seventy-some years, Hugo went to the library and read every yellowed USGS report. He collared a young Superior geologist, Mike Waldman, and together they soon filled a whole room with copies of every old report, jour-nal article, or newspaper clipping. To these they added newer gradu-ate theses and modern publications on exploration technology. Through assiduous searches, Dummett discovered several American geology professors who had maintained an interest. He visited them, pumped them, and interviewed their students to see if they had, or were interested in, diamond-related projects. In his prodigious head he memorized every fact, theory, and conversation so completely that he was on his way to becoming a latter-day Kunz. Then he found out something scary: He was not alone.

Suddenly everyone wanted back in. The 1970s had seen a record runup in diamond prices and demand, and everyone could see that two-thirds of the gems now came from diminishing pre–World War II pipes. Hugo estimated that without new supplies, Kimberley itself would be exhausted in ten years, all known mines by 1999. Interest had increased in the United States in 1975, when the first diamondiferous kimberlite since Crater of Diamonds surfaced, in Colorado. It was very poor, but enough to make sleeping eyelids flut-ter. Around the same time an amateur mineralogist, Al Falster, found a diamond in a sandbank in Antigo, Wisconsin. Falster, who managed the Peli-Clean Carwash nearby, converted it to wash sand over a plas-tic gasoline-price sign smeared with margarine, having read that dia-monds adhere to grease. He found six more. The same year at Crater of Diamonds a female tourist from Amarillo, Texas, spotted a spec-tacular specimen: 16.37 carats, worth $80,000. Newly refined aero-magnetics and other techniques showed promise for spotting pipes that may have been missed before. Old diamond locales suddenly looked ripe for reinvestigation.

As Superior got going on U.S. diamonds, so did Gulf, ARCO, Shell, Mobil, Occidental Petroleum, and Bethlehem Steel, among others. Big Canadians including Kennecott, Ashton, and Cominco worked both sides of the border. In all, some eighty outfits scoured the United States. One big surge was in the Great Lakes, after the USGS wrote a report on a new pipe on Michigan's Upper Peninsula. Dow Chemical got hold of it before publication, staked out the pipe, and started a rush. Mousseau Tremblay was still active as a consultant; Oklahoma City oilman A. J. Magness hired him to advise whether more pipes could be found in Arkansas. With high 1970s oil prices, there was plenty of money. Tremblay visited Magness's mansion, entering a glass-floored foyer where exotic fish swam underfoot; he noted that there were so many bathrooms, every once in a while Magness would find a new one.

Only De Beers seemed shut out of the U.S. rush; in 1974 it had been indicted again, this time on criminal charges, for fixing the U.S. industrial diamond market. It pled no contest, signed a paper saying it would no longer act like the monopoly it was, and agreed to let the U.S. Attorney General rifle through any company records found in the States, without a search warrant. In other words, they were not welcome. If they wanted to do anything, they had better keep it quiet.

In 1979 a brand-new outfit sprang up seemingly from nowhere with the biggest operation of all. This was Exmin Corporation, with a staff of forty and expensive new quarters in Bloomington, Indiana. Exmin spooked Hugo. According to a dossier he compiled, it was a subsidiary of Sibinter, S.A. Sibinter was incorporated in Luxembourg and was in turn owned by a Belgian company, Sibeka. And Sibeka was 19.6 percent owned by De Beers. Sibeka shared two board members with the De Beers board and ran joint ventures with De Beers in Ireland and Sweden. Exmin president Derek Fullerton practically growled and bared his teeth when anyone—in fact nearly everyone—suggested Exmin *was* De Beers. According to résumés Hugo got hold of, Fullerton and his two lieutenants, by the names of Hawk and Kramers, had worked for the Oppenheimers most of their lives. It was not long before he saw signs that they were spying on him—or so he thought.

Hugo picked a large number of targets, including the old dikes under Green Street, Syracuse; known kimberlites under Kansas cattle ponds; a train of indicators found in the Texas panhandle;

unidentified circular structures in New Mexico. To research property titles and negotiate mineral rights with landowners, he employed a full-time "land man," Robley Berry, whose business it was to figure out such matters.

They went first to the historic little backwoods kimberlites at Isom's Creek, Kentucky. When Berry went to the county courthouse to check ownership, the clerk looked coldly at him. "Your competition was in here last week," he said. Berry asked who that might be. The clerk kept his mouth shut and glared.

Hugo, Berry, and Mike Waldman hiked up a dirt road dotted with swayback wooden shacks, where people sat idly on porches, all drinking out of their own personal half-gallon bottles of Pepsi. Everyone waved as if expecting them. When they seemed near the pipes, Berry picked a porch and negotiated for access with a ninety-year-old man. During a half hour or so, he gradually realized the man was senile; if he owned anything around here, he probably did not remember. The elder gripped Berry's arm and drifted from one bizarre story to another, gradually tightening the grasp. Berry tried prying off the fingers. Behind the porch door, he could hear an unseen relative snickering. Finally they escaped without concluding any business and found a pig path through the woods. Up on a little knoll was one of the pipes. It contained four big, fresh holes, made by heavy equipment. They looked around, hurriedly hacked off chunks with their geologists' hammers into five-gallon plastic paint buckets, and ran like hell.

In search of obscure local nineteenth-century newspaper accounts from the Great Lakes, Hugo dispatched Waldman to the town library in River Falls, Wisconsin. "Oh, you want the diamond stories," said the librarian. "Right over here. The other fellows were in here yesterday."

"What other fellows?" asked Waldman.

"Oh, I don't know," she said. "They were looking for diamond stories, too."

They visited the California state geologist's archives. There a clerk told them they were early.

"Early?" said Berry. "We didn't make an appointment."

"Oh." she said, "Sorry. That wasn't you who called about the diamonds?"

Amid these mysterious close encounters, Chris Jennings came from Africa for a few weeks to take samples and consult. Hugo was not necessarily enthusiastic about the help; he wondered if Jennings might be here to somehow steal his show. Jennings's pipes in the Kalahari were still being evaluated, and thus Keck had no proof Jennings had really done his job there.

One sunny summer day Jennings went to check a kimberlite in rural Rockbridge County, Virginia. Found by academic geologists, it was mentioned in a journal article. Jennings located it on a deserted country road, cropping from a creekbed next to the wooden Mount Horeb Church. As he walked toward it, something seemed out of place. When he got close, he realized it was a row of aluminum pie pans on the church lawn. In them, wet, colorful sand was drying— obviously kimberlitic minerals panned from the creek. Next to them was a big, new Samsonite suitcase embossed with gold letters. Jennings read them twice: DR. ARNOLD ELZEY WATERS, JR. A tag on the handle showed a Vermont address. There was no car nearby. Jennings peered into the silent woods. He pricked his ears. No movement. Perhaps Waters had gone for lunch? He bent with a magnifier over the tins and saw gorgeous big grains of dark garnet. Then he took out a pair of tweezers and some small vials, went down the line, and lifted the most interesting-looking ones. He looked around once more for Waters and fled in his rented car.

Hugo now became convinced that spies were everywhere. The staff drove to a Tucson hilltop and held a meeting there because they feared the office was bugged. Strange clicking noises plagued the phone lines. Hugo told everyone to note license numbers of any vehicles that seemed to be following them. Hugo took pictures of some of the suspect vehicles; in addition to his other pursuits, he was an avid photographer. They could never find out the owners, though, because no one knew how to get around state motor-vehicle clerks, who were forbidden to give out that information. Now, whenever they got a book out of any library, they copied down the names of everyone who had checked it out before, off the card. Hugo had read of this investigative technique in *All the President's Men*, which described how newspaper reporters had recently uncovered perfidy in the Nixon White House, partly by discovering who checked out certain library books.

Nothing led anywhere. Not the counterintelligence, not the prospecting. Their kimberlite samples from Isom's Creek, Virginia, Kansas, and New York only confirmed their status as barren. Indicators in the Texas panhandle and elsewhere petered out to dead ends. There was, however, one place that Hugo was convinced no one had investigated deeply enough: Crater of Diamonds.

Wolfhard and Hugo invited Chuck to Tucson, where he gave a rambling presentation about heavy-mineral separation. Chuck had no idea they were looking for diamonds; Superior was also hunting gold, and he figured that was what they wanted from him. His lecture was full of stutters, tangents, and inconsequential details. Hugo loved it; he recognized a kindred prospecting soul. They hired Chuck to process some gold samples from Nevada, and he did a good job. In fact, if Superior had moved a little faster, the samples would have led them to a minor deposit; instead, a competing company staked it out. Shortly, Wolfhard called Kelowna. "Can you do concentrations for diamond indicators?" Chuck thought fast. "Sure. Sure, I can do that. No problem." He still had the jar from Kimberley to practice on. They advised him this would be a major, long-term project. Superior advanced him money on contract, and Chuck started building his dream lab.

The lab was situated, and always would be, in a low, unmarked cinder-block building off Kelowna's now growing and charmless strip-mall row, between a Quonset hut dealer and a maker of curtains for recreational vehicles. Conveniently down the road was an amusement park called Original Flintstones Bedrock City. The first employee was Chuck's teenage sister, Carol, who rode there five miles on her bike each day. Chuck was used to being poor and was exceedingly stingy. Carol at first accepted donations of clothing in lieu of wages. Chuck needed someone with engineering expertise, so through a government small-business program hooked up with a retired Canadian Navy engineer willing to work for $3 an hour.

Together they figured out how to build an operation of sheet metal, custom plastic and glassware, and off-the-shelf machine components that would separate minerals on an industrial scale. Chuck started traveling back and forth to Tucson. Marlene kept the books, paid the bills, ordered supplies, and made travel arrangements, as well as packing clothes, lunch, cash, and credit cards.

Chuck worked 9 A.M. to 2 A.M., testing and refining the lab, and before long the little walled-in parking lot in back was piled with twenty-pound plastic sample bags of sand from the United States. From exactly where, Hugo would not say; Chuck's job was to sort, not ask questions. Depending on the particles' fineness, each bag contained between 5 million and 20 million grains.

Inside the garage bay by the parking lot Chuck tipped the contents of each bag into a deafening, yellow baby dinosaur–size apparatus, which shook and water-sprayed the stuff violently down through a series of ever-finer circular sieves. It soaked anyone running it but sorted each sample nicely by size. He selected smaller size fractions, most likely to hold indicator minerals, dumped each in its own stainless-steel kitchen pan, and set the pans to dry in a giant homemade oven eight feet high.

The next room over was heavy liquids. Most geology labs used the nasty substances by the cup; Chuck ordered fifty-five-gallon drums. Under a big fume hood he had twenty deep glass funnels filled with dark-brown tetrabromoethane (specific gravity: 2.96) or methylene iodide (specific gravity: 3.27) and attached to mazes of copper and glass tubing that filled or drained them. He dumped the grains into the funnels and watched the fun begin. Everything lighter than the liquids floated—usually 90 percent or more. Once the heavies settled to the bottom, he drained them out through a spigot, washed off the poisonous liquids with equally poisonous acetone solvent, and recycled the heavy liquid for the next batch, all in a closed system.

The magnets were next door. Chuck bought eight identical sewing machine–size devices that took the particles distilled by the liquids and shook them down a sloping chute toward a magnetized wheel. Anything magnetic—say, ilmenite—would stick to the wheel. The rest—say, garnets or chrome diopsides—spilled into a separate pile. Then there was a big glassed-in box. Chuck dumped the grains onto a metal chute, turned on the juice, and a giant Frankenstein-like arc of electricity crackled through. This gave each particle a greater or lesser electromagnetic charge, depending on what it was. Again the particles slid down a chute and ended up sorted in different piles.

By this time the twenty-pound bag was generally pared down and sorted into seven or so tiny paper envelopes of heavy-mineral dust, each with different weights or magnetic qualities. Then the moment of truth: He poured the grains into a petri dish, put them under a microscope, and picked through rapidly with fine tweezers. Under strong light and fifty times magnification, they came alive on the dish's glass bottom. Each was a world in itself, with its own map of eroded angles, pits, valleys, ridges, continents, and oceans of various colors and shapes. Some were like geodes, cracked open to reveal crystals within crystals. Others were transparent, with fine inclusions and veins of other substances visible inside. He was good at identifying many by color, fracture, transparency, or texture. From the jar of Kimberley minerals he saw that pyrope garnets and chrome diopsides were generally identifiable by look alone; ilmenites nearly so. He hired a few poorly paid workers and showed them how to work the equipment and, when they got good enough, to pick samples.

It emerged that Chuck Fipke was a genius at this. In the first batch of thirty sample bags, he concentrated almost all the grains and found nothing. He had a microscope at the house, so he took the last bit with him one midnight. While he picked, Marlene called from the kitchen for the second time that supper was getting cold. Ignoring her, he flipped over a harmless-looking black sand grain. Stuck to its underside was a single fleck he recognized straight off for its striking green shade and the perfect way it cleaved on broken edges: chrome diopside. He picked it off and stared; it swam under his eyes with the intense coloration of a planet with a poisonous atmosphere. It was the only grain from approximately 300 million—a trial run from the old goldfields of California, where Hugo had sent some samplers.

Hugo was impressed. "If this guy works in the field like he works in his lab, we've got it made," he told Wolfhard. Shortly Hugo called Chuck. "Would you like to take some samples for us?" he asked.

Chuck had not seen Stew Blusson for years. He never forgot him, though. When he had time, he called to say hello and get back in touch. They struck up a new relationship, and eventually Chuck came down to asking Stew for some help organizing a sampling crew.

The timing was good. Still mapping the Mackenzie Mountains in the Northwest Territories, Stew was also having enough success on the stock market that he now had his own small helicopter. He continued to ignore Mother's pleas to marry. However, things were changing. By now the GSC had been nearly everywhere once. Stew hated the prospect of merely refining maps he had already made. Pioneer days were over; it was time to do something new. Some GSC colleagues were now irked to see him increasingly take on outside consulting jobs looking for metals and, they suspected, prospecting on his own account. He was not quite the same man.

As Stew was distancing himself from the GSC, he came to see the lab. He was impressed, too. Together he and Chuck made a short prospecting trip to the southern Canadian Rockies to help a Kelowna prospector look for copper. Stew flew them in, then watched Chuck plunge into an icy creek up to his waist, nearly disappear while digging a hole, sieve the material down in record time, and come out smiling, as if stepping out of a phone booth. This was the guy he remembered. Stew knew nothing about diamonds but figured he did not need to; he envisioned the two of them running a small GSC-like wilderness operation. Chuck enthused over Stew to Hugo, and Stew started as part-time crew manager, for a handsome per diem.

Brent Carr, an unemployed Kelowna truck driver, saw an ad in the *Kelowna Courier*—NEEDED: EXPERIENCED SAMPLERS. REMOTE AREAS. There was a phone number. Carr, a tall, dignified man who loved being outside, had no idea what samplers sampled, but he liked remote areas. His call was answered by a high, childlike voice. "Is your father at home?" Carr asked politely. "My father?" said Chuck, fumbling for his father's number on the farm. Carr realized his mistake. When told he would have to hike, camp, and fly for a living, he begged. Chuck also hired Paul Derkson, a lean outdoor-sports fanatic whom Stew had met on an airplane, and Dan Tomlin, the hyperenergetic, underemployed teenage son of a lab worker. "Dynamite Dan," as they called him, sported a heron's expansive frame, a plierslike handshake, and one of the continent's first male nipple rings. Others would come and go, but these were the core, the faithfuls who would dig the holes like John Williamson's Issa and

Ibrahim. For thrifty Chuck, they were ideal; they viewed this as a permanent campout with lots of beer, and reveled in using their resilient young bodies so extremely, genuine slaves would have been a poor bargain by comparison. The pay was $50 a day, Canadian; hours, unlimited.

One spring morning Chuck and Stew picked up Brent Carr for the first expedition. Chuck was driving a battered black 1968 Oldsmobile sedan with a souped-up engine. His head did not reach far above the steering wheel, and his teen driving habits were intact. As soon as Brent got in back, Chuck peeled out and ground destructively up through all the gears. Heading into the mountains along a twisty, narrow road, Chuck said straight out they were looking for diamonds. He got so excited talking about it that he forgot where he was going and veered toward the shoulder. At the last second, he bounced like a pinball off the invisible edge of disaster and headed for the center line. They came up behind a black Porsche. Brent gasped as Chuck veered around and offhandedly drag-raced the Porsche, right at a blind curve and a cliff. Stew sat next to Chuck hard-eyed and unflinching like a copilot and calmly told Chuck to take his foot off the gas, but Chuck did not seem to hear. He just kept talking and driving. Carr began to reconsider his employment choice, but it was too late. Chuck would not even say where they were going.

This became standard practice. On Hugo's orders, no one beyond Stew and Chuck was to know destinations in advance. That way there could be no leaks. The crew was generally just told to show up at an airport at a certain time, how long they would be away, and what kind of clothes to pack.

When Chuck had eight men, including some temporary hires, he told everyone to show up at the Vancouver airport. At the departure gates, he handed out tickets. They read: ALBUQUERQUE, NEW MEXICO. Chuck saw them off and said he would catch up later; he had to go back to the lab. Stew and the samplers got on the plane, stopping over in Calgary, where they passed through an outpost of U.S. Customs. One teenage sampler had chosen to pack an ounce of marijuana in his sock. The ever-vigilant Americans detected him instantly; they called the RCMP and had him dragged away in handcuffs, crying. The crew was one short already.

At Albuquerque airport they rented four identical white Ford Broncos with New Mexico license plates. Then the samplers learned they were not going to Albuquerque. They were driving to southeast Arkansas, 900 miles off. Dynamite Dan filled one truck with vast quantities of gear he'd brought to cover all contingencies—boots for wading, boots for snow, boots for logging, a large assortment of specialized outdoor clothing, big hunting knives, boxes of nuts and bolts, spare wires for his stereo, and the stereo itself, which he plugged in and cranked to 10. "Think Pink" was his favorite recording at the moment. Dynamite Dan drove much like Chuck, and when he discovered that in some American states you could buy minor explosives at truck stops—illegal everywhere in Canada—he emptied his wallet and threw firecrackers out of the moving vehicle onto Interstate 40. He was not called Dynamite Dan for nothing. Thus they cruised toward Crater of Diamonds.

The pipe was a public venture now. In 1972 the Arkansas state Department of Tourism had bought it for $750,000 and turned it into Crater of Diamonds State Park, complete with campsites, a nature trail, and rangers. Otherwise it was unchanged; anyone could hand-dig for $2 a day and keep diamonds. Weathered wooden signs dotted the wide, muddy field where major ones had turned up— cumulatively, quite a few. The half-dozen smaller outliers, still privately owned, remained abandoned in the woods.

To check things initially, Hugo and Waldman had simply bought tickets. Once inside they blinked; it was like 1870s South Africa. With the current high diamond prices, hundreds of diggers were there, many regulars—retirees, out-of-work loggers and such, with $75 season passes. Deep, dangerous holes were everywhere, bordered by pans, rockers, sieves, coolers, and folding lawn chairs. Some diggers carried flat plastic boxes in their back pockets with finds for sale at tourist-trap prices—$100 for tiny, brown, eaten specimens that looked like rotted teeth, worth maybe $10. On Howard Keck's money, Hugo and Waldman bought dozens for study. They came across one stooped old man who was prospecting by poking the ground with the tip of his white cane. From his pocket the old man pulled a clear, beautiful stone. He wanted $17,000 for it, and Hugo saw it was worth every cent. His heart raced. He briefly considered buying it for his wife, Nora, but could not afford it.

After this, they hung around quite a lot. Hugo read about the conspiracy theories. He wondered if they had some truth; to him, the pipe's historical yield seemed respectable, given all the obstacles. He made intricate calculations of possible diamond grades, qualities, and profits for Superior, given different scenarios. It seemed promising, but he needed a bulk sample of the ore—not permitted in the park. Several times they filled their backpacks, then sweated like drug smugglers as they walked out through reception. This was not working. Finally Waldman made a deal with a local digger. Little by little the man unobtrusively hauled 2,500 pounds in 100-pound bags through rattlesnake-infested brush to the perimeter. It took days to get it all over the chain-link fence into a waiting truck. When they processed it, it contained a fair number of microscopic diamond chips—enough, Hugo believed, to suggest the presence of many big ones farther below. The harmless old codgers with their lawn chairs now looked to him like deadly competitors. He conceived the idea of seizing Crater of Diamonds from them.

Hugo soon became a regular visitor to Richard Davies, head of the Arkansas tourism commission. He told Davies that if Arkansas would let Superior dig a big, deep hole, instead of having all these little, shallow ones, the state might make millions in royalties. Davies was affable but firm: State parks were not for sale. Hugo persisted. He commissioned a firm to do a statewide poll to see what the public thought of strip-mining a state park. The public hated the idea. He got the tourism commission to put it to a vote. He lost 12 to 1. Those who hated the idea most of all were the residents of Murfreesboro, which thrived on tourism and where the mayor and the editor of the weekly *Murfreesboro Diamond* violently denounced Hugo's idea. He just kept coming back with more and better proposals, including one for an underground shaft that would let the diggers stay.

The governor of Arkansas at this time was a relatively young man and indirect beneficiary of De Beers: William Jefferson Clinton, who had studied at Oxford on a Rhodes scholarship. When Clinton heard of the proposal and the possible financial figures, he appointed a separate commission to short-circuit the tourism commission and examine the facts. Hugo was impressed with his education and open-mindedness. "You know, Clinton will be president some day," he said to Robley Berry. Berry, a northerner, snorted in derision.

Meanwhile Waldman opened a one-room office off the Murfreesboro town square. To deflect attention in the hostile atmosphere, it was located inside the Farmers Insurance office, behind an unmarked door. They were certain no one even knew it was there until the bomb exploded.

It was a fairly powerful device, detonated by an adjacent bank's drive-in window. It was late on a Friday night, so fortunately no one was around to get killed, but it blew out a lot of windows, including the one to Waldman's office. By morning, the town was alive with rumors that a competing company was telling Superior to get out of town. Others said it was a local giving the same message. Either way, most folks approved. Waldman's wife started keeping the doors locked, and Waldman was careful where he went.

A few days went by. Then police arrested a local drunk and well-known incompetent. It turned out there was no terrorist plot; he had been trying to break into the bank.

In any case, Superior was now convinced Crater of Diamonds had to go on the back burner; instead, they would reconnoiter nearby. They had to move fast, though, for by now competitors had the same idea. Among them were ARCO, the British Selection Trust, and Chuck's old employer Cominco, all clogging area motels. Hugo hired an airplane, loaded it with magnetic instruments, and sent it buzzing over 900 square miles of surrounding countryside to detect a great many barns, backyard appliance dumps, chicken coops, and rusting pickup trucks, but no new pipes. So Hugo called in Chuck's newly recruited foot soldiers to cover 1,500 square miles—four rural counties—with sampling for heavy minerals.

On their first evening in Arkansas, the sky outside the motel darkened strangely long before night. Stew turned on the TV just in time to hear a tornado warning. Neither he nor Brent, who had never been farther south than Washington state, had ever seen a tornado, so they went on the balcony to watch. Only later did they realize that everyone else was hiding in the basement. Just then the electricity popped out. A psychedelic roaring started, and trees bent double. Down the road part of a Baptist church ascended to heaven, then changed its mind and crashed in a heap. Boats in a nearby marina rafted from the water and cruised briefly through the air. That was it. In a few minutes it was over. They were not in Canada anymore.

In the morning, while firemen cleaned up wreckage, Stew bought a pair of denim overalls and a straw hat, which he noticed a lot of the locals wore. He began practicing what he imagined to be an Arkansas drawl. Stew had once read the James Dickey novel *Deliverance*, in which a couple of city fellows on a rural river trip get ambushed and sodomized by gun-toting southern hillbillies. He hoped that maybe if he looked and acted right, no one would notice he was not from around here.

Chuck, still absent, had drilled everyone carefully on how to sample. Over the years, he had developed a finely tuned sense of how hydrologic forces deposited particles. Usually he marked Xs on topographic maps at regular intervals, then headed out with plastic sample bags. The ones he used now held twenty pounds each. Rivers or creekbeds were preferred; stream junctions were ideal because you got two drainages at once. At the X you sought a spot where water slowed and dropped suspended particles—often the inside of a turn, or head of a riffle, where a gravel bar formed. Standing in shallow water, you dug about three feet. The hole's bottom was where you actually took the sample because heavies tend to sink. Under small boulders was good because particles wormed in and got trapped. While leaning into the watery hole, it was important, though awkward on the back, to lift the shovel straight out and level with the bottom each time, so as not to lose any material or touch the sides of the hole, which might introduce material from other layers.

The stuff went into a large sieve, which in turn sat in a bucket of water. You squeezed dish detergent into the water to break up fine particles, then shook the sieve forcefully to get sand through the fine holes—usually 400 per square inch. It took considerable upper-body strength. Pebbles and cobbles, to be discarded, got tossed over your shoulder as you went. Sieved sand settled on the bottom of the bucket. You poured off the water, then scraped the sand into your bag, being careful to get every grain.

The whole process was repeated until you had the twenty pounds. It all took about an hour of backbreaking labor—three or four, if you were less of a man than Chuck or Stew. Samplers were expected to take up to eight samples a day.

As for less technical details: Unlike Canada, where most land is open, there were "No Trespassing" signs everywhere in the States. Stew and Chuck both despised such signs—rocks existed before anyone presumed to own them—so it was understood that everyone could wipe their asses on the signs if they liked. If challenged, samplers were to be nice and say they were engaged in a harmless U.S. government "mineral survey." Chuck had a line: "Let me tell you about rules. Rules are made to be broken, and I break them all the time!" Then he would laugh and laugh and laugh.

The first day, Stew and Brent emerged from sampling a roadside creek on private land to find a police cruiser sitting behind their Bronco. The officer was friendly. He had noticed they were newcomers. "How can you tell?" drawled Blusson. They were taking a lot of chances by going out in the woods alone, advised the cop. Especially on posted property. Be careful. That was all he had to say. Stew was given to understand that the police might not be able to help them if they got in trouble. The other three Ford Broncos received identical visits that day.

The more immediate threat, as far as they could apprehend, was not landowners but the land. After cool, piney British Columbia, Arkansas seemed about equal parts mud, brambles, and snakes. Counting roadside rattlers became a sport; Stew jokingly announced that less than one per hour meant you were shirking. A water moccasin zipped between Paul Derkson's legs and was gone before he could yell. Another arrived when Derkson was halfway in a hole, and together they inspected each other at a range of four feet until the sample was taken. Getting into creeks often required belly-crawling through humid, ground-level tunnels in the underbrush in which briars and ticks drew blood. Within three days one man from British Columbia previously unacquainted with poison ivy went to the hospital, converted into a massive blister. Only Dynamite Dan, whose wardrobe included a heavy, full-body canvas suit, escaped all harm. It was hard even to find the theoretical nice gravel bar at a stream junction; everything was a swamp.

Chuck kept in daily contact by phone from the lab, but when it became obvious work was not proceeding at the desired pace, he flew down and threatened to fire any laggards. Brent told him he was crazy; fire away. They went out, and Chuck crawled rapidly forward through one brush tunnel to show them how it was done. Then he came crawling out backwards even faster with his butt in the brambles, eye to eye with a moving water moccasin. When Chuck got thirsty, he stuck his face in a creek and drank. Within days he had such bad dysentery, he could not drink coffee for the next four months. After a week or so he went back to Kelowna to run the lab and left the employees to their own devices.

Toward the end of a sweaty day, Stew and Brent climbed over a barbed-wire pasture fence and fanned out in waist-high grass along a stream to dig. Stew was almost done when he heard a rustle behind him. He turned to see a middle-aged man with cloudy eyes and a swollen red nose wearing a crushed-up felt hat. He held a double-barrel shotgun against his shoulder, with both hammers drawn. Stew had never noticed before how big and bottomless the holes of a .12 gauge shotgun barrel seem when they are pointed in your eyes. The irritating banjo-guitar duet of *Deliverance* (the movie version) involuntarily started in his head.

"What are you doing? What are you doing in my creek, boy?" the man hollered.

"Oh, *hi*," said Stew in his friendliest voice, making sure to keep both hands in sight. "We're just doing a mineral survey . . ."

"Mineral survey. Mineral survey," said the man. He did not want to hear it. He was convinced they were stealing gold from his creek; otherwise, why dig? Stew started to tell him there was no gold, then decided not to correct the man. "No, no, actually we're looking for diamonds," he said. He instantly saw that was worse.

"I mean, no, we're not exactly looking for *diamonds*. We just want to see if there's minerals here that might tell where diamonds *come from*. It's just a sample. You know, that state park over the hill . . ."

The man looked at him as if he were dangerously mad. He cornered Brent also and marched them both with the nearly full sample bags through the pasture and over the fence, where he pointed to the "No Trespassing" sign. Then up the embankment to their truck. The man's own pickup was pulled up in back. Stew tried appeasing him by showing off their equipment and explaining its uses. He gave the man a shiny new shovel and pan as a gift. The man warmed a little, uncocked the shotgun, and laid it across the hood of his pickup.

Then he saw the license plate.

"Mexico? You're from Mexico? You're *Mexicans?*" He fumbled behind him for the gun. Brent started clearing his throat to point out that the plate said *New* Mexico. Stew kicked him in the ankle and smiled, meaning: Shut the fuck up, Brent.

Just then another pickup pulled up, and another man with a shotgun got out.

"These here boys are from Mexico," said the landowner. The other fellow seemed to be a brother, or cousin. The whole round of accusations started up again.

Finally the Arkansans made the Canadians dump out the samples from their creek into the roadside grass. Then, for good measure, they made them drag the entire day's samples from a half-dozen other creeks from the back of the Bronco and dump those too. They made a sad, wet little pile. They were free to go.

Some of the half-dozen other companies now prowling bought rights to woodlands near the park. Selection Trust got hold of a small known pipe, surrounded it with eight-foot chain-link and razor-ribbon fence, and started digging.

Hugo could not bear not to know what was going on in there. One afternoon he and Waldman parked on a nearby dirt logging track, crept through a quarter-mile of thick timber and bushes, and peered through the fence. Hugo snapped pictures with a telephoto lens. There was heavy equipment at work, and a big rotary pan. Men in coveralls were moving about. Also an armed security guard in a blue uniform—who looked their way at just the wrong moment. The guard yelled, and they took off. Hugo, who loved to eat and got little exercise these days, was in poor physical shape, so his bearlike frame was impeded by bushes. By the time they got to the car, they were scratched and breathless, and the Selection Trust headman was waiting.

"Who are you?" he demanded.

"We don't have to tell you who we are," said Hugo. "This is a publicly maintained road."

"No, it is not a publicly maintained road. It is leased to private parties. Who are you?"

"It really is none of your business," said Hugo indignantly, beginning to catch his breath.

"This is private property. I'm calling the sheriff. I'll have you arrested."

"You can't have us arrested. We haven't taken anything."

The man looked in the car and saw a shovel.

"You're geologists!"

A big screaming fight ensued. The man kept demanding their identities. Hugo bellowed about American laws and constitutional rights and tried outbullying his opponent by presenting his sheer bulk. The argument got nowhere. They climbed in the car; Waldman started the engine and began to drive. Hugo made him stop a few dozen yards on. He was not about to be chased.

By this time the uniformed guard was coming up at a run with a rifle in his arms. "I'll have you shot!" screamed the Selection Trust man. Waldman hit the gas.

To be fair, there were some friendly people. One sunny day Stew bought the $2 ticket to Crater of Diamonds and found himself surrounded by the usual gaggle of semiprofessional prospectors, all eager to stop and chat. Many had shiny, shallow pans nearly the size of bicycle wheels—*sirucas*, much like those used by the *garimpeiros*. Some world traveler had introduced them here, and locals discovered they worked in the right hands. They showed Stew how to throw a few shovelfuls in, then deftly bounce and twirl the *siruca* in a bit of water. This centrifuged lightweight elements to the sides and heavies toward the middle and bottom. Then with a fast motion you flipped the pan upside down. If you did it right, everything inside cohered and landed in the inverted shape of the *siruca*, like a well-turned flapjack. Halfway to the shallow mound's apex was where you would find a diamond, if there was one.

Stew watched carefully, and they handed him a pan. He knelt, filled it, and worked vigorously while they cheered him on. When it came time for the flip, he performed it perfectly, and everyone fell silent. "Oh, there's something," said one digger. Stew saw it at the same time. It caught a ray of sun. He picked it up between thumb and forefinger, calmly unfolded a little magnifier hanging from his neck and looked. It was a limpid half-carat diamond, shaped like a teardrop. Turned the other way, it was a heart. He could not believe it. His first try. Everyone patted him on the back and congratulated him on his beginner's luck.

Stew would never forget the day he found his first diamond: Sunday, May 11—Mother's Day. When he got home, he took it to Burke's, a fancy Vancouver jeweler. He wanted it preserved rough. He had it wrapped around by a gold locket and mounted on a chain, hanging in the heart shape. When the jeweler finished, Stew took it over to his mother's nearby apartment. Now in her mid-seventies and widowed, Edith Blusson seemed reconciled to his lifestyle, though she worried. She knew he flew a lot. Every time she heard of an air crash, she fretted until she was sure it did not involve Stew. On infrequent visits, she listened hungrily for tales of where he had been. She especially liked hearing about cute, young animals he saw, so he emphasized those; the ones about bears scared the hell out of her. Edith Blusson said many Hail Marys for her son. Stew helped his mother fasten the locket around her neck, and she kept the diamond as a treasure until the day she died, many years later.

In Kelowna Chuck sorted the Arkansas samples and quickly isolated a few that held indicators. From creeks near Crater of Diamonds, the minerals did not seem to emanate from Crater of Diamonds itself. Waldman took a topographic map, bushwacked upslope through the woods, and came upon an oval area where vegetation thinned and the soil looked subtly greenish—a previously unnoticed pipe not a mile from the park. Beginner's luck.

The land was owned by a paper company. After some wrangling, they leased the land, cut a road in, and started test-digging. Hugo was sure they had their diamond mine. He called in a backhoe to run fifteen-foot-deep trenches through the pipe and extracted 170 tons. They put up a washing plant and ran the ore, and when the results came back, they knew they were in trouble. There were only fifteen diamonds, weighing a quarter-carat all together, brown and rotten.

With so much new prospecting underway, samples from Murfreesboro and elsewhere were circulating among universities and exploration companies. Armed with new instruments, they were drawing ever-finer distinctions among the rocks. Up to now, everyone had considered Murfreesboro classic kimberlite; now they began to reconsider. The rocks appeared to carry a subtly different chemical signature. Researchers had identified a related rock called lamproite, thought to come from less diamond friendly edges of cratons. It could carry diamonds, too—but generally even less than kimberlite, and usually small and blasted, like the ones here. The Arkansas rocks, including Crater of Diamonds, were soon reclassified as lamproites.

Around this time, an increasing flow of papers reclassified other pipes as well. Only a petrologist—a professional rock-namer—could love the fine distinctions. The dikes around Ithaca, New York, were redubbed alnöites. Certain odd plugs in Montana became ultramafic lamprophyres. The Diamond Hoaxers' garnet sources in Navajoland became serpentinites. Then you had your carbonatites, syenites, aillikites, nepheline syenites, nephelinites, picritic monchiquites, and mugerarites. Not to be confused with ijolites, melilitites, melnoites, dolerites, altered lamproites, feldspathoidal lamprophyres, uncompahgrites, ouachitites, alpine peridotites, ophiolitic peridotites, monticellite peridotites, minettes, limburgites, trachybasalts, katungites, kamafugites, olivine leucitites, vogesites, or even meteorites. All with poor diamond potential.

Suddenly everything seemed more complicated. Hugo sent some of the Arkansas minerals to John Gurney in Cape Town. Gurney was supposed to be telling them what they were looking for, but to Hugo's and Jennings's annoyance, he was late with his report.

Chuck and Stew had no problem with any of it. They got paid whether diamonds were found or not. And Hugo already had another job for them: two more prospects, both in the American West, both strangely close to the site of the century-old Diamond Hoax.

CHAPTER 11

The Secret of the Anthill

In a way, it all started the day Tom McCandless lost the use of his legs forever. It was summer 1975, and twenty-year-old McCandless was hell-bent on racing his dirt bike across the western desert. One day he was traveling at high speed when he came over a rise and found himself confronted with a choice: He could either hit a desk-size rock in front of him and die, or vault it. McCandless gunned the engine, ramped onto a hump, flew twenty-five feet in the air, and went flying—just like Superman, he would always remember. When he got out of the hospital just in time for fall semester at the University of Utah, he had the new experience of meeting with a counselor for the handicapped.

"What do you want to do?" asked the man.

McCandless's father was an off-and-on gold prospector, and McCandless had always liked rocks himself. Without thinking much, he said: "I want to be a geologist."

The counselor tried to dissuade him. "How are you going to do the field stuff?" he asked.

"I don't know. I can just figure that out when it comes along," was all he could think of.

After some time, the counselor quit arguing and pulled open a desk drawer. "Well. Then take a look at this." He fumbled under some papers and removed a plastic aspirin bottle. It was filled with tiny, colorful candies, dark red, orange, and green.

"I used to know an old sheep rancher," said the counselor. "About twenty years ago he picked these up. Found them on some anthills. I'm not sure what they are."

McCandless recognized the dark pyropes immediately because his birthstone was garnet. The orange ones, he later learned, were almandine garnets; the greens, chrome diopside. He learned this when he started taking geology courses and found a textbook picturing the minerals—alongside a few sentences describing Gavin Lamont's Kalahari anthills. He figured he had the ultimate student project: finding a diamond mine.

Heber Campbell, the sheep rancher who knew the location of the anthills, was long dead. In the vast badlands where Utah, Wyoming, and Colorado all meet, McCandless and the counselor spent a year crisscrossing his haunts in a van. They had no luck until one day in a remote valley the van had a flat. The counselor got out to fix it, and there by the dirt track was a volcanolike ant abode six feet across, strewn with red, orange, and green grains shining like dew. It was maybe forty miles west of the Great Hoax site, though no one knew for sure; after 100 years the exact location was lost. McCandless figured out a way to do the field stuff. He drove a four-wheel-drive truck with hand controls to the anthill valley, set his wheelchair out the door, dropped in, and got going across the hard-packed, rolling surface.

In this place, anthills sparkled all around, some made of 10 to 15 percent indicator minerals. Not only were the garnets huge, but under the microscope many also had kelphytic rims—soft, brain-textured ridges on the surfaces formed by chemical reactions during the kimberlite eruption. Such rims quickly erode to nothing when exposed or transported any distance, so it appeared to McCandless that the ants had bored straight to the source.

For a long time the only others who knew the secret were his academic adviser, his girlfriend, and a couple of fellow students, whom he paid in beer to help dig up anthills. This infuriated the ants, which poured out and stung everyone. The group also traversed the area with hand-held magnetometers and bored as far as eighteen feet with a small auger. They never could find any kimberlite. Over a couple of seasons McCandless drank a lot of beer out there and wondered what to do next.

Eventually his adviser mentioned the project to a Wyoming Geological Survey man. Shortly after that the man received a visit from Hugo—just making the rounds, looking for useful information. The geologist did not know the anthills' location and was unsure of their potential, but he casually brought them up. If Hugo had been a German shepherd, his ears would have stood straight up. "How interesting," he said coolly. "Can you get me in touch with this young man?"

McCandless figured big mining companies would cheat him, but when Hugo phoned he immediately trusted him. Hugo spoke with too much erudition and passion to be a pure businessman. Then he heard Hugo's deal: Show me your anthills, and I'll give you a million dollars for every diamond mine we find. To get a fast overview, Hugo wanted to fly over in a helicopter. Negotiations got serious, and McCandless figured there was something he had better mention. "Well, you know, I don't know if it makes any difference. But I'm in a wheelchair. I mean, I don't know if you can fit a wheelchair in a helicopter."

"We'll get a bigger helicopter," said Hugo without missing a beat.

This is the guy I need to work for, thought McCandless. He has no limits on the way he thinks.

When they landed in the secret valley, Hugo lifted McCandless out of the aircraft, set him in his chair, and side by side they traveled. Hugo was staggered; he was in the Valley of the Indicator Mineral.

Rob Berry confirmed it was federal property, run by the Bureau of Land Management. And so, owing to the provision in the 1872 General Mining Law engineered by the Diamond Hoaxers, it was open for staking. Hugo called in crews and instantly dotted the badlands off to the horizon with upright white plastic pipes as claim markers. It was time to call in the Canadians again.

Hugo had something else to check: a dozen true kimberlites 240 miles east along the Wyoming-Colorado border. The first were spotted by Colorado university students in the 1960s but were not publicly identified until 1971. In 1975 a USGS technician wrecked a carborundum wheel while sectioning the rock for study; he had hit a microscopic diamond. It turned out several of the pipes had them. They seemed too tiny and sparse to be worth much, but within months the first of twenty-some companies were acquiring rights from landowners.

The biggest pipe, in a mountain pasture on the old Sloan Ranch outside Fort Collins, Colorado, was still available. The owner was a tough, leathery sometime prospector named Frank Yaussi. Yaussi had long sold the rock, not knowing what it was, for calcium-rich fertilizer and decorative building stone. Made into polished terrazzo, among other places it graced the floor of the men's room in the Cheyenne, Wyoming, airport. After the identification, a wiry, emaciated-looking man in his seventies had shown up and tried to buy it. He was gentlemanly and honest about his identity: Dr. Arnold Elzey Waters Jr. Yaussi liked him, for despite his age and underfed appearance, Waters stuck his arms up to his elbows in a cold stream to pan for minerals. Waters and Yaussi could never agree to a price, though. Later, Exmin also bid without success. Hugo now bested their offers: $300,000 just for permission to dig and a huge percentage of anything they found. Some Superior executives thought he had lost his mind, but Howard Keck approved, and in 1979 Hugo set up a $2-million ore-processing plant on Sloan Ranch and called in heavy equipment.

Superior made a terrible neighbor. Scared of spies, and now thieves if there were diamonds, Hugo had an ex–U.S. Army officer put up razor wire around the plant. The man donned fatigues and mock-charged it himself to make sure no one could get over. Hugo hired armed guards. Rumor somehow got out in Fort Collins that they had Uzis—illegal and untrue, but a good rumor to have going. No one knew its source. The guards regularly shot up tin cans on a nearby hill with pistols and shotguns and took to stationing themselves on the gravel county road nearby. If anyone dared stop or even slow down on the road, the guards raced over and told them to move. They shadowed the next-door rancher as he checked on livestock until the rancher threatened them serious harm. In short order, though, they did chase several night intruders apparently trying to steal samples.

Meanwhile, Superior seemed to do its own spying. The main competitor was Cominco, which had leased a number of pipes nearby, but did not guard them as well. One morning the Cominco men returned to one pipe to find that someone had drilled a big hole in it during the night. Their main suspect: Hugo, though Hugo denied it. They were sometimes followed by people who resembled Superior people. Cominco geologist Howard Coopersmith discovered someone other than himself had entered other Cominco lands, bearing official-looking business cards reading HOWARD COOPERSMITH. The real Coopersmith was a shy, slight, ponytailed youngster; this other Coopersmith was described as pushy, six-foot-six, and topped with a shock of flaming red hair—much like one of Hugo's geologists. Cominco hated them. Of an evening, when Cominco geologists saw Superior geologists emerge from a bar and come walking down the street, they would cross to the other side and refuse to look. The atmosphere worsened when Cominco threatened to sue Chuck on the theory that he had stolen their proprietary mineral-separation techniques. The action never materialized, partly because no one clearly owned the techniques.

The Sloan Pipe turned out thousands of diamonds—though most were dust-speck size. At least the numbers sounded better than those from Arkansas. Howard Keck himself came to see them, along with Jennings's boss at Falconbridge. When Hugo met them at the Denver airport, Keck insisted on driving Hugo alone in a rental car so they could talk diamonds. Keck got so excited about the Sloan, he ran a red light, sat to receive a ticket from the state police for the offense, then drove far over the speed limit the rest of the way. At the plant, he picked greedily through the few tiny stones he had paid millions to find and held the biggest to the light between his fingers.

"This is the way you gotta do things in Africa," growled Keck at Jennings's boss; unfortunately, the Kalahari pipes were still showing conflicting results. Keck then beamed sideways at Hugo. Hugo later heard the Falconbridge boss had collared Jennings and humiliated him without mercy. It was sweet victory but temporary; secretly, Hugo knew he had better find something else fast, before Keck realized the miniature Sloan stones would never pay anywhere near what it cost to dig them up.

The sampling team charged across the Alberta-Montana border in Chuck's 1968 Olds and a C. F. Minerals Chevy Blazer, already rebuilt several times after Chuck had rolled it into snowdrifts, mudholes, and other roadside obstacles. A cold early winter wind was shrieking on the Great Plains, and Dynamite Dan was blasting "On the Road Again" on the stereo. The truck cab was freezing, so they constructed a nest of sleeping bags, blankets, and clothing in back, where everyone piled in like a family of squirrels. It was here, on the fenced-in former frontier, that they perfected the fine art of trespassing. It took several seasons off and on, but they finally got it right.

Stew tried to be civilized, given his past bad experience. If he saw a house within sight of an X on the sampling map, he would drive over, drink coffee with the ranchers, and tell them he had a college geology project. Most would not let him dig no matter what; they feared he would hit oil or some damn thing to upset the cows. With all the coffee eating his kidneys, Stew took to digging by moonlight, courteously backfilling his holes so that steers would not step in and break legs. The others were bolder; they just looked both ways in broad daylight across the open landscape and scooted under a fence. As they got better at it, some samplers packed mountain bikes for quick, soundless incursions. Then, in Colorado, Brent got an idea: The South Platte River ran through much targeted territory, away from the gauntlet of houses and fences on roads. He bought a used rubber raft, put in, and rode the current like Huck Finn. Every once in a while he landed to dig. Nobody saw him, and he had a splendid few days.

Dynamite Dan had no patience for such finesse. Confronted with a fence and no gate in sight, he would veer off the road at a good clip, crash the wire, go down a bank into the ditch and up the other side— sometimes with a quick stop to reattach a radiator or fender. Chuck was equally ruthless, and he expected everyone else to be. In Colorado one sampler saw a rancher driving toward him and dropped his filled bags in a creek to run. Chuck ordered him to retrieve them that night, but the man feared ambush. Chuck cursed, called him a pussy, and crawled through the dark drainage himself at 4 A.M. Later Chuck sent a more courageous sampler on a night mission to a U.S. military base, where the guards almost certainly had real machine guns. Another was a wide-open area from which a rancher had chased them repeatedly. There was early snow, so Dynamite Dan waited for a moonless night, donned a pure-white snowsuit, and walked a long distance undetected. At the desired creekbed he swathed himself in mosquito netting to camouflage his tools and quietly chipped through a thin coating of ice for the sample. The only impenetrable place was a spread in western Wyoming owned by the entertainer Bob Hope. There were electronic alarms all over. Chuck figured they had enough samples anyway, so they left Bob Hope alone.

It was not a bad life out there. Sometimes after a dawn-to-dusk workday they just lay down next to their trucks and slept under the stars. Occasionally they would find a hot spring and take a bath. Some nights they convened in back-road cowboy bars, often filled with single, available girls. Stew was happy to go to sleep early, but Chuck led the charge, buying rounds of beer, chatting up everyone in sight, doing backflips on the floor when he'd drunk enough, and making outrageous comments to or about the waitresses.

Naturally everyone soon had at least a few gun stories. In Canada gun ownership was strictly controlled, but in rural America they learned to identify .357 Magnums, .45s, .30-.30s, and all sorts of shotguns and revolvers by manufacturer and model, sometimes just by the kind of light, heavy, or medium click they made when irate landowners who caught them sampling cocked and pointed them. This was nowhere more frequent than in the largely roadless region around McCandless's anthills, ringed with particularly hostile ranchers. However, it was imperative to get in there; the mineral train, if there was one, seemed to run right across their land. Stew hit on the natural solution: a hired helicopter and pilot to lift himself and another sampler over miles of fences, in and out before anyone knew.

Early one morning Stew had himself dropped into an empty valley, and the pilot departed to drop the other sampler. Stew was barely finished when the aircraft came over a hill ahead of schedule and roared in violently with a fast turn. Before it touched ground, the door was open, and the sampler was frantically waving him in. Stew barely had his foot on the skid when they lifted off, and he turned to see what they were running from: a pickup truck careening around the hillside trailing a big dust cloud. In back were four men with rifles. They began firing from the moving vehicle.

They escaped, shaken that anyone would be crazy enough to try and shoot down an aircraft. However, the next day a rancher tracked down the pilot by the big registration number on the bottom of the craft, and asked politely if it was his. When the pilot said yes, the rancher knocked him to the ground with one punch in the eye and walked away. Stew had gotten the sample, though.

The closest miss came when Chuck was in a wooded area that they later learned was frequented by a violent biker gang. He heard a whoosh behind him and turned to see a broad-headed hunting arrow quivering in the ground. Chuck looked at the arrow and pulled it out; he thought it would make a neat souvenir to take home for Mark and kept it. There being no further movement, he went about his business. Later that day Stew saw the arrow on Chuck's dashboard and heard the story. It was evident Chuck had not considered what the arrow meant.

"Chuck! Chuck! You know where that came from? They're trying to kill you! For digging a hole! Don't you see?" Chuck looked at him blankly. Stew slapped his own forehead hard and groaned. Chuck was still coloring outside the lines.

"Oh. Oh, you think so? You really think so?" said Chuck in his customary meek voice.

After they had spent enough time together, Stew and Chuck became like their own little tribe, where each person knew his ritual place. They were yin and yang. Stew represented the intellectual, Chuck the irrational. Stew was the generalist, good at spotting rock formations and inventing ideas; Chuck was the specialist, zeroing in on sample sites and doing detail work. To the samplers, they were good cop and bad cop. If a sampler told Stew something was impossible, Stew just smiled. Chuck never accepted an excuse from someone who came back without the sample. Between the two of them, somehow everything got done. Over two years, Hugo sent them all over to check out leads—Montana, Arizona, New Mexico. Stew and Chuck ended up alone in the old California goldfields trying to hit the last, thorniest locations.

One of these was within an elaborate open-pit gold and mercury mine north of San Francisco. They decided to enter in disguise, wearing brown coveralls like the drillers wore. Stew took a clipboard to scribble on so he would look like a supervisor while Chuck dug. They were nearly done when a couple of men shouted at them. Stew maintained character and kept scribbling, but Chuck panicked, grabbed his samples, and ran. The men pursued. When Chuck's short legs would not carry him fast enough, he dropped the samples and hid in a dip. Here they discovered him cowering behind a rock. To his terror, one called him by name, as in a bad dream. "Chuck! Chuck! Don't you know me? It's me, Dick." They were fellow small-timers from Kelowna. They were snooping here, too, except they were interested in gold.

They needed another sample from near the mine, on a spread that happened to be owned by George Gamble of Procter & Gamble, maker of cosmetics and soap. Chuck had tried driving in once, but Gamble had personally collared him. Chuck must have looked hungry, though, for the magnate softened, took him back to the house, and fed him supper before kicking him off. It was a tough mission; the spot they wanted was in a small creek right next to Gamble's house. It was patrolled by guards similar to Hugo's own goons, who chased anyone who stopped on the public road. They drew up a plan, rehearsed carefully, and waited for nightfall.

It was dark and raining when Chuck drove a rental car up the road with Stew slouched low in the right-hand seat. There was a big culvert running under the road across from the house, and here Chuck slowed down. Stew cracked the door and rolled onto the wet pavement, over the shoulder, and down the slope. Chuck just kept driving.

Stew entered the culvert in water up to his chest and emerged on the other side to see Gamble's house perched on the bank. The good life was underway in the living room above; warm yellow light, people drinking wine, talking, and eating hors d'oeuvres—the kind of dinner party to which Stew Blusson would never be invited. He spent the next forty minutes furiously digging and sieving, the noise covered by rain. When he was done, he crawled back through the culvert to wait. On schedule, Chuck came down the road the other way, rear left door ajar, and slowed. Stew scrambled crablike over the pavement, threw himself in, and slammed the door. He lay on the floor panting. When they got around a curve, they could see the soft light of George Gamble's windows. They began to snicker. Then to cackle, and wheeze, with uncontrollable laughter, until they both could barely breathe.

Back in Kelowna, when not on the road, the samplers doubled as lab technicians, running the machinery and fixing equipment. Few samples yielded much, except the anthill region. Here indicators were spread over 800 square miles. McCandless could see no pattern, but he figured one thing out by doing some rudimentary entomology. The ants were *Pogonomyrmex occidentalis*—western harvester ants. He realized why they were called harvesters when he noticed that the fragments of a beer bottle he and his friends had smashed into many pieces on a rock had all been carefully moved; now they were atop an anthill twenty feet away. It was one of those stupid accidents of science. The ants did not dig indicator minerals; they gathered them off the surface.

This did not tell them where the minerals came from, but it was a clue. He and Hugo drove across the desert looking for more. When they got to terrain too steep or rough for a wheelchair, they disembarked and Hugo picked up the 120-pound McCandless in his arms and carried him like a child. For longer treks Hugo hoisted McCandless onto his back, and off they went with their lunch, chatting away, pointing things out to each other, stopping once in a while to look at rocks.

One day McCandless was down in a draw when Hugo came trotting down off a steep hillside, out of breath.

"Come on, I have to show you something."

He carried McCandless up the 30-degree slope and set him on a boulder.

"Look at that."

A little stream of chrome diopside crystals ran downhill along the sunbaked surface for six feet. At the apex they dug and pulled out a small rock. It was not kimberlite; it was conglomerate, loaded with indicator minerals and little pieces of granite. They both groaned. It meant the pipes had erupted in a granite; then both pipe and granite had eroded, cushioned in a fine-grained ancient mudflow, which hardened into this conglomerate. That was why the garnets still had brainlike kelphytic rims; the mud protected them. The conglomerate, scattered all over the place, had to be ancient, and it was now dissolving, dropping minerals out, which the ants gathered for some unknown purpose. Which meant they were screwed; the pipes had been gone for a billion years, the contents scattered everywhere.

Then they made a very strange discovery. A retired USGS geologist, Lowell Hilpert, came upon another trove of mineral-rich anthills. He mentioned them to Dan Hausel, of the Wyoming Geological Survey. Hausel told McCandless, and McCandless went there. The anthills were ranged around a big reddish sandstone ledge, and nearby were sieved-out piles of gravel, indicating that prospectors had visited. McCandless dug into the anthills and sifted; he quickly found seventeen rubies, thirty-four extraordinarily beautiful pyrope garnets—and seven small diamonds. Problem was, rubies are not normally found around here. And the garnets had distinctive orange-peelly kelphytic rims, unlike the brain-textured ones he was used to—a sign they had come from some other place, too. As for the diamonds: They were small and industrial grade. Some trained eyes might have said they had been transported by a mysterious geologic force all the way from Africa. They had rediscovered the site of the Great Diamond Hoax.

Great news came from South Africa. John Gurney had found the secret indicator-mineral formula.

Gurney was a blunt, pugnacious man with the face of a dockworker and the gritty accent of his native Liverpool. Once a harddriving athlete, he had planned a professional soccer career. Then his family moved to Cape Town, and a couple of flying tackles left him with permanent hobbles in both legs. Needing a sedentary job, he instead went into geochemistry. He turned out to have a special talent for it, and in Cape Town kimberlite was the main game. Just as he was getting started, he got a lucky break.

There was a new way to analyze mineral chemistry: the electron microprobe. This focuses a beam of electrons on a spot as small as $1/1,000$ of a millimeter. Atoms get excited by the beam, and each element emits a different, identifiable wavelength of X-ray. Detectors pick up the X-rays, a computer sorts them out, and seconds later a screen displays the percentage of any element you wish, out to three decimal places. It was revolutionary for chemistry in general and indicator chemistry in particular. It was now possible to analyze the tiniest diamond inclusion or indicator grain.

In 1965, while Gurney was still a student, the Carnegie Institution of Washington, D.C., got one of the world's first microprobes. Researcher Henry O. Meyer put flawed diamonds into a tiny vise, cracked out the inclusions like nutmeats, and microprobed them. He confirmed that pyrope garnets and chrome diopsides are common inside diamonds. Meyer also identified inclusions of shiny black chromite, previously mistaken for ilmenite. Ilmenite itself, long considered an important indicator, was weirdly absent. Meyer then started building chemical profiles of typical inclusions, which obviously formed with or near diamonds. The unstated implication: If you could build a good database of inclusion chemistry, then find mineral grains outside diamonds with compositions matching the ones inside, you would be on the trail of a diamond-rich pipe.

Gurney heard about Meyer, and in the early 1970s went to Washington to do postdoctoral work at the neighboring Smithsonian. There, in a back room, he discovered that mineral curator George Switzer had a small box of garnets he had collected from De Beers's Finsch Pipe—from which Chuck had received his own souvenir jar. The Smithsonian also had a gift from Gardner Williams, sitting there since the turn of the century: a huge collection of assorted minerals from the nineteenth-century De Beers mines. They lay untouched in their dusty boxes. From both collections, Gurney picked out minerals in shades that looked like Meyer's inclusions and put them under the Smithsonian's own brand-new microprobe.

After a year, he returned home to a research post at the University of Cape Town and microprobed more minerals. At first he got samples from an unexpected source: Barry Hawthorne. Hawthorne was willing to supply material to outsiders because he figured that whatever information he got back would outweigh any leakage. He gave Gurney diamond inclusions to study, helped finance assorted projects for Gurney's students, and, most importantly, let Gurney come with a garden trowel and fill plastic bags with concentrated heavy minerals off the mines' dreg piles. Apparently Hawthorne did not fully grasp what

Gurney intended to do with them. Then one of Gurney's students wrote a thesis implying that microprobe analyses could help prospectors. Around the same time, Gurney went to work for Superior/ Falconbridge. Hawthorne had the dreg piles ringed with security.

Hawthorne, still friends with Gurney, politely begged him to drop this line of inquiry. But Gurney said he could not compromise academic freedom. Hawthorne lamented the fact and backed off. After some silence, Gurney received a visit from a highly placed De Beers geologist who said that if he would stop the work, De Beers would discreetly deposit, tax-free, in a foreign bank account in any country Gurney chose, $30,000 a year. Gurney was indignant. He thought it was worth more; but mainly, he said, he was serious about scientific research. He later took some students on a field trip to Kimberley and was called into the office of another geologist, where both the carrot and the stick were applied. The man upped the offer to $100,000—and told him that if he did not accept, the company would find a way to break his career. A nasty shouting match ensued. Gurney said no price would buy him off, spat out some choice Liverpool slang, and stumped out the door on his two bad legs. Some De Beers lab people, working along the same lines as Gurney was, overheard the ruckus and later marched in to demand why they were not being offered $100,000 a year for *their* work.

With funds from Superior and Falconbridge, Gurney did huge numbers of microprobe analyses, getting colleagues to send mineral grains from both diamondiferous and nondiamondiferous pipes everywhere—Russia, China, Brazil, the United States, southern Africa. When he got enough data, he personally delivered one copy of his report for Jennings, one for Hugo, and kept a third in his own safe. It ran only twenty-five pages, including graphs and footnotes, and was ridiculously simple.

Garnets had highly variable formulas, some as long as Gurney's arm. But every diamond mine everywhere had mixed in it somewhere among its vast assemblage of garnets some that were anomalously high in chromium—over 6 percent—and low in calcium— under 4 percent. These matched the compositions of the garnets inside the diamonds. By comparing the chemical compositions of the garnets with the known grades of certain pipes, he saw that the more such garnets in a pipe, the more diamonds; and the more subcalcic and superchromic they were, the better. Pipes lacking such garnets never held diamonds.

The peculiar compositions might be tied to an abundance of heavy chromium and paucity of light calcium in diamond regions, or maybe with the way molecules fit together in particular temperature-pressure conditions. But the reason was not his concern: Only the empirical observation mattered. He plotted his answer on a high school algebra-type graph, the x axis quantifying calcium, the y chromium, and drew a sloping line; all those to the left were good garnets. He called them G10s, after Group 10 pyropes, one of twelve cruder classes assembled by an earlier researcher. Color was not necessarily a reliable corollary. G10s were generally some shade of deep red, but depending on the part of the world, they could also be purplish, pinkish-purple, blue-violet, rose, lilac, or a bottomless blood-red—almost black. Non-diamond-indicating pyropes, often in similar colors, were classed G9, G12, and so on. Only the microprobe could tell.

At the same time, Gurney cracked the secrets of other indicator minerals. Chromites worked in much the same way; high chrome indicated diamond. Chrome diopside had no telling formula—but since it was soft, it was still a sign you were close to a pipe if you saw one. As for iron-rich kimberlitic ilmenites, they did not appear to form directly with diamond at all—but there was still a link. They formed either above or below diamonds and came up with every-thing else. When the probe showed their iron was unoxidized, there were usually diamonds, as long as there were G10s; but if the iron was highly oxidized, there were no diamonds no matter how many G10s. Apparently, ilmenites mirrored the heat and available oxygen in the erupting kimberlite. If they got oxidized on the way up, so did diamonds—vaporized, gone. In sum: G10s or chromites were like a trail of blood; they led toward the scene of a crime. A nice big green chrome diopside was like seeing the victim's shirt; you were probably near the body. Ilmenites suggested whether the victim was lightly wounded, badly wounded, or dead.

In this system Superior had a secret diamond-detecting weapon. Gurney's work, refined, consolidated with others' studies, and pub-lished years later, was key to the now-accepted theory of diamonds:

They do not form in kimberlite, but rather are picked up from certain garnet-rich rocks. Most of the rock disaggregates on the way up, leaving only constituent minerals. A species of peridotite rich in G10s, garnet harzburgite, supplies most mines' diamonds, but old T. G. Bonney was at least partly right about orange-garnet-studded eclogite as the "parent rock." Eclogite also supplied diamonds to some mines, though usually fewer. Gurney and others later discovered that eclogitic garnets, anomalously high in sodium, could be used as indicators in some cases, too.

Meanwhile, De Beers's scientists had trouble sorting through the understandable confusion over which minerals to analyze and what elements to look for. Simultaneously, they came up with a different formula, the so-called "R-Factor." It was based wholly, and erroneously, on ilmenite grains. According to their observations, those high in chromium and magnesium and low in iron signaled diamondiferous pipes—true at least of the pipes they investigated. They missed the garnet formula altogether and drew up an equation in which one multiplied and divided ilmenite elements to calculate diamond grade. They developed a whole prospecting system based on it. By 1980 they were slowly realizing it did not work—a disastrous setback. One of the lab men responsible for the R-Factor, an old college chum of Gurney's named Harold Fesq, later committed suicide.

Chuck was now inundated with samples and chronically behind. As his lab's reputation grew, other corporate clients were jostling to have him sort for gold, zinc, and copper indicators. He was doing well without finding a thing, so Superior samples often waited. Partly because of this, McCandless opened a lab in Tucson, where a half-dozen workers did nothing but microscope-pick Chuck's heavy concentrates. When pickers found apparent indicators, McCandless had them mounted on metal discs with double-sided tape and shipped to Gurney's microprobe. To supplement this, Superior set up another microprobe in a Houston office park around which Howard Keck's gigantic prize steers were pastured.

Early on, Hugo called McCandless into his office and let him in on the G10 secret. On a scrap of paper he drew the x/y axes showing the low calcium, the high chromium, and the sloping line. Fearing the room could be bugged, he took care to explain it by pointing to the elements, but not speaking their names.

"Do you think you can memorize that?" asked Hugo.

"Sure. It's simple," said McCandless.

"Good," said Hugo. He ripped the paper to shreds, piled them in his ashtray, and set them on fire.

They made a point of not telling Chuck anything. He was asking increasingly nosy questions about things he had no need to know. Chuck was excluded from meetings in which G10s might come up; any reports he received were expunged of G10 references; they took care not to even say the "G-word" around him.

McCandless and Chuck had to spend large amounts of time on the phone coordinating their activities, but whether in the field or in the lab, Chuck often did not return phone calls for weeks at a time. His workers claimed they could not find him. Then he would call out of the blue late at night, and you could not get rid of him. Chuck talked easily nonstop for two hours about some minuscule aspect of a single mineral sample. He could insert an hour-long digression about beer, or Marlene, then return to the minerals. This addled McCandless's mind. During one call, McCandless got so tired that when Chuck asked about a particular sample, McCandless blurted, "Yeah, there were a couple of G10s in there."

"Oh. Huh. What's that?"

McCandless thought: Shit. How did he get me to say that? "Nothing. It's just a garnet," he said, and hung up fast.

McCandless went to Hugo and confessed his slip.

"You know, Chuck can get your wife's underwear size off you, if you don't watch out," said Hugo. McCandless had never thought of Chuck that way. But Hugo had begun to perceive that perhaps Chuck was craftier than people thought.

To cover themselves, they sat down and concocted a lie. McCandless called Chuck back. He said G10 garnets were indicators of diamond. The tipoff was high titanium—as far as McCandless knew, a horrible sign. He gave Chuck figures on how much titanium was good—8 percent or so—and even supplied Chuck with phony charts showing how it worked on particular samples. Chuck swallowed it completely and swore he would tell no one. Around this time, he had an idea to turn north.

CHAPTER 12

A Nunatak

East of Kelowna on the Trans-Canada Highway lay the high Canadian Rockies. In 1976, a bulldozer cutting a logging road across a piney slope had accidentally exposed a small kimberlite fifty miles north of the U.S. border, outside a town called—truly by coincidence—Kimberley, British Columbia. Cominco and others quickly found a few other pipelike formations, but none held diamonds. The search was hindered by the mountains: near-vertical, mostly roadless, covered with twelve feet of avalanche-prone snow most of the year. Trees stopped growing about a mile and a half in the sky, and beyond this, alpine glaciers reigned among the clouds in isolated, deadly splendor. This was where Chuck wanted Superior to go. Since it was his own idea, he added, he should be a partner this time.

Hugo agreed; he said Superior would front a modest budget; Chuck would get a per diem, plus 10 percent of any mine they found. It was a good deal for Chuck, so Hugo had a hard time keeping his temper later when Chuck signed some papers, then spent months trying to rework every clause to make it even better for himself. "If I knew what I was signing when I signed it, I wouldn't have signed it, hey?" he objected. Hugo just shook his head. Stew stayed out of the royalties business but agreed to work for wages. The deal was slightly complicated by the fact that Superior was now out of its bailiwick; its sister company, the Canadian Falconbridge, which employed Jennings, was supposed to work this side of the border. But Falconbridge showed little interest, and it became merely a silent partner.

They first sampled the lower reaches of mountain creeks crossed by roads or logging tracks. Just before each trip, Marlene would get up before 7 A.M. and make the whole crew Chuck's favorite breakfast—bacon, eggs, hash browns, and massive quantities of coffee. Afterward, she knocked egg white from Chuck's goatee and turned him outside. Mark, now past thirteen, was old enough to earn his keep, Chuck figured; he brought him when school schedules permitted. Marlene had learned how to do many lab operations herself, so often when they were gone for a few days and the younger kids were in bed, she would sit and look through a microscope at the latest concentrates.

During one mountain trip, in September 1980, Chuck had eight people including himself and Stew crammed into a single cheap motel room to save money. They were all dead asleep in shared beds and on the floor when the phone rang around midnight. It was Marlene.

"I found a sample with a whole bunch of the green ones in it."

"Chrome diopside? Are you sure?" asked Chuck.

She was sure. She had seen enough to know, even if she forgot the technical name.

The grains came from an alpine creek above the town of Golden, at the northern end of their planned range, 180 miles north of the U.S. border. Chuck rousted everyone before sunrise and chartered a helicopter to fly himself and a sampler up the dizzying canyon from which the creek flowed. Above the tree line was a cirque—a glacier-carved, armchair-shaped hollow, its backwall ending against the bare mountain. It was partly filled with the ice that fed the creek. They flew back to the tree line, hacked down a stunted pine with an ax, and returned to plant it in the snow as a claim stake.

Chuck was in a quandary now. The helicopter cost upward of $500 an hour, more than his budget permitted. But that was only way to reach such country without climbing on foot for days. There was a ready solution: Stew still had his own helicopter. It was parked down in Calgary at the moment. It was a tiny Hughes 300, little more than a bubble with a half-dozen exposed fanbelts, a small motor, and a slim tail striped red and white like a barber's pole. It was not designed for high-altitude use; this usually required a more powerful machine to whip up lift in the thinned-out air. On the other hand, there was nothing Stew was better at, or loved more, than flying in the mountains. He rushed down for it, and a day later set down by the Esso gas station in Golden, alongside the Trans-Canada.

In the first light of frosty morning, Stew and Chuck got ready to take off for the high ground. Stew knew he might lack rotor power, so to save weight, he unscrewed the doors and tossed them off. He piled survival gear on the ground. He filled the gas tank only part-way; an aircraft's fuel weight matters. As Chuck watched these preparations, he grew nervous. He was almost relieved when the engine would not start; something had killed the battery in the night, and the gas station had no helicopter batteries.

"What about the dump?" Stew asked the attendant.

There was a pile of junked auto parts out back.

"Help yourself."

Stew found three mostly dead twelve-volt batteries and lined them up in series. Lacking jumper cables, he used baling wire to hook them to the motor and turned the key. The engine cranked but did not quite turn over. Chuck tried dissuading Stew from continuing, but Stew was already rummaging in the dump again, pulling out a length of rope. He looped it loosely over the main rotor and left one end hanging, then positioned a sampler to hold the dangling end. He hooked up the batteries again and turned the key while the sampler jerked hard in the direction the rotor was supposed to go. To Chuck's horror, the rotor came around hard, sending the rope sailing, and the engine belched to life. He had never seen anyone jump-start a helicopter before.

"Come on, get in," yelled Stew.

Chuck hung back. "It won't run without a battery, will it?"

Stew looked heavenward with annoyance. "It *is* running Chuck," he said. They would be in trouble only if the engine stopped somewhere—

say, some isolated peak or in midair—because there was no way to start it again. "Get *in*, Chuck," bellowed Stew. Chuck obeyed.

They lifted off and followed the highway a few miles, then turned up a narrow, forested river valley, toward the spectacular crags of Banff National Park. It was a clear day, and sun glistened off the glaciers and peaks. Stew headed for a big precipice where the sun was beginning to bake the rock walls. Suddenly they were lofting upward, caught on the updraft like raptors. At the top of the cliff Stew moved in against the mountain again, caught another draft, and repeated the operation—a trick he had seen his ex–Korean War pilots do many times. Soon they were soaring above the tree line, in open space. Without the doors, it was freezing; there was nothing around them for miles but thin air and nothing holding them in but their seatbelts.

After hopping several ranges, they caught sight of the cirque where the chrome diopside had come from, and Stew descended in a spiral pattern.

"There's the pipe!" he hollered.

"Where? Where?" said Chuck, bewildered.

"Right there! Look!"

Stew pointed. In the steep rock wall just below the glacier's snout was a spot where the weathered beige Paleozoic limestone looked subtly green. It was nearly imperceptible, but the more Chuck looked, the more he saw. It was nearly round, 120 feet across, flush with the limestone like a raisin in a slice of fruitcake. Only Stew could have seen it so easily; his eyes had long been filled by the mountains' continuous layers of sedimentary and metamorphic rocks, so the little intrusion had jumped out almost before he consciously knew.

Elated, they turned up the adjoining ridge. At the upper edge of the ice, Stew shouted. "There's another one! Look! Look!" There was another subtle round spot, this one slightly reddish, like a chalked-on bull's-eye, clinging to the very edge of a high spine. Chuck was marking the position on a map when Stew cried again. "Chuck! There's another one! . . . And another!"

The things were all over the place. They turned to adjoining ridges. Even as they rounded a corner, Stew would sense a disturbance, something nonstratigraphic, a small shift in color, or a different texture. In the excitement Chuck several times turned the map upside-down and marked the wrong spot. Stew kept grabbing it out of his hand and turning it right-side up. Soon Chuck calmed down and spotted a few pipes himself.

Most of the places were too vertical to land on, but Stew found a teetery spot near the reddish bull's-eye, and they jumped out at 11,500 feet with the engine still running. They pounded out some of the rock as fast as they could, then got back in, and Stew revved up. They bounced up a little, but the rotor could not beat up enough air to really get off. Stew had seen the war vets do another trick. He told Chuck to hang on, and he gunned the engine a few times. The machine bounced a few feet up each time. With one last bounce, Stew let the helicopter fall sideways over the cliff. They went down and forward fast. Chuck thought they were going to die, but the speed of the fall itself gave Stew enough airspeed to steer. Eventually they hit thicker, lower-elevation air and slowed.

Farther on, they found the biggest, scariest pipe of all. They both spotted it at the same time, sticking out from a vast horizontal ice field within a basin framed by peaks, icefalls, and masses of dark cloud. Lying at nearly 10,000 feet, it formed a nunatak—a bare island of rock totally surrounded by ice. They saw it was different from the gray limestone edging the glacier—a series of dark, knifey ridges splashed ochre and yellow from minerals, crumbly and crudely layered with xenoliths. The glacier itself was combed with crevasses plunging into bluish darkness. The pipe appeared to run under it, for a sheer cliff of the same rock reared from the glacier's snout a quarter-mile away. A meltwater creek cascaded over this isolated chimney and plunged down the precipice into a canyon. The glacier was receding; a few years before the pipe must have been under ice, invisible.

With an eye to Superior corporate politics, they dubbed the nunatak Jack Pipe, after Hugo's boss, Jack Langton. Others they named Hugo, Marlene, Stew, Chuck, Mark. They used up the names of Chuck's other kids and various samplers and by lunchtime had seventeen. They returned to Golden, refueled, and found seven more before dark along a thirty-mile line of high peaks. That night Chuck phoned Hugo to babble ecstatically, then went off to the bar to get drunk. Stew went to bed early and instantly dropped into a deep sleep.

They did not want to risk landing on the Jack Pipe, so they took a sampler up in the helicopter and tossed claim stakes out the door. Most clattered down crevasses and disappeared. On the way out, Stew flew over the creek fed by the snout, and they spotted what looked like small kimberlite boulders in the bed. Stew hovered, then set down in the streambed so they could pick up a few. The craft was now heavily laden, and he had trouble getting off again, but he managed it. Just then, as they were a few feet up, a cold katabatic wind gust swept off the glacier's surface and onto them like a flyswatter. They slammed back into the bouldery creek. The skids spread apart and the tail rotor broke. The sampler, fearing fire, instantly leapt out and scrambled away over the rocks. Then he looked back. Stew was still strapped in, flicking switches and assessing damage. Chuck sat next to him, apparently unwilling to abandon his buddy. The sampler came back looking sheepish.

They pulled out the radio and hiked down to the tree line. They had not planned very well; they lacked food or heavy clothes, but they were at least able to build a fire, thanks to some wooden matches Marlene had put in Chuck's survival vest. Night came on fast. The sky grew black, and masses of bright stars and nebulae swam above.

It would be tough to hike out even in day; it was too rough, the distance too far. The radio worked—but only on line-of-sight transmission, and they were boxed in by peaks. In the glow of the fire Stew lay back and scanned the sky. One star blinked and moved perceptibly—a commercial jetliner, miles above. He grabbed the radio and called out a Mayday with their position. He thought he heard an incoherent response before the radio died. That night they dozed, curled practically into the fire, which baked their frontsides while their backsides collected frost. At dawn, a heavy helicopter motor broke the stillness. The jet had heard.

A few days later Stew managed to patch up his little machine and fly it back out, but after this he let Superior charter a bigger machine with a hired pilot. Hugo was now game to put up the money.

It was fall. After a bit more staking, it started snowing. It kept up for months; the work season was over. When it let up a bit toward the spring of 1981, they came back and proceeded to stake the other pipes. They did this from a huddle of drafty summer tents Marlene got on sale at Zeller's discount department store. At least her intentions were good. The samplers got so cold, they built a creditable igloo to sleep in.

Working in the lower altitudes involved slogging in waist-high snow all day, often amid avalanche chutes—steep, raked-off mountainside gouges where frequent slides had sheared away trees. At the end of one day, Dynamite Dan found himself at the top of a chute on a high ridge. Base camp was visible far below, but he was too tired to walk down. He felled a tree, tossed it onto the chute, and watched it spin more or less straight down the 2,000-foot drop with its lower trunk forward, branches back. Satisfied it would work for him, he felled another, jumped on, and shoved off like a one-man luge team, just ahead of various mini-avalanches he was starting. Nothing ever caught up with him, and he disembarked in front of his comrades in time for supper.

Things got more dangerous in late spring, when much snow melted off. This reexposed the pipes themselves, and Chuck wanted samples. Many were barely knobs sticking out of sheer faces, reachable only with ropes. Chuck tried rappelling to a few himself with the aid of a hired mountain guide, but he grew impatient; this could take forever. His solution was to load the helicopter with samplers. The pilot pulled as close as he dared to a ledge, knob, or slope, and the sampler was told to jump off and have a nice day. Chuck disembarked during these runs with the aplomb of a commuter getting off a bus. Everyone else wore parkas, but he wore a white short-sleeved knit golf shirt and a red many-pocketed vest—a sort of uniform he had adopted for himself. With a heart-stopping drop just feet away, he would step off the helicopter skid, then sometimes lean back into the hovering machine to rummage for his hammer or rock bucket, mumbling to himself all the while.

However, there were some spots even Chuck feared. To these he sent Dynamite Dan or, more preferably, Mark. Chuck wanted Mark to follow in his footsteps as a geologist and all-around tough person, and he insisted that Mark work even harder than the rest. He said Mark needed to grow up fast, as he himself was forced to—he came right out and said it—by Mark's untimely birth. Mark did whatever he was told. He competently set up camps, fixed machinery, and dug—all for a bargain $10 a day. It was rarely good enough. Chuck bullied him for the slightest error or sign of fear. Even the cynical samplers were sometimes horrified to see fourteen-year-old Mark scrambling up some impossible ridge on Chuck's orders, with no safety rope. They felt sorry for him, but they were also often annoyed. Mark drank beer with them at night but could not handle it. He always lost the brawls he started by mouthing off. During one, Dynamite Dan grabbed Mark's ax, held Mark down, and sawed at his neck, pointing out that Mark would be dead now if he had bothered to sharpen his ax like a grown-up. When the men went to bars to pick up women, Mark had to stay in the motel room.

One spring day they were scheduled to investigate a cliffside where the helicopter could not get close enough for anyone to jump down onto the outcrop; someone would have to jump *across* several feet of high, empty airspace and land on the other side. They flew up next to the spot and hovered. Chuck, sitting next to Mark, coaxed him onto the narrow metal skid and told him to go. Mark launched off, landed safely, and waved them off nonchalantly. However, later he looked down and knew he could never do the leap in reverse—not with seventy-some pounds of supposed diamond ore on his back. Chuck and the pilot returned an hour later to find him gone. They spent five frantic minutes searching the cliff face before spotting him safe in a creekbed 1,200 feet below. He had somehow managed to climb down with the ore on his back. When they got to him, Chuck cursed Mark violently for not staying put and fined him the five minutes' helicopter time—close to a week's pay.

Later that day Mark and another sampler were turned out simultaneously into a sandy creekbed to dig while the machine ran, eating up more dollars. Mark apparently wanted to redeem himself by being fastest. He took the sample and was sprinting back when he saw the other man running ahead by ten feet. As they neared the machine, his competitor dodged around a fallen tree trunk lying under the wash of the blade. Mark saw his opening: directly over the log and straight for the door. He jumped onto the fallen tree with both knees bent and sprang into the air. Chuck was inside, watching, screaming, and panting. In Canada, there is a law requiring every helicopter passenger be drilled over and over, never, ever to stand up straight near a low-swinging helicopter rotor. Especially on uneven ground. The reason is obvious.

The eyes of father and son met just as Mark Fipke ascended. Mark saw an expression on his father's face he had never seen before. When he caught that gaze, he cringed, just as his head entered the zone of dead air that lies an inch or two under a moving helicopter rotor. His body dropped down. His head was still attached; the last-second tuck had saved him.

Sometimes at night when he thought his tentmates were asleep, Mark would lie awake in his sleeping bag and think. Then he would begin to choke, and tears would come. He would cry as quietly as he could, terrified the men would hear him. He wished he were back in Kelowna. Free of Chuck. Free of this search. Just hanging out with his friends and being a normal junior high school kid.

The word *nunatak*, from the language of the Inuit, means "lonely hill." Applied to the Jack Pipe, it was an understatement. However, it was the biggest prospect. The outcrops tearing up through the ice suggested it measured 140 acres or more underneath. Early on, a sample of the rock taken from creekbed had been broken up, and to Chuck and Hugo's delight, it contained a single diamond—a microscopic speck, actually, which Chuck grandiloquently described in a report as "$37,320 \times 10^{-8}$ carats." Chuck pushed hard to prove up the Jack.

They got on to the central nunatak by landing on a relatively solid part of the glacier and picking their way across in knee-deep snow. Crevasses could be hard to spot because snow bridges formed over them. The first time, Chuck was walking along when one gave way and he went down. Lucky for him it was only ten feet deep and buffered with snow at the bottom. When they got to the nunatak, there was almost no place to walk, or even stand; it was mostly a jumble of sharp, crumbly rocks that threatened at every moment to break and hurl them off into a deep, wind-carved moat ringing the formation.

Early in the summer, Chuck ordered Brent and Paul Derkson to bulk-sample the nunatak's apparent extension, the chimney of rock at the glacial toe. Before leaving for the job, they got one hour of sleep, having worked at another site until 3 in the morning. Chuck and the pilot picked them up at 4:30, and shortly they found themselves dropped on the inescapable rock. Water roared from the end of the ice, over part of the chimney forming a creekbed, then plunged off. Behind them the glacier steamed in the rising sun; ahead lay an unbelievable view of the rest of world.

They spent the day moving some 5,000 pounds of material from the meltwater creek and sieving it down to 700 pounds, alternately suffering drizzle from fast-moving clouds and enjoying brisk sun. Later it drizzled continuously, and clouds threatened to close in. It was nearly 10 P.M. but still light when they heard the comforting clatter of the helicopter.

Chuck jumped out and informed them the helicopter could not take them and the rocks at the same time, so the rocks were going out first. They protested; they had been working a day and a half straight. The weather did not look good. What if the clouds closed in and the helicopter could not return?

"Leave the samples, will you? Who's going to steal them?" said Brent.

"I gotta get off of here. I'm getting hypothermic," said Derkson.

They yelled at him for a while, but Chuck won. As a concession, he said he would stay, too. They threw the bags into a net under the helicopter and cursed as it lifted off.

A few minutes later the clouds closed in and lowered close to the rock from both above and below, leaving them in an opening surrounded by mist. The rain and wind picked up. No one was dressed for it; Brent had a down jacket, useless when wet. Chuck had his golf shirt and vest. Derkson had nothing.

"Shit. What if this keeps up, Chuck? What are we gonna do then?" said Brent. "Fuck, we don't even have a boulder big enough to hide behind."

The wind now shifted direction and blew down directly off the glacier. It was freezing. Rain changed to wet snow. As big flakes coated them, they huddled shoulder to shoulder like a herd of horses on the winter range. The two tall samplers had the diminutive Fipke between them.

"What if we die out here, Chuck? Will it all be worth it to you then?" yelled Brent. "How about the diamonds then? Fuck. Shit!"

"Yeah. Yeah. I guess you're right," mumbled Chuck. "We should've gone back."

"A little late for that, isn't it, Chuck?" said Brent.

"We won't do this again," promised Chuck.

"We might not get the *chance* to do this again," said Derkson. "Shit."

It was at times like this that Chuck repeated his big promise. "Listen. Listen. When I make it big, hey, you'll make it big, too, hey? I'll make good. When we, when we hit it, you guys will be taken care of. We're all gonna benefit from this, you know. Come on, guys."

He always sounded so earnest when he said this, and he said it on many occasions. They believed him and swore they would hold him to it.

Looking into the darkening sky, they stopped cursing and shouted to the gods of the mountains for a hole in the mist, stamping at the snow piling up around their ankles. All at once they heard something imperceptible that seemed not to be wind. They all shut up and strained to hear. It was the helicopter, below. The drone grew rapidly. Suddenly the machine shot straight up like a tropical sunrise, feet from the pinnacle wall through a slight aperture in the mist. The pilot had been watching every second for an opening. He brought the machine down, and they ran for it. As they took off and were descending, they looked upward and saw the hole close back up. After that, it stormed heavily all night in the high mountains. No one would have survived.

After this, most days the samplers were still thrilled for the adventure, but resentments started surfacing. The long hours, low pay, danger, and lack of rudimentary safety equipment were getting on their nerves. Chuck was no manager; he would never even remember to pack food unless Marlene told him to, and he rarely thought to thank anyone. Busy running the lab much of the time, he was often not even around. Stew sometimes filled in as chief, wading into icy mountain streams in his currently preferred footwear—suede Hush Puppies and thin black dress socks—but he never yelled at them if they did not do the same. However, Stew was often off on his own projects, and they were left on their own.

Mutiny broke out when Chuck returned to camp. One evening Derkson straggled back at nightfall, looking half dead. On a lower slope, he had just spent two days hacking a half-mile line through a steep tangle of slide alder so magnetic instruments could be carried through. Chuck had already sent him back once to make the line straighter; then again, to make it neater. Derkson dropped down and started eating dinner. Chuck, dinner plate in hand, told Derkson now that the line was not long enough. "You're not working as hard as everyone else here," he whined. Derkson stood up.

"I don't care if you're the fucking queen of England! You can't talk to me that way."

He walked over and booted Chuck's plate out of his hand. It went clattering into the rocks. There was a moment of silence. Then Chuck sprang up on his compact frame and punched Derkson on the chin.

Derkson was still standing. "Go ahead, Chuck. Beat me up. Go ahead." He knew Chuck could do it. "I'll quit."

"I'll fire you."

"No, I quit."

"You bloody fucker, you're fired." Chuck held his oversize fists up in front of his goatee. "Come on."

"Fuck you, Chuck."

"Come on, you fucker."

Brent rose. Disorder and violence deeply upset him. But if it came down to it, he was going to help Derkson kick the shit out of the boss, and they could all quit, diamonds or no diamonds. It did not come down to it. The combatants backed to opposite corners and turned away from each other. Things simmered down.

After a few minutes Derkson approached Chuck and said tersely: "I'm sorry I got angry."

Chuck glanced at him shyly. "It's OK," he replied. Dinner was concluded.

Later, in the dark, when most everyone was asleep, Chuck and Derkson were left alone by the fire.

Chuck spoke up. "When I was a kid, you know, I used to fight a lot with my younger brother. I suppose I was a bit of a bully to him."

That was all. It seemed like some sort of apology, though Derkson was not sure what sort. He did not know what to say back, so they just sat side by side and looked at the fire together without saying anything more.

Aside from the promise of wealth, Chuck's secret to maintaining loyalty was simple: He was fun. He always bought beer and pizza when they went out, kept the jokes coming, and eventually everyone forgot to be mad. When visitors came to Kelowna, Chuck regaled them with drink and banter, traveling a regular route of bars that inevitably ended at the Willow, a strip club down near lakeside. When Hugo paid a visit, Chuck proudly took him straight to the Willow, but Hugo was a bit of a prude and was devoted to Nora, his wife. Once he realized where they were, he turned red and scraped his chair awkwardly around so his back faced the stage. Marlene hated the Willow; Chuck seemed to have a fixation on it. She was astoundingly flexible about it, though. She not only let him go; she went around mornings and collected the credit cards he forgot on various counters. She knew his route because she instructed him to take a matchbox from each place so she would know where to go.

As for Stew: He was becoming more socialized now, in the outside world. When not on the diamond project, his services as a consulting geologist were in high demand. He hunted gold on his own account and for various clients. He also discovered that, for some odd reason, a great many unattached women swooned over a single, hardworking, craggy-looking Ph.D. in his early forties who pounded rocks, invested, and flew aircraft for a living. One day a pilot friend introduced him to a beautiful Air Canada flight attendant, Marilyn Ballantyne. Before anyone knew it, she was Marilyn Blusson. Mother was ecstatic.

They bought themselves a little house near Vancouver and settled in. On the surface, it was a mismatch. Marilyn hated dirt, wild animals, and helicopters. Stew tried taking her camping, but after a few trips she begged off. They bought themselves a big four-poster bed with a canopy and tried making a baby, though with no results. Marilyn kept her flight attendant job, but Stew tried to stay home as much as possible, which was quite unlike him. Whenever she was flying, Stew never slept in the four-poster. He unrolled his sleeping bag on the living room floor, climbed in, and dropped off there alone.

The Blussons often visited Kelowna, skied with the Fipkes, and just generally hung about. Stew and Marilyn loved the Fipkes' small old house, where they always had a giggly Fipke child or two on their laps. There were three cute little daughters. When another boy, Ryan, was born, Stew and Marilyn were named godparents. Chuck never fixed anything, so the house was a wreck, but no one minded. Chuck worked too hard to be home much, but when he was, he made everything an adventure. If a light burned out, he lit a candle. The little kids were delighted when overnight guests came, because that meant they were kicked out of their shared bedroom and allowed to camp in a living room fort Chuck made. Stew and Marilyn barely fit in a child's bed; Hugo's feet and calves hung well off the end. Once when the Blussons were there, a solar eclipse occasioned an all-night pre-eclipse beer party to build up excitement. In the morning Chuck set up a telescope in the middle of the tree-lined street and outfitted everyone with dark glasses, including the family dog. Cars just swerved around them.

In late summer John Gurney arrived in the States for a series of meetings with McCandless and Hugo regarding indicator minerals and the terrains they were working. It was Gurney's strong opinion that the Jack Pipe and the other mountain pipes looked bad. There was only one diamond and few garnets—few, if any, of which plotted as G10s. He felt the setting was wrong, too. His ongoing research and others' all bolstered the long-held idea that you needed deep, old, stable cratonic rocks for diamond formation. There was no evidence the tormented young thrusts and folds of the Rockies were underlain by such a place, or had ever been. Hugo was not ready to give up, though. He insisted Gurney and McCandless come up for a look.

Hugo had been there before, but Gurney and McCandless had never seen such spectacular mountain landscapes. The first day they were alternately scared and exhilarated when the helicopter pilot threaded his way through scattered clouds, asking them to watch out their side for cliffs. Chuck did his ledge-hopping act, and near sunset left Gurney standing half an hour on a mountaintop in his shirt-sleeves so some rocks could be ferried out first.

The next day they landed on the glacier surrounding the Jack Pipe. Hugo insisted on better-than-normal safety precautions, so they were roped together behind a mountain guide who led the way, probing for crevasses with a long pole. Chuck was right behind. Hugo hoisted McCandless on his back and plowed through the snow; Gurney limped through on his own power. Some 200 yards on, Hugo sat McCandless down on the sharp rocks and they gazed at the fearsome beauty assaulting them on all sides.

Gurney was quiet for a while. "Are you all insane?" he said finally.

Chuck and Hugo grinned wickedly. McCandless just stared around, grateful to see it.

"This is terrible!" shouted Gurney. "You fuckers are not listening to me! This is totally the wrong place for diamonds! Maybe you found one diamond, but you're not going to find any more. It's wrong. It's wrong. How many times do I have to say it?" Gurney surveyed the crevasses. "If you keep coming up here, you fuckers are all going to get killed."

After this, McCandless was ready for a vacation. He had been working nonstop for Superior the past two years, and it had been super-tense. He never allowed himself to ease up; if he did, he might make a mistake. When the others left, Chuck offered to drive him down to Kelowna for the weekend, and he accepted. McCandless and Chuck got along well, partly because they were both westerners, and both hated what they considered citified easterners. McCandless had never driven with Chuck, though, and was not sure he had made the most relaxing decision when Chuck floored the gas and passed every car on the road until they screeched into his driveway a few hours later. It was with great relief he met the family. Then he discovered the Fipkes were planning a weekend trout-fishing trip partway back up the mountains they had just come from. McCandless said he would stay at the house.

"No, hey, Tom, you're part of the family now," insisted Chuck. "We're not leaving you behind. You got to come." McCandless saw it was no use. They all piled into the ancient C. F. Minerals Chevy Blazer. McCandless noted that Mark had packed his dirt bike in back.

When they reached the fishing spot a few hours later, Mark unloaded the bike and went roaring off on a hillside. The rest clambered down the embankment. Chuck insisted McCandless had to fish. In great excitement he carried the wheelchair down and carefully leveled it in the rocky shallows. Then he went back for McCandless, carried him down, and set him in. Then he scurried off to cut a long willow branch and attached a line and a small hook. He gave McCandless a bunch of live worms.

"Now, you're fishing, hey?" said Chuck. "Catch some big ones. See you later, Tom." He splashed off to find the little kids.

McCandless had never fished with a simple stick before. He tossed his line out. Among the rounded rocks, he could see six-inch trout leaping from the surface, almost within grasp. Without trying, he forgot all about kimberlites and microprobes and G10s, and everything that had filled his mind for so long. He just thought about leaping trout and listened to the rush of water. He could hear Chuck roughhousing with the girls around the bend. For the first time in years, a feeling of deep peace came over him.

CHAPTER 13

Blackwater Lake

The move 1,000 miles north was swift.

Chuck always had an ear out for other opportunities, and one day on a Vancouver street he ran into an old college acquaintance, Colin Godwin. Godwin, now a professor, had two students with summer jobs looking for gold in the Sayunei Range, near where Chuck had met Stew thirteen years before. In one valley up there, they ran across a pipelike formation. The company they worked for declined to stake it, so Godwin and a friend flew up and claimed it themselves. They dubbed it Mountain Diatreme, diatreme being a fancy name for a pipe. Situated in a barren labyrinth of worn sedimentary pinnacles, cirques, and arêtes, it was a modest cylinder of greenish refuse tumbling from a wall into a scree-filled chasm. It showed no visible diamonds and was too remote for mining, so De Beers and others turned down options. However, nothing was too far for Chuck or Hugo. Chuck easily persuaded Hugo to have Superior buy the option, and again had himself cut in as a partner. They arrived with a fourteen-man crew in July 1981.

Stew felt ineffable joy to be back on home ground. In fact he remembered this valley; eleven years before, in the GSC, he had photographed it from the air. He never noticed the pipe, but he was not looking for it. Among his meticulously filed photos, he dug up a shot of a nearby ridge, and there in high sidewall was pictured a round, rusty spot—another one.

To supplement a chartered helicopter, Stew brought up his little Hughes, and they camped fifteen miles off on Palmer Lake, one of the few big enough to land a plane on and low enough in elevation to have trees. They commuted to the diatreme. It had come up between two layers of gray dolomite, once under an ancient sea, and was loaded with conodonts, tooth-shaped marine fossils. This allowed them to date the eruption: Ordovician, 440 million years, more or less. Part of the crew hammered out samples while the rest were airlifted in all directions to search surrounding drainages for indicator minerals and more pipes. According to a plan Chuck drew up, they worked in two shifts in the twenty-four-hour daylight, each man dropped by helicopter six or eight times a day. Mark, now fifteen, worked his shift as hard as ever, but Chuck went through both shifts, stopping only to throw a frozen steak directly into the campfire, pulling it out bloody and covered with ashes to stuff in his mouth.

The only serious new hazard was grizzlies. They were common in British Columbia, but not like this. Brent was about to be dropped one day when he looked down and saw the biggest bear anyone there had ever seen, ambling toward the sample spot. They diverted to another site, but as soon as the helicopter was gone, Brent turned to see a different bear a stone's throw away. It looked at him, then walked away. At Broken Skull River he was making his way through head-high alders—a favorite bear hideout—when he heard a crunch, crunch, crunch, paralleling him ten feet away. Unarmed, he was about to wet his pants when he saw moving caribou antlers sticking above the brush. He later followed what looked like a caribou path to a river. It turned out to be a bear highway. He heard footsteps and turned to see one jogging at him from behind. Dropping his pack, he leapt into the stream and floated 300 feet to a sandbar. He watched the bear sniff his pack for a while, then wander off. Brent emerged shivering; luckily there were no snow flurries that day.

No one had guns. Chuck's line always was: "Nah, you don't need a gun. The animals won't be scared of you if you're not scared of them." Back at camp Stew listened to Brent's story and grinned. "Oh yeah, you see bears all the time here," he said. "Look." He pulled up a plaid shirtsleeve to show the old grizzly claw mark in his bicep. Everyone gaped; they had never heard that one before. After this trip, everyone but Stew decided that maybe the Americans were not so crazy after all, and they acquired their own small arsenals.

A yellow Twin Otter supplied food and fuel every couple of days from Norman Wells, an oil settlement in the forested Mackenzie River valley, 120 miles east. Hugo showed up on the supply flight one day, bearing a wooden case with twelve industrial diamonds he'd acquired in New York. He passed it around so that if anyone saw a diamond, they would know what it was. It seemed wildly optimistic; in all their travels, Stew was still the only person who had ever seen one in the field, and he had had to buy the $2 Crater of Diamonds ticket to find it.

When the Otter returned, Hugo went out on it. Since he was the only passenger, he chatted with the pilot and worked in a question he asked anywhere he went.

"Are you serving other camps around here?"

"Oh yeah, we have a big contract with something called Dia, Dia, Dia something."

"Errr . . . Diapros?"

"That's it. They got a camp down on Blackwater Lake. Contract for 1,000 helicopter hours this summer. They got money, I'll tell you. Last week we flew in a load of ice for their scotch."

"Really?"

"Yeah. A few of them have an accent sort of like yours."

When they landed at Norman Wells—a huddle of oil rigs, trailers, and a big airstrip—Hugo bought the map sheet showing Blackwater Lake. It was in hilly, roadless forest 170 miles east of Palmer Lake, on the far side of the Mackenzie. He marked an X and neatly penned on the map: "Pilot tells me De Beers has got a big camp here. Why not go out there and find out what they're doing?" He sealed it in a manila envelope addressed to both Chuck and Stew and asked the Otter pilot to take it on his next run.

Hugo figured: a contract for 1,000 hours of helicopter time at, say, $500 an hour, plus aviation fuel, charged separately. Diapros was spending a million dollars this summer—just for helicopters. Something had to be up.

Something was up. De Beers had been in the far north for eight years now, largely undetected. About 1,000 miles on the other side of the Barrens from Blackwater Lake was Somerset Island, part of the archipelago above the coast where Sir John Franklin had disappeared. GSC bedrock mappers had done basic reconnaissance on Somerset by aircraft and found what they called a "basic intrusion." In 1973 two Canadian geology professors flew up for a closer look and proved it was kimberlite—the North American Arctic's first. On the December day their report came out in the *Canadian Journal of Earth Sciences*, Dr. Arnold Elzey Waters Jr. saw it.

The sane thing would have been to wait until spring, but Waters, now up in years, probably wondered how much time he had left. With one assistant he caught a plane to Resolute, a bleak Inuit community and military outpost on an island, and hired a ski plane. Since he had no arctic experience, it was thanks only to a hired Inuit guide that he survived. On Somerset, the guide discarded the tent Waters had brought and built an igloo. When the guide returned after scouting, he found Waters in the process of melting the igloo by cranking his Coleman camp stove too high. For provisions, the Doctor had brought canned goods, jars of ketchup and pickles, and other items that invariably freeze and explode on arctic campouts. Waters learned the curious fact that once a chicken egg has frozen, it will not flatten properly no matter how much you fry it. However, the old man survived. He staked out miles of land and ordered crews back in the spring. When the snow cleared, they found several more pipes. They dug for several seasons with equipment barged in during the few summer weeks when pack ice abated.

It turned out that Somerset Island was a bust—Diapros found only a half-dozen microscopic diamonds—but it opened the north for them. In 1975 they started moving secretly along the upper Barrens coast, sampling major river mouths, including the Coppermine, plus parts of Banks, Cornwallis, Prince of Wales, Devon, and Bathurst islands—all equally grim and unpromising. Then they decided to go inland along the north's only big waterway: the mighty Mackenzie, which meets the ocean far west of the Coppermine. This was more hospitable, since it lay conveniently within the upsloping tree line and was more or less navigable all the way to Great Slave Lake. In classic fashion they worked their way upstream looking for indicator minerals. In 1979 they hit a batch at a tributary 1,200 miles south—the Blackwater River, flowing from the direction of the Barrens, which lay 300 miles east. Diapros made a few recon flights to the edge of the tree line, but never passed; just twenty-some miles upstream in the low Richardson Mountains near Blackwater Lake, they hit glacial till brimming with large garnets, ilmenites, and chrome diopsides.

Diapros was sure they had hit it. At the time, they were still using the faulty R-Factor, and the readings suggested diamonds. They took out prospecting concessions on forty-five square miles of steep hills combed with stunted spruces interspersed with bouldery plains and wet muskeg, and camped near the apparent mineral epicenter—a sandy creek beach in a U-shaped valley.

They did everything they could think of. Numerous samples were poured into portable cement mixers to break up clods and float off clays, then shipped in drums to South Africa for more separation. They ordered cement mixers by the dozen. They hired Dene and Europeans from faraway Yellowknife to chainsaw 240 miles of straight, parallel lines through the wilderness, then carry ground magnetometers along them in search of kimberlite—a method usually used only to outline pipes once you have found them. Labor and equipment were supplied by a Yellowknife subcontractor, Brian Weir, one of several geologists there who had smartly given up looking for gold themselves and instead made handsome profits organizing crews and equipment for visiting southerners.

The Dene, not told what they were looking for, were mystified. To avoid messing up the subtle signature of kimberlite, magnetometer-carrying Indians were ordered to remove watches, coins, metal buttons, and even the zippers on their pants. Later they were sent to walk the cut lines with scissors, snipping spruce twigs and sealing them into letter-size envelopes. The twigs were secretly shipped to South Africa and analyzed for traces of nickel, cesium, strontium, and, most particularly, niobium—elements rarely found in the crust, but often detectable in kimberlite or other deep-origin rocks. Theoretically, if a pipe was underfoot the elements might be taken up by plants. Diapros brought in a glacial geologist to study the till lying everywhere. Finally, they brought heavy artillery—drills nearly the size of cabins to auger to bedrock, 1,000 feet under the till in places. By this time De Beers was even supporting research to find out whether certain kinds of bacteria grew over kimberlite, but this was not used—about the only method not applied here. The only other thing missing was Dr. Arnold Elzey Waters Jr.; Somerset Island had been his last hurrah. Faded now, he retired and lived on until August 1985 in Vermont.

The campaign was mostly invisible. Permits and other official papers were filed in the names of minor employees to keep perusers of public files off track. Required government reports were written in well-turned gibberish, so that no one would know what they were looking for. Pilots in Norman Wells were asked to keep quiet. At least most did. Out in the woods, the camp lay unprotected except by a big golden retriever named Duke, who was trained to bark at bears.

A few days passed before they could get away from Mountain Diatreme. With work wrapping up, Chuck sent word to have a small plane pick him up, making sure it was not from Norman Wells, but a more southerly settlement, Fort Simpson. He would not tell the pilot exactly where they were going—just go southeast, he said. After taking off from Palmer Lake, they traversed nearly 100 miles of peaks. Then suddenly the mountains ended, decisive as a tabletop, and they were over the boggy alluvial plains of the Mackenzie, with its broad central channel surrounded by braided subsidiary streams, cut-off oxbows, and mucky ponds. On the other side, stunted mountains picked up again—the Richardsons. Chuck was just guessing where he was supposed to go: on the map, crescent-shaped Blackwater Lake was twenty-five miles long and quite wide in spots. A few minutes later he saw a giveaway: the parallel lines the Dene had cut through the stunted trees, 300 feet apart, running over hills and down swales as if someone had run a rake through. He also saw cleared squares with pipes sticking out of the centers—drill pads. "The bastards are right on top of something," muttered Chuck.

They came over a ridge at an altitude of 900 feet, and there suddenly there was the De Beers camp.

"Don't slow down! Don't slow down!" yelled Chuck. Don't *look* down!" He did not want anyone to think they were spies; they were just passing through. Trying not to incline his head too much, he scanned the layout. A big cluster of orange-roofed tents and bright yellow fuel drums. Off to one side, cement mixers, lined up in sharp military formation. He saw a couple of people emerge from the tents and look up. One, wearing khakis head to foot, had a pair of binoculars and was checking Chuck out.

"Keep bearing straight ahead," he said. He marked the spot on his map. Minutes later they were over the horizon, where their engine would be inaudible at De Beers. He had the pilot do a wide U-turn, taking care to go nowhere near the camp again, and they headed back to Palmer Lake.

This, Stew agreed, called for an all-out spy mission, complete with sampling from around Diapros ground. Two days later, in the early evening, they departed in Stew's helicopter without telling anyone in camp where they were going. It was by now early August, and there was still a dusky half-light after midnight. They went to Norman Wells, refueled, and waited at the airport. Even De Beers men had to go to bed sometime. After 12 they took off again. It was 2 in the morning when they came in fast over the Mackenzie, then dropped and hugged the rising ground on the far side. Stew stayed inside valleys so the enemy would not hear them.

"Jeez, kinda like Vietnam, huh, Chuck?" grinned Stew.

"Yeah, yeah, Vietnam," giggled Fipke.

They found the creek leading to the camp and followed low, the water roiling under them. About a mile off they touched down on a gravel bar, hidden by a ridge. They were both giddy and a little scared; they had never claim-jumped De Beers. Chuck leaped out and furiously shoveled a three-foot-deep sample hole at the head of the bar while Stew kept the machine running in case they needed to leave fast. At the same time he scanned the hills, trying to identify rocks.

Chuck jumped back in and they moved to another spot to repeat the exercise. Then they flew a safer distance, probably beyond

De Beers's claim bounds. Stew shut down, and they listened for air-craft in pursuit. Nothing. While taking a turn digging, Stew looked down in the streambed and was shocked to see so much glacial debris; everything had obviously been ground up by ice around here.

After a couple of hours, they had a half-dozen samples. On the way out they looked from the air for striations or other clues about where all the glacial junk might have come from, but the light was too feeble. They flew back to Norman Wells, where they had break-fast and put the samples into air freight to Kelowna.

When Chuck got home, he personally rushed the bags from Blackwater Lake through the lab and did the microscope picking himself. As soon as he got the first petri dish under the lenses, his heart leaped. It was loaded with garnets, chrome diop-sides, and black grains that looked like ilmenites, all relatively big and unabraded. These were nice-looking. He shipped the grains to McCandless in Tucson, who photographed them and sent them to the Houston microprobe.

A week or two later, Chuck got a jubilant call from Hugo.

"You know those samples you took at Blackwater Lake?"

"Yeah?"

"They are loaded with G10 garnets. I mean loaded."

Each sample had a dozen or more—fabulous, on the scale of indi-cators. Hugo was still lying about the composition of the G10, but Chuck knew they had something. They were both so excited, they almost forgot the prospective ground belonged to De Beers. Or did it?

Stew, in the meantime, had been gathering intelligence. He dis-covered some of his GSC buddies had recently worked the Black-water area, and he got hold of their maps. From what he could see, Diapros had staked the wrong spot.

Diapros's ground, on the north side of the lake, was deeply buried in division after division of glacial till, making bedrock outcrops hard to find. The few exposures the GSC could find, though, indicated the till was underlain by relatively young rocks. South across the lake, on unstaked ground, was older rock—in fact, the same conjunction of 440-million-year-old rock through which the Mountain Diatreme had emerged. Stew was willing to bet the Blackwater pipes, wherever they were, had come up at the same time. And in this particular area, the glaciers had moved south to north, meaning the indicators could have been moved to the other side of the lake, to mislead De Beers. Stew argued they should stake that open ground, and Hugo agreed. Chuck told the samplers they were leaving again for a few weeks, and to bring warm clothing. Mountain Diatreme, Jack Pipe, and their other prospects were instantly forgotten.

In October five samplers found themselves camped in a single tent with Chuck and a helicopter pilot along the Blackwater River. Stew was off on an unconnected consulting job.

It being October, the crew was already hemmed in by temperatures well below zero in the day and nights that fell dark around 5 P.M. Snowshoes came in handy, for drifts around the twelve-foot spruces already lay armpit deep. Diapros was nowhere in sight; they had apparently gone south for the season.

Every morning each man was dropped alone to stake a pre-planned route with map, compass, ax, and lunch, then picked up at the end of the day by helicopter. To minimize carrying stakes, when possible they used living spruces as claim posts, cutting a blaze in the trunk. The downside: Branches were loaded with snow, so every ax stroke brought a cascade onto head and neck. The upside: Now that there was snow, the bears were hibernating.

With daylight waning daily, Chuck pushed the pilot to stay out as long as possible, but finally this backfired. On one of the last evenings snow began falling and the aircraft had trouble spotting Derkson. It wasted precious fuel before it set down on a frozen riverbed and Derkson dashed in to find Chuck sitting up front. Brent was still out, but with darkness gathering and fuel running low, the pilot told Chuck they had to go back to camp.

"You can't leave him out here," said Chuck.

"We don't have a choice," said the pilot. Brent was experienced; it was assumed he would burrow under the snow, wait for morning, and be OK.

"No, no, you can't leave him out," repeated Chuck. "Five more minutes. Just five minutes, OK, then we'll go. OK? OK?"

"OK. Five minutes," said the pilot.

Brent was wondering where they were. As darkness fell, he tried building a fire by a grove of trees, but all the dead branches were under the snow. He got a smoldery little affair going with green ones, and was kneeling and puffing on the flames when he caught a move-ment from the side: a single wolf, 100 feet away, color untellable in the snow and dusk. The wolf froze, then melted into the storm. Just then Brent heard a *too too too too* sound. It turned out to be his fire, sputtering moisture.

A few minutes later he heard it again: *too too too too*. This time it was the helicopter; he could see its safety light blinking a quarter-mile away. Stupidly, he had forgotten to ready a signal flare—the one piece of safety equipment he had. He ripped off his mitts to fumble with the flare's plastic wrapper, got it into its gun, and launched it amid the falling snow. The machine kept searching; they had not seen it. Hands numbing, he got another one in as fast as he could and shot it. This time they turned straight for him.

Brent's five minutes had been more than up; Chuck had just kept arguing with the pilot, refusing to retreat without his man. Now that they had him, it was truly dark and stormy. Just then one of the aircraft's headlamps burned out. The pilot looked at his compass and started fly-ing by dead reckoning in the direction of camp, supposedly forty miles distant. No one was sure if they were going the right way; it was hard to see more than 100 feet. "If you recognize anything at all, let me know," said the pilot. Chuck navigated as best he could with the map while they went twenty miles an hour so they would not hit anything. A sheer rock wall suddenly loomed out in a place where no rock wall was supposed to be. "Oh, oh, oh, wait a minute," said Chuck with a big smirk, and swiveled the map 180 degrees. They veered and headed off in another distant, eerie direction. Brent sighed and pulled his mitts back on. "What are you doing?" said Derkson. "Getting ready to go back out," said Brent. "Have you read the gas gauge lately?"

A few minutes later they came upon a broad, flat road—the frozen Blackwater River, a sure landmark. Confident now, the pilot turned, followed at high speed, and made it to camp with three minutes of fuel left.

Two days later, a single-engine Otter landed on skis to take them out. They threw in the equipment, cramming the tent into the cargo door unfolded. Overloaded as usual, they could hear the tops of miniature spruces scraping the bottoms of the skis before they lofted into clear air. It was Halloween eve, 1981.

C huck and Stew adopted Diapros's tactics and filed the claim papers at the Northwest Territories Mining Recorder in Yellowknife under the name "Harley E. Pyett," a person who had never staked a claim in his life. Harley E. Pyett was Marlene's father; he agreed to provide his name as a cover and secretly sign the claims over to Superior. They were greatly pleased with their own cleverness, and they could tell from their many phone conversations with Hugo that he was practically drooling in anticipation for snowmelt. Thus the calls they separately got from Hugo on the same day shortly after New Year's Day 1982 were all the more shocking. He called Chuck first.

"I have bad news."

"What's that?"

"We have to get out."

"What do you mean?"

"Chuck. This is not easy to explain. We have to drop this whole thing. Superior is getting out of this joint-venture project with you. We're getting out of Canada."

"Huh?"

"I'm really sorry, because I like what we're doing together. It's not my decision. Those are orders from the top. We have to pull out. It's over. I'm truly sorry."

For once, Chuck Fipke was speechless.

T here were three explanations, depending on whom you asked: the official reason; the unofficial reason; and the conspiracy theory.

The official reason: Superior's territory was the United States, not Canada. It was in effect competing with its sister company, Falconbridge. Falconbridge had in fact put up some money for the Canadian program, but then ignored it. It was inefficient, confusing.

The unofficial reason: Howard Keck's many enemies had finally caught up with him. Howard had been bullying and ignoring stockholders for decades, most particularly his sister Willametta, the last survivor of his five quarrelsome siblings and a major share owner. In turn, Willametta and others had been forging behind-the-scenes alliances against Howard. Just now, they managed to gang up, and they forced Howard out as chairman. The new CEO, ex-Exxon executive

Fred Ackman, was intent on making his mark right away, and his first target was Howard's pet mineral-prospecting programs—especially this nutty North American diamond venture. It had cost over $11 million and netted nothing. The budget was slashed, Canada cut off; only a vestige of U.S. exploration was left.

Then there was the conspiracy theory. Howard was now sixty-seven. Just before his unseating, rumor had it that he wanted to sell out any diamond finds to Harry Oppenheimer at an outrageous price. He had only one; among Jennings's sixty-some Kalahari pipes was a single formation called Gope-25 that, extensive testing finally showed, would make a modest diamond mine. Jennings never got to enjoy it. Just as his prize proved up, Howard ordered him moved to Toronto to manage Falconbridge's worldwide explorations. It was a gigantic promotion, but Jennings found himself an immigrant, suitcase in hand, far from his beloved desert discovery, shivering with cold and surrounded by a whole country of unfamiliar rocks. Then Howard ordered Jennings to fly back with him to Johannesburg. Howard met with Harry while Jennings was made to sit alone for days in his hotel room like a bad boy, waiting to answer "technical questions" regarding the pipe. Then they went home. In the plane, Jennings noticed Howard's wife was wearing a brand-new ring with a huge, ostentatious diamond—a De Beers diamond. Shortly, De Beers took over Jennings's pipe. The cartel had won again. He was devastated.

The connection: It was rumored that Harry had agreed to pay the ridiculous asking price for Gope-25—if Howard would stay out of Harry's face in Canada. It seems Diapros knew very well that Harley E. Pyett had not filed those papers claiming the ground across Blackwater Lake. They kept good track of competition and knew the names Hugo Dummett and Chuck Fipke. Few people pass unnoticed through tiny Norman Wells or Yellowknife. The Superior pullout from Canada took place in January 1982, almost simultaneous with Jennings's exile.

Superior's U.S. program languished on, allowing Hugo and McCandless to keep their jobs a little longer. The mostly retired Howard Keck now occupied himself mainly with running the

W. M. Keck Foundation, named for his father. Howard became fascinated with telescopes. He funded two of the world's largest, Keck I and Keck II, on top of Mauna Kea, a volcano on the island of Hawaii.

The week of the pullout, Chuck and Stew spent many hours on the phone with each other. Finally Stew called Hugo. He wanted to know: If Superior was getting out, was it possible he and Chuck could be assigned the Blackwater claims? Hugo brightened; maybe someone could carry on. They discussed how it might be arranged. If it led to something, suggested Stew, they would need a joint-venture partner again for financial backing; Hugo would get first option, whoever he worked for. It was not a legal, but a moral, promise; they owed him.

Hugo asked for a personal audience with Fred Ackman, which took guts; Ackman had just been listed by *Forbes* magazine as one of the world's ten toughest bosses. Ackman was nice about it, though. He said he saw no reason to hold on to the claims. When the details were settled, Chuck's company, C. F. Minerals, got title to the Jack Pipe, Mountain Diatreme, and the other alpine properties; that was in his contract. Blackwater Lake was a separate deal; it went to Chuck and Stew jointly. Chuck and Stew wrote up a one-page typed agreement calling themselves the Blackwater Group and signed it at the bottom, not bothering with legal witnesses. Starting here, and wherever it might lead, they promised to go 50-50, friends and partners. They were on their own.

CHAPTER 14

Tree Line

The first thing they needed was money. They each agreed to raise half.

Chuck went to a craggy Kelowna mechanic named Stan Emerson, a plainspoken ex-prospector who had once owned a gold property outside Yellowknife. The ingenious Emerson now had a job at a factory fabricating plastic objects for which no designs existed, including Chuck's heavy-liquids system and other parts of the lab. The two had worked together for years, and Chuck trusted Emerson completely. Chuck showed him the Blackwater indicators. Emerson knew nothing about diamonds, but he loved anything prospecting-related. Having no money, he went to the owner of the plastics factory, a businessman named Bill Shemley. Shemley knew nothing about diamonds or even geology, but he viewed Chuck as a kind of mad professor, handicapped by his stutter and strange habits—and thus probably too odd to be fake, smarter than normal people. Shemley knew Marlene well; she kept a steady hand on the family business and always paid C. F. Mineral's bills on time. He put up $15,000 and lent Emerson an equal amount so he could get in, too.

For his half, Stew typed up a prospectus explaining the "rare volcanic rock known as 'KIMBERLITE'" and their "lucky break" at Blackwater Lake. "I am confident we will find a Kimberlite, pipe . . . and at the very least can expect to make a favorable deal with debeers," he wrote. He need not have troubled. He was boarding one of Marilyn's Air Canada runs one day when she introduced him to the captain, Eric Cartmell. Cartmell, like many Canadians, was addicted to penny mining stocks. When he learned Stew was a fellow pilot, investor, and geologist all in one, he pumped him for stock tips, then phoned his broker from the gate to make some trades while the passengers waited a good twenty minutes. Stew saw his opening. Within days he had $10,000 from Cartmell, who in turn corralled a stockbroker and another investor in Toronto to put in the rest.

With an initial backing of $60,000, the five outsiders got 49 percent of any discovery, while Chuck and Stew, as "discoverers," kept 51—fairly standard. Chuck's contribution was the use of his lab and his expertise in mineral grains; Stew's was his helicopter and expertise in larger geologic units.

The only thing holding them back now was winter; prospecting work in the north usually stopped now. In the meantime Chuck and Stew traveled off and on together on freelance consulting jobs. Since they both needed to earn a living, they were still at it by summer. In July, while flying from a lead/zinc prospect, they were assigned widely separated seats in a crowded commercial plane cabin. Stew was about to settle into some reading when he detected a spookily familiar sound behind him: Hugo's soft South African accent, discussing diamond prospecting in impressive scientific detail. Except that it was not Hugo. The man's companion sounded Canadian. Stew adjusted his seat back slightly, cocked one ear aft, unfolded a newspaper low on his lap, feigned sleep, and started taking notes with one eye half-open.

The African was obviously a top-ranking De Beers geologist on tour—in all probability Barry Hawthorne. Stew learned De Beers was still haunting at least one Arctic island and now also working the west coast of Greenland. Interestingly, it was hunting diamonds on Africa's Skeleton Coast seabottom, thought to be washed down the Orange River. Then the Canadian pronounced the magic words: "Blackwater Lake." As of July 2, Stew learned, the staff agreed the minerals around camp must have been moved by glaciers from the south—Stew's exact diagnosis. However, they still believed the source lay within their own ground. Line 61, whatever that was, looked best. They had four drill holes underway into glacial overburden, two already showing significant indicator pockets. Stew noted the grain numbers and types. The cement mixers were being readied to process actual kimberlite. "We should have something soon," said the voice. Stew heard no mention of their own claims next door.

When they made an intermediate stop, Chuck's seatmate got off, and Chuck stood in the aisle and yelled.

"Stew! Hey, Stew! Hey, come on, Stew, we can sit together now! There's a seat up here, come on!"

Stew made as if stupefied by exhaustion. He snortled, wheezed, and winked groggily. He raised one hand weakly and nodded to signal he could not possibly manage to speak, never mind get up. Chuck looked on without the slightest comprehension and sat back down.

For the rest of the flight, the De Beers men dozed; there was nothing more. On the tarmac at Calgary, where everyone walked toward the terminal, Stew grabbed Chuck's arm and hissed what had just happened. Chuck's mouth fell open. The De Beers men were hurrying off ahead of them and did not look back.

"De Beers is out to lunch," said Stew. "They didn't go far enough south. We've got them."

Sixty thousand dollars goes fast in this business, so they had to keep it simple. In August they decided that Chuck would stay in Kelowna to run the lab, while Stew went with the helicopter and Brent. Stew had supplies and a cache of fuel drums flown in by float plane, and they set up a small pup tent on a western arm of Blackwater Lake. Once again, De Beers was nowhere to be seen.

The center of the Blackwater Group's staked-out ground was a wide valley flanked by ridges, exposing cross sections of the 440-million-year-old dolomites. They concentrated here. Stew dropped Brent off a half-dozen times a day to dig, then went off to look for kimberlite float in streambeds and scan the dolomites for visible pipes. It was not quite as dangerous as mountain flying, but in this relatively open country, the wind blew hard much of the time, pushing the little heliicopter around. Every time a big gust came along they went side-ways, sometimes clear into another drainage. Then they would have to fiddle with the map and figure out where they were. Each morn-ing Stew got up to check the exposed fanbelts for cracks and proper tension. This made Brent nervous. When Stew was not looking, Brent checked for tension and cracks himself. Some mornings it was too stormy to fly, but Stew rejoiced; he loved fishing, so they spent the day at that instead. At night they sat around the campfire and talked about their childhoods—Stew about Powell River, his mother, his poverty. Brent shivered, unable to sleep from the damp and cold rising under the tent. Not Stew; he slept like a child.

Panning turned their hands numb in the cold water, and they used boulders as stepping-stones to keep their feet out. About a week in, Stew was crossing a stream in this manner when his eye stopped at the next stone. It was no local bedrock, nor was it kimberlite, but rather a terrible omen: a sharp, fresh-looking boulder of multibillion-year-old pink granite. He recognized it instantly. Identical boulders had been found by the GSC weirdly stranded a mile up in the moun-tains across the Mackenzie, where no such bedrock existed either. They were glacial erratics, ripped from the Barren Lands—the very same stuff seen in much greater quantity by John Richardson in 1825. Stew had never been out there himself, but he knew it was utterly characteristic of the Slave Province. Known exposures were shown by new GSC maps. The exposures did not start for another 200 miles. Now, the more he looked, the more of it he saw in the stream—softball-size cobbles, numerous, sharp, and uneroded. If those rocks had traveled so handily, so far, kimberlitic minerals could, too. Which meant they might not be at the start of the glacial dis-persal at all, but far downstream—down-ice, as glacial geologists say.

He prayed his fellow bedrock mappers had simply missed some granite outcrops poking from the cover of younger rocks closer by. He left Brent for a day and flew beyond their ground, southeast, up-ice along the apparent glacial path, looking for them. From the air he could see tons of pink granite boulders scattered along in lines, indicating the direction, and also boulders of ancient gray gneisses and schists—additional signs of the Slave—but no bedrock outcrops. He landed among the boulders and dug a few mineral samples.

A few days later he decided to cut their trip short; if everyone was in the totally wrong region, he and Chuck had better figure it out before Diapros did. They left a cache of fuel drums at their beach camp and buried all the shovels, pans, sieves, and canned food in the sand, planning to come back. Brent hung a pair of hip waders from a small spruce to mark the spot. To avoid spies in Norman Wells and to have the helicopter ready for a return, Stew parked the helicopter at their former mountain camp on Palmer Lake and got a lift out from another pilot down to Fort Simpson.

In Kelowna, Chuck rushed the samples marked "B" for Blackwater through the lab. They were laced with indicators all over their staked ground—and beyond. At best, they had staked only some pipes; at worst, none.

Stew wanted to go back immediately to search for the end of the indicator train. If he went east far enough, and somewhere the minerals disappeared, he would know he had probably gone too far and could back up. He planned on widely spaced samples in a 200-mile arc peaking seventy miles from Blackwater. In his discussions with Chuck, Stew understood that he would do the reconnaissance, and after Chuck analyzed the samples, they would consider restaking.

It was late September when Stew had himself dropped alone at Palmer Lake. After a freezing night, he got up while it was still black to scrape ice off the helicopter air intake. He turned the key; the engine cranked but would not catch. Stew cursed. The oil must be congealed solid in this temperature; it had to be warmed. There were several old bush-pilot tricks for this, one of which involved running a plumber's blowtorch around the engine block to warm the oil; but he had no blowtorch. Thinking fast, he ran through the dark to gather dry inner spruce limbs, then moss for tinder, and piled them under the aircraft. He got some canvas for a wind shield. Then he lit a fire under the helicopter. Not a big one—a controlled one, just enough to make the oil viscous. He blew on the fire and draped the canvas around to hold in heat, taking took care not to get the flames too close to the gas tank.

A half-hour later, just as morning was taking shape, he had turned the engine over and was moving through the mist over the thin new ice forming on Palmer Lake. He kept his radio off. Except for the pilot who had dropped him, and who was sworn to secrecy, no one knew where he was. It meant no one could follow him; it also meant that if anything happened, there would be no rescue. He grinned; the secrecy added relish.

A couple of hours later he dropped through the clearing mist into their Blackwater Lake camp and refueled. Brent's waders were still hanging from the tree, already tattered by wind. After refueling, he took off to make his big arc. It was not easy. Unlike the hilly ground they had claimed, most of the country east was low-lying and swampy. He had a tough time finding good spots. At least there were no mosquitoes or blackflies; it was too chilly. Over two days he took twenty samples. In the middle of the second day it hit him all at once: He was completely alone, just like the old days. No bosses. No sampling crew. No annoyances from Chuck. Just himself, making a map, trying to solve a puzzle. He had forgotten how much fun it was. He had to hit himself to make sure he wasn't dreaming. He could not stop grinning.

His last, farthest-out sample was from a sandy spit on a stream called River Between Two Mountains, toward dusk. Now with 240 pounds of gravel tied in bags to the skids, he had just enough fuel to get out, and a bit over an hour to dark. He planned to fly straight, coming out along the Mackenzie at Fort Simpson, about 100 miles south. Since the map showed no obvious landmarks, he had to go by the crudest dead reckoning: Take a compass heading, keep an eye on the speed indicator, and watch. At eighty miles an hour, he figured to hit Fort Simpson in seventy-two minutes, just around darkness.

It was a complete swamp—a featureless tangle of scum-covered ponds and meandering water channels that went nowhere. Mossy hummocks and dead trees protruded through the muck like infections; living trees grew at drunken angles because there was nothing solid to root into. Everything everywhere appeared to be sinking and rotting— no hills, no identifiable lakes, not even a rock. A while later Stew looked at his watch: forty minutes to Fort Simpson, and forty minutes to dark.

He could never remember getting lost anywhere, and he tried not to think about it now. But he could not help looking down and wondering what would happen if he missed Fort Simpson. The answer was obvious: He could not land. To the west, the setting sun was already turning deep orange and reflecting off the waters. He wanted to believe that some of the moss hummocks would be all right in a pinch, but knew they were traps; one or both skids might sink in, and he would tip over. Spiky trees, alive and dead, were sticking up every-where—there was hardly even any clear water. If his tail rotor even touched one, it would fly apart. There was no way to walk or canoe in or out. This land was totally defended.

The sun dropped, and as its last glimmerings subsided into after-glow, he knew Fort Simpson should be in plain view dead ahead. It was not; there was only more swamp. He felt an unfamiliar emotion, but recognized it: panic. He was sorry he had done this. He wished he were home, in the four-poster bed.

As darkness took over, he glanced at the fuel gauge: ten minutes. It was pitch black. He was going to have to land, or crash; either way, there was no difference. He was dead. A single light appeared directly ahead, though no light had been there a second before—a hallucina-tion, he was sure.

Fort Simpson sits on an island below a low bluff along the Macken-zie. If you come in low from the north, you cannot see it until you are practically on top of it. He was on top of it; in a moment, dozens of lit-tle lights sparkled below him. Five minutes later he was on the ground, trying to keep from shaking. He had lost count of which of his nine lives he had lost this time. Stew was sure only that he did not want to lose any more of them, because he might be on number eight.

I n Fort Simpson, most inhabitants were Dene, visitors few. Stew checked into a pack of trailers that passed for a motel and fell fast asleep. In the morning he was getting ready to leave when two out-siders walked in—Chuck and Brent. Stew was dumbfounded.

"What the heck are you doing here?" said Stew.

They all stepped outside.

"We came to get more ground, like we agreed," whispered Chuck.

"What do you mean more ground? I just got the samples. We don't need more ground. You're supposed to look at the samples. Then we can talk about more ground."

"I thought we said we were going to take more ground," said Chuck in a stage whisper.

They did not seem to be communicating properly. Stew fumed and protested that Chuck was supposed to be back at the lab now.

Chuck looked sympathetic. "Jeez, Stew. You look terrible. Did you have a hard trip? Are you upset about something?"

They went back and forth, but Stew could see it was no use. Chuck was being dense again. Chuck had his mind set on some course that only Chuck knew. Stew took the samples with him so Chuck could process them when he came back, and left town in an agitated state. The investors' money was almost gone.

Chuck and Brent checked into the trailers and made ready to stake ground south and east with a chartered helicopter. That was September 28. From September 29, when Chuck planned to start, until October 6, when he planned to stop, fog, sleet, and snow prevailed completely. No aircraft could leave.

Fortunately the motel had Cinemax, HBO, and other cable TV channels. Each morning their routine was the same. They rose early in their shared room, checked the weather, and went to breakfast. Then they turned on the television and sat side by side with their stocking feet sticking forward on their twin beds and watched movies. At midmorning they checked on weather, waited for the previews that came around lunchtime, then decided what to watch next. They ate and took in movies until supper, when they checked the next day's weather.

Chuck rarely watched TV—he was always working—but after a few days of enforced idleness, he confessed there was one show that he did always watch at home.

"What's that?" asked Brent idly.

"*Dallas.*"

Brent roared. The then-enormously popular American import about a fictional Texas family was so like the real-life Kecks, magazines compared their heartless internecine intrigues side by side. Chuck had worked for Howard but had never been important enough to meet him; however, Chuck's hero was *Dallas* villain J. R. Ewing, a Howard clone who ran the family business with unabated meanness. During Chuck's enthusiastic discussion of *Dallas*, it became clear to Brent that Chuck viewed himself as a sort of prospective J. R., founder of a great dynasty, if only he could find the diamonds. The constant victim of J. R.'s schemes and bullying was his brother, Bobby; Chuck said Wayne was his Bobby. He knew how badly he had once treated his brother. Chuck claimed that now, whenever he was in the field and unable to watch *Dallas*, Wayne dutifully videotaped it for him—apparently a sort of Fipke family therapy. Brent almost rolled off his bed and onto the carpet with laughter; Chuck seemed perfectly serious.

Evenings they spent in the bar, where they were the only non-Dene. The first night they sat down and immediately began flirting with two of the many uncommonly pretty local women. All male eyes immediately turned poisonous and the place fell silent. Finally a 250-pound man rose steadily from his table, walked over, and held out his hand. He introduced himself in a loud, welcoming voice as Jonas Norwegian, a welder and distant descendant of at least one Norwegian. He sat down and explained in a more confidential tone that if someone did not take it upon himself to be friendly at this juncture, Chuck and Brent would currently be having their brains kicked in by six or more customers; drunken beatings are a form of nightly entertainment in this part of the world. He bought them a round and made them promise to be more careful.

During that week the weather lifted enough for them to fly one hour—enough to glimpse the ground Chuck wanted staked—then be driven back by a hailstorm. Time was up. They had to go back and process Stew's samples.

Every sample was impregnated with indicators. They lay in no particular pattern; it was a regional dispersion, with no clear source. Stew and Chuck tried to put the best face on it for Emerson, Shemley, Cartmell, and the other investors.

There was some good news, though. They were still in sporadic touch with Hugo, and he made an offer: If they would send the indicators, he would have them microprobed and tell them whether they signaled diamonds. Hugo figured to do this for free by hiding the work in one of his ongoing exploration budgets—a huge bargain for Chuck and Stew, as microprobing one grain could cost upward of

$1,000, and no one else knew how to analyze diamond indicators anyway. It was also good for Hugo because it kept his hand in. Hugo continued protecting Gurney's formulas, though.

Chuck sent scores of garnets and ilmenites, and by the end of November 1982, results started coming in: Many garnets were G10s, and the ilmenites suggested a high degree of diamond preservation. McCandless inspected the surface textures and saw many garnets had fresh kelphytic rims, like those from his anthills. Either they had not traveled far; or they had traveled far, protected in ice.

Chuck and Stew had to decide whether to keep casting about Blackwater Lake or move up-ice. Stew pulled out the geologic maps and tried assembling a plan, but Chuck brushed him aside. "It's not brain surgery," he said. "The ice is coming from the east, hey, we have to follow that." Actually it was more complicated than that. The ice probably had moved back and forth many times, and by the signs Stew had seen, even the latest ice sheet in various locales had turned and meandered from the north, south or east. But Chuck was basically right: Maps showed the dominant direction was from the east, starting with the Keewatin ice dome on the other side of the Barren Lands.

Nothing more was done with Blackwater Lake; they let the claims lapse. The equipment was left buried under the beach, the waders hanging from the spruce.

Over the long cold season of 1983 they waited again for snowmelt and looked at maps. The most useful was a wide-scale aeronautical chart used by bush pilots. It clearly showed set upon set of long, thin parallel lakes—glacial fluting, carved by ice. Together, they formed a giant J, tilted hard on its side to the right, extending 200 miles beyond their last sample. That was obviously the ice path. They would trace back south, southeast, then northeast along the sideways J. They convinced the investors to ante up another $30,000 and waited for early fall; mosquitoes and blackflies would abate by then. In early September they headed out with no assistants.

They figured that instead of having Stew fly, it would be faster to have a hired helicopter to drop them in tandem. To foil spies, they met their pilot in the mountains, alongside the Alaska Highway below the Yukon mountain town of Whitehorse. Only when they were in the air did they tell him where they were going. He never asked what they were looking for.

Starting in the appalling swamps that Stew had escaped the previous year, they quickly ran up a huge bill just looking for running water. Streams shown on the map were actually scummy, stagnant channels wandering through the moss. The best they could find were a few islands in lakes big enough to generate waves along beaches. Since there were so few safe landings, they took turns having the pilot come in over a foot or two of water, then stepping off in hip waders and sinking knee deep into organic goo. They sloshed up to the shorelines on the lee sides of the islands to collect whatever little sandy material the waves had exposed. Most was useless peat; at some stops Chuck sieved 500 pounds to fill a twenty-pound sample bag. And the insects had not abated; the year was unusually warm, so they attacked in force as the men sweated.

Their base camp was near the bottom of the J near the Willow-lake River, a rare true stream. Every morning they woke to the same daunting sight: hordes of mosquitoes gathered at the tent door netting, waiting. Chuck got up, tore open the door, and fought through to make a fire for breakfast. The smoke helped a little, but whenever he stopped moving more than a few seconds, hordes convened on him, top to bottom. Most people faced with this wore a bug jacket— a head-to-foot mesh suit—but it made you sweat, was hard to see through, and imposed a hateful encumbrance through which bugs always penetrated anyway. Both Chuck and Stew refused to wear one, and neither of them liked caustic 100 percent DEET bug repellent either; it did not work any better, and they figured it probably gave you cancer. They braved the torture bare-skinned.

Something merciful seemed to watch over them. One day, as Chuck was kneeling to fry eggs, a great flock of sparrows appeared from nowhere; they surrounded him and began gorging on his envelope of mosquitoes. For several minutes the birds flittered through the air a foot or two away, hopping rapidly on the ground to gorge on insects. When the mosquitoes were all devoured, the birds flew off, and the bugs did not return for some time. That same morning a mosquito-eating dragonfly alit on each of Stew's shoulders and performed the same service with great effect.

A twenty-year-old GSC map of glacial features put a name to much of this bug-ridden morass: Glacial Lake McConnell, named after the GSC man who first saw that the bogs must be the soggy last remains of a vast meltwater lake that had once sat in front of the receding ice. Fine suspended sediments had dropped to its bed for thousands of years, deeply burying most minerals and rocks with clay and organic material. A few days of it was enough to make them both admit they were probably not going to find anything. They decided to vault farther up-ice and hope there was something there.

They had the pilot fly northeast to Lac la Martre, a big, open lake near the end of their planned sampling pattern. On a bit of high ground was a peninsula with an isolated Dene community with the same name as the lake. Lac la Martre was about 200 miles from Blackwater Lake, 100 miles from the tree line.

People were friendly here. The Dene chatted, then invited them to camp in the home economics room of their tiny schoolhouse. Chuck provided them with a laugh when he accidentally dropped his pocketknife off a lakeside dock, then dove after it into the freezing water wearing only his blue jeans. He emerged dripping and triumphant with the knife, while a bunch of children applauded and poked fun. Stew snapped a picture of the saturated Chuck standing on the dock holding up the knife with a big smile on his face. He kept it for years to prove that they really did have fun on that trip.

Lac la Martre provided a better selection of sampling sites, including short lateral moraines left by the glaciers, lying parallel to one another like bunches in a carpet. There were also creeks running into the lake from the east. Across a flat expanse easterly sat a lone ridge that wound continuously off into the horizon—the foot of the great esker system traversing the Barrens. This looked good, but they could not land because it was too peppered with little trees. Finally they turned back and found a detached esker segment sloping down into the lake, all bare sand. Chuck was delighted; here at last was beautifully sorted, well-drained material, perfect for their purposes. The digging was easy and fast. He hoped for more eskers.

Some twenty miles beyond Lac la Martre, they came to another new sight: scattered low domes of bare rock rising out of the monotonous greens and browns of the muskeg—ice-polished pink granite. They had arrived at the first outcrops of the crystalline rock seen lying in the streambeds of Blackwater Lake. With a few minutes of eastward travel, the pink domes erupted everywhere.

They sampled a few sluggish creeks around them. Amid the high marsh grass, Chuck was awed to see a rare sight: a large group of black-masked trumpeter swans, nearly as tall as himself. He had not seen the great birds on their previous travels. Chuck stayed still as they strutted along the shore and gave off great *oh-oh*! bugle cries. For once, he almost forgot about the samples. The wildlife was changing in other ways, too. From the air Stew spied a brownish area that looked like a solid landing spot. Suddenly the ground wavered chaotically, then began rushing toward them. It was a well-camouflaged herd of caribou, migrating from the tundra for winter. The pilot swore and pulled up just in time to avoid crashing into them.

From here, the closest town was no longer Norman Wells, but Yellowknife, about an hour's flying time south. Having reached the edge of the Slave Province, it seemed like the logical place to stop, go home, and see what they had. When they were done, they asked the pilot to drop them at Yellowknife airport.

The flight was routine. At the airport the pilot made a phone call, then came back to bid them goodbye. He told them he had just learned he had to return north, near where they had just been. "My company just got a big job up there," he explained. "Flying some other geologists." He mentioned the company: Selection Trust. "Sounds like they're doing something like what you guys have been doing," he said.

Chuck and Stew simultaneously flinched in opposite directions, then straightened back up to look at each other. They're on to us, thought Stew. The logbook, was all he could think. The pilot's logbook. They'll only have to look at his logbook to know exactly where we've been. Holy Christ, how did this happen?

The pilot did not seem to notice their reaction; they must have recovered well. To him, it was just another job. It was not De Beers that was following them. Worse: It was Jennings.

Now based in Toronto, Jennings had been deeply hurt by Howard Keck's betrayal. He felt thrown away, displaced, useless. As head of worldwide exploration at Falconbridge, he had paid little attention to Superior's work in Canada, even though Falconbridge helped fund it. It was only after Superior cut off the project in 1981 that he had looked at a summary report regarding the very last samples, containing G10 garnets, and realized they were on to something. He called Hugo to ask for more data, including maps, but Hugo politely refused: The rights had been signed away to Chuck and Stew. Privately, Hugo was glad: He had little desire to help Jennings. To Jennings, this was highly unfair. In his eyes he was the grandfather of it all—the man who had gotten the diamond program started; the man who had helped Hugo; the man who had found John Gurney, and thus the secret of the G10 garnets. He argued, but Hugo had a legal point; since the data no longer belonged to Superior, he could not give it to Falconbridge. Jennings was locked out again.

He was not, however, finished. The report did not mention Blackwater Lake itself, but he knew the samples had come from somewhere on the east side of the Mackenzie below Norman Wells. Still new to Canada, he was not familiar with the geography or geology, but he started learning. In 1983, while Chuck and Stew were planning their move east, Jennings quit Falconbridge to became head of exploration for Selection Trust, which was actively prospecting diamonds in both Canada and the States—in fact, the same company whose Arkansas geologist had threatened to have Hugo and Waldman shot for trespassing. Officially, diamonds were only a minor part of Jennings's job—he was supposed to find gold, nickel, and uranium, too—but diamonds were his real desire. Jennings immediately set aside a half-million dollars and sent crews east of the Mackenzie.

Meanwhile, unknown to anyone, the threat from De Beers had momentarily receded. The cartel saw Blackwater Lake as a dead end; there were no discoverable kimberlites. Back in Kimberley, Barry Hawthorne had gathered around himself a school of geologists who proposed that they head east to look at the really old cratonic rocks— the granites, schists, and gneisses of the Slave Province. De Beers had never been out there, but Hawthorne had the same GSC bedrock maps as everyone else, showing unit after unit of ancient rock, the obvious center of a craton like the one in South Africa. The old hypothesis that cratons were good for diamonds had gained ever more credence, codified in 1967 as a dictum called "Clifford's Rule." Hawthorne also had a few GSC glacial maps, suggesting ice could have moved the Blackwater indicators from the Barrens. However, one look at a different map—the road map—revealed no way in. Even Somerset Island had been more accessible; at least there was navigable sea a few weeks a year. In the tundra, there was nothing; analysts said that even if they found a mine, they could not afford to run it.

Thus in the boardrooms of Johannesburg Hawthorne's pleas were rejected. By the end of 1981 De Beers had ordered a full retreat south. It was cheaper to operate there, and many still-unexplained indicator anomalies lay about—in the Ontario woods, the prairies of Saskatchewan and Alberta, the lower Hudson Bay. In other words, they were to round up the usual suspects and interrogate them once again. Both Hawthorne and the Diapros geologists were disappointed. From rumors, they knew Fipke and his partner were still skulking around up there somewhere, but there was nothing they could do about it except keep their ears to the ground for their footsteps.

When Chuck got home from Lac la Martre, he was overrun with clients' samples flooding his lab. He was forced to process their own samples piecemeal, so it was not until mid-1984 that the picture started coming clear. The bogs of Glacial Lake McConnell contained only a few indicators, as expected. He was elated to find, though, that farther east the trail picked up. Once they hit the crystalline shield beyond Lac la Martre, there were fifteen or twenty garnets in each bag—far more than before—and the Superior microprobe showed many were G10s. This included the sample they had taken from the foot of the esker at Lac la Martre. The increasing numbers signaled they were closer.

This was not good enough for their small-time investors; there was too much wilderness out there. Emerson, Shemley, and the others retained an interest if something was found, but they refused to put up any more cash. Chuck and Stew were stymied; neither of them was poor, but they did not have enough to finance the venture. Chuck worked away at the lab, Stew with consulting projects, and as 1984 wore on, little got done.

Hugo had been thinking all along, and he now raised a disturbing possibility. If there once had been, say, a cluster of twenty average-size pipes, and you added up the density of minerals spread over the area thus far prospected, it was possible—even probable—that there no longer were any pipes. They might have been totally eaten by ice—the reason there were so many indicators lying around for them to find. Stew took this in stride; he just wanted to know where the minerals came from. He told Chuck that if the pipes were wiped out, they could write a fascinating paper proving they had once existed. This drove Chuck crazy; how could Stew think of pure science at a time like this? He only wanted to find the diamonds. Secretly, he began to suspect Stew was losing heart. Maybe Stew was afraid to risk his own property and savings—the likely next step. Or maybe Stew feared his number was up; next time, he was finally going to get killed. Stew, for his part, thought Chuck was getting too worked up; they were on a trail that might never lead anywhere. Better to learn something and let go of the monetary results. He thought that maybe Chuck was the one losing heart.

Growing pressure was indeed getting to Chuck. His lab had grown to thirty-five employees, but it was boom or bust; unless samples poured in, he was faced with laying off staff and borrowing. Mark, now seventeen, was in full rebellion. He showed no desire to keep working for Chuck, become a geologist, or even finish high school. They fought all the time. During one pitched battle, Chuck laid down the standard father-of-a-teenager tenet: As long as Mark lived at home, Mark would follow orders. Mark accepted the challenge and promptly moved out. He dropped out of high school. Shortly his girlfriend, Leslie, was pregnant, and he was forced to marry. Chuck saw his own painful youth repeated before his eyes. He had worked so hard, so long, and could not understand what was happening. Mark seemed to hate him; the four younger children barely knew him because he was never around. Marlene constantly badgered him to be more of a father. His hair was starting to thin, and deep, permanent creases were forming in his brow, which seemed to jut farther out all the time, like it was overfull. He was thirty-six years old. Where were the diamonds?

Chuck looked back to the old mountain claims, owned by C. F. Mineral. Nothing more had been found in the Jack Pipe, which mineralogy tests now suggested was a lamproite. Mountain Diatreme appeared to be a melilitite or alkali basalt, a step farther down; 800 samples from surrounding drainages contained not one indicator mineral. Nevertheless, he decided to form a public company to reexplore them. He dubbed it Dia Met, for diamonds and metals. He figured that if no diamonds turned up, they might find gold or copper. That was his personal hope; the Blackwater Group's secret indicator trail was a separate outfit, a different domain, Chuck and Stew's own cabal.

Dia Met appeared to be just another crazy idea of Chuck's, but one person believed enough to help get it started: his brother Wayne. Wayne had never ceased to be amazed by Chuck's tales of adventure, nor to be Chuck's admiring little brother. Now managing a theater in Edmonton and dabbling in playwriting and acting, he lived vicariously on stories from Chuck, and sometimes Stew, who occasionally stayed at Wayne's house after field trips. Stew had once arrived after weeks outside, failing to take a hint from Wayne's wife that he needed a bath even after she opened every window in the house. Wayne saw the prospector's life as quite exotic.

Chuck needed investors for Dia Met. Wayne had little cash himself but knew somebody who did: his old college buddy Dave Mackenzie, who had earlier gone drinking with him and Chuck. Like Wayne, Mackenzie was a bit of a dreamer. Even before finishing college, Mackenzie got himself a pilot's license, and later did a bit of bush flying for oil companies and the government in the north woods. Actually, he did not need to work: He was heir to a shipping fortune. However, Mackenzie was embarrassed about being rich, so even Wayne did not even realize it until Mackenzie one day bought a huge house in Edmonton and a couple of airplanes. Wayne told Chuck about this. Together, they instantly set on Mackenzie. Mackenzie toured Chuck's impressive lab and consented to put up $50,000 in startup expenses. Wayne bought $2,500 in shares—all he could afford. Chuck retained the majority of shares, in return for deeding his claims to Dia Met. Stew got a big chunk, too, and an automatic seat on the board; Chuck honestly acknowledged the big part he had played in finding the mountain claims.

Dia Met went public in October 1984 on the Vancouver Stock Exchange (VSE). The VSE was the center of Canada's penny mining stock world—a place where people routinely threw away money, hoping maybe one investment in twenty would pan out. However, even here most accredited brokers would not touch Dia Met—especially after Chuck made a round of their glass-tower offices wearing the single rumpled blue suit he owned, tie and eyeglasses askew, spouting an incoherent pitch about diamonds. Finally one man who knew and trusted Stew agreed to sponsor an offering, and others fell in line. Venture capital funds and several well-heeled Kelownans bought 50¢ shares—mainly because a federal tax law encouraging high-risk mining ventures guaranteed them a $1.33 writeoff for every dollar they lost. Like Bill Shemley and Stan Emerson, they saw at least that Chuck was the real thing—a prospector who actually went out looking for stuff, not one of the slimy promoters usually plaguing their doorsteps. After this, shares rarely changed hands and plummeted as low as 9¢. Dia Met had suffered apparent crib death.

Just then, more bad news: Superior Oil was put up for sale—a move long resisted by Howard Keck, who no longer had any say. Mobil Oil agreed to a takeover and swallowed Superior whole. As soon as they had it, Mobil took one look at Hugo's truncated U.S. diamond program and axed it. This ended not only Hugo's and McCandless's jobs but also Chuck and Stew's free microprobes. In February 1985, Hugo and McCandless split up the diamond library and collected their severance checks. McCandless moved to South Africa to work for John Gurney. Hugo went to an American gold and copper outfit that had no interest in diamonds. Mobil donated the stones from the failed Sloan property, about 300 carats, to the Denver Museum of Natural History. The Blackwater Group now had no claims, no funds, and no help.

Chuck and Stew talked about it and agreed that they had enough evidence to justify one more shot. One hundred miles beyond the last samples lay the Barrens. The only thing now was to go there without delay and cover the entire ice path to its point of origin, all at once. They would have to pay for it themselves and do it fast; the more time that passed, the more chance the competition would pick up the trail. They had never been in the tundra, but they had survived everything else from Arkansas to the Yukon. This did not look any worse.

That winter Stew returned to the familiar GSC headquarters in Ottawa and spent many hours in the attached photo library of the Topographic Survey. By now there was a complete set of black-and-white aerials of the Barrens from decades of government flying. Stew considered these better than maps, recently published on a crude 1:250,000 scale. The maps, in fact, were made from the photos. And on the photos, one feature was etched in better detail than ever: the great central esker system, at the foot of which they had found indicators. The Barrens had little more river drainage than Glacial Lake McConnell, but the eskers were a paleodrainage, right down the middle. Box after box of big contact prints clearly showed the channels standing out pure white amid the chaotic grays of swamp and the blacks of lakes. Stew saw they had only to follow, and dig. Their end point would be the figurative headwaters of the esker, on the far east side by Hudson Bay—the Keewatin ice divide, the ice spreading center. Logically, the pipes must lie somewhere in front of it.

Chuck and Stew ran up huge phone bills over the cold months of 1984 and 1985. Stew wanted to start southwest of a big, caribou antler–shaped lake he saw on the map, Lac de Gras, and work systematically east. To save money, he would double as pilot. Instead of a relatively slow, expensive-to-run helicopter, they could rent a float plane in Yellowknife to use. Out in the open land, they could travel fast and set down on lakes, the old-fashioned way. They awaited summer eagerly.

Then a series of seemingly unaccountable things happened.

It began when Stew started his own miniature public company around the same time Chuck started Dia Met. It was called Pioneer Metals. Like Chuck, Stew was organizer, chairman, and chief geologist. Like Dia Met, Pioneer had no ostensible connection with the Blackwater Group. Its main property was a gold prospect in Manitoba. Stew was able to get a few investors to put up money for exploratory drilling, and by spring he was on site, running the project.

They had agreed to fly into the Barren Lands on July 1, 1985, shortly after usual ice breakup. Just beforehand, Stew called Chuck to say he was still in Manitoba; could they instead leave in the middle of the month? No problem, said Chuck. When mid-July rolled around, Stew called; he was still trying to wrap up. Could they go at the start of August? Chuck agreed, but this time was annoyed. He could not shake the feeling that Stew was losing his nerve—afraid of financial ruin or death. Whatever it was, by this time they had not been in the field for almost two years. Summer was waning.

As August approached, Stew called once again; he was still busy at the gold site but would be free by mid-August. He promised they would definitely go then.

Chuck now believed Stew had no intention of really making this trip. They were going to lose another year, and they could not afford that. And indeed, Stew did not reach Chuck on the phone in time for the mid-August departure—Chuck was out somewhere. But finally at month's end, Stew finished and went home to Vancouver to see Marilyn. From home, he called Chuck to tell him he was ready to go to the Barren Lands.

When Chuck picked up the phone, he tried to speak, but stuttered more than usual. That usually meant he was nervous. "I uh, I uh, oh, oh, oh. Well . . ."

Stew waited patiently.

"I just got back," said Chuck finally.

For a moment Stew was speechless.

"What do you mean you just got back?"

"I got it done, Stew. I just got back." And that was not all.

CHAPTER 15

The Sandman

When Chuck failed to hear from Stew, he headed for Yellowknife alone.

The town had changed little since the 1950s. Its main business was still the ceaseless search for minerals, mainly gold and copper. Some years migrating caribou on the outskirts outnumbered the 10,000 people and twenty-four bars. A planned "New Town" of modern buildings had recently risen on bare rocks along the optimistically named 50th Avenue, the intended center of future Avenues 1 through 100. So far only 49th to 54th existed. Aside from the mining recorder, New Town's main attraction was the Gold Range bar, a box with windows permanently plywooded over to intercept flying objects, such as humans, and towels draped on tables to absorb spilled beer or blood, whichever came first. Dene, Inuit, and Europeans mixed daily to hunt members of the opposite sex, drink, brawl, brawl, drink, and perhaps drink a little more, before being herded out at the 1:45 A.M. closing to tip over trash cans in front, hang off the guy wire to the utility pole, and sleep on the sidewalk. In cold weather people were known to freeze to death this way. Quiet Old Town, on an adjoining island connected by a small bridge, was still the spiritual center, focused on a winding lane called Ragged Ass Road. Weathered wooden trading posts and log cabins abutted the float-plane docks nearby, where small bush aircraft buzzed in and out all day.

Yellowknife was chronically infected by mining-town malaise: an excess of cash from occasional gold strikes and the lack thereof whenever they ran out. In either case, little filtered down to native people living on the outskirts. In 1961 a gravel road was finally pushed through the woods from the south, bringing fresh vegetables, a few tourists, and a sign at the edge of town: EDMONTON 1482 KM. It was a rough fifteen-hour drive if you did not stop, including ferries across unbridged rivers, which were impassable during fall freezeup, spring breakup, and assorted frequent bad weather.

Bush planes were expensive, so most prospectors still confined themselves to a 100-mile radius of town. Their finds formed a ring of scattered, worked-out pits, often filled with toxic waste. Bearded locals in baseball caps gathered each morning in the steamy, cigarette smoke–filled diner next to the Range to trade information, gossip, and lies about their wilderness exploits and to speculate where the next strike would be. After breakfast, many spent days loafing near the float docks, noting the directions in which planes took off and timing departures and arrivals—useful in calculating where competitors were landing. Whenever copper or gold prices spiked, some large companies came up and reventured to the long-abandoned Barrens—they had the major backing to fly such long distances—but the results ranged from bad to disastrous. In the Northwest Territories, the leading cause of death for young men was not homicide or AIDS; it was boating accidents. The tundra saw them in abundance, as well as a variety of other ways to die. Air crashes were right up there, followed by snowstorms and pure disappearances.

In June 1983 an aging, nearsighted Yellowknife prospector called Shorty (his real name was Alan Reid) staked a gold claim out near the edge of trees, about 100 miles from town. A few weeks later a friend flew in and found only his eyeglasses—nothing else. Shorty was never seen again. Two years later a southern outfit called Connors Drilling was making a gold-exploration hole in the lower tundra during cold weather when there was a whiteout, a common condition in which blowing snow cuts visibility essentially to zero. An assistant driller from Newfoundland walked 1,500 feet from the drill to a lakeside to refuel a water pump but apparently could not find his way back. Remnants of his tracks were later spotted heading south. Tracking dogs and airplanes never found him.

In July 1989, two geology students employed for the summer by the Australian company Broken Hill Proprietary—the company Chuck had once visited in his wandering youth—were dropped by helicopter in the far eastern Barrens to search a lakeshore for gold ore. When the helicopter left, a grizzly came over a rise. Unarmed, they retreated into the freezing water, and once they were in there, the bear would not let them out. Even in summer, the water in these lakes is too cold for people; fatal hypothermia often comes in minutes. One of the students was David Forget, age nineteen. It is not clear whether David stepped off an underwater ledge and his heavy clothes dragged him down or if his heart just stopped beating from cold or fear. His companion, nineteen-year-old Al Rennie, survived.

To the west that same summer, another student, Brandon McWhinnie, was out with geologist Valerie Arthur checking a ferric gossan on a high rock. It was a glorious, sunny day marred only by blackflies. A brisk wind blew above. Suddenly a huge, dark cloud mass came over the horizon. It blotted out the sun, and everything grew still. When the mass was directly overhead, it paused, then shot a bolt of lightning into McWhinnie's head and out his leg into the gossan. Valerie Arthur was knocked down by the force. When she came to, she climbed atop him, tore off her bug-jacket hood, and did cardiopulmonary resuscitation for nearly two hours while blackflies attacked her naked face without mercy. Long before she stopped working over the body, the cloud moved off to expose blue skies again, and a beautiful rainbow appeared. Brandon McWhinnie was not even a geology student; he was studying business, and took the summer job because he thought it would be an adventure.

Not all such expeditions were failures. As a result of modern aerial prospecting, a minor gold mine did manage to open just past the tree line, 150 miles north of Yellowknife, but it lasted only from 1964 to 1968. The so-called Tundra Mine was supplied by air and, in the coldest three months when ice froze deepest, a "winter road"—a widened Dene dogsled trail plied by bulldozers hauling sleds over frozen lakes. The trip took about two weeks. By 1984, the winter road had been extended to two other mines, Lupin and Salmita, and smoothed so trucks could pass. But this narrow fissure through the snow did not go anywhere except the isolated gold holes, posed grave hazards on the way, and disappeared most of the year. Salmita closed shortly after Chuck's visit, and Lupin did not last, either. The world still ended at Yellowknife.

Down at the Old Town float docks, Chuck engaged a small charter outfit, Raycom Air, to take him out sampling in the Barrens. The fliers were used to secretive prospectors, but Chuck was beyond the usual. Most geologists going out that far worked for major companies and paid by check. Chuck told them he had no company and would be paying from a wad of anonymous $100 bills. He wanted receipts—but with no name written on them, apparently to avoid leaving carbon copies. Cash, of course, was no problem; Raycom set him up for a couple of weeks with a rotating cast of pilots to take him wherever he wanted.

Stew's intellectual plan to sample the esker system was discarded. After a brief reconnaissance flight, Chuck decided to just crisscross the tundra in north-south lines. He figured he would look down, and when he saw a good spot with a safe landing adjacent, drop in. His guide was the GSC's "Glacial Map of Canada," a 1:5,000,000-scale chart of the entire nation, showing only the most general ice-flow features and useless for navigation. His plan did not even involve camping out; with enough extra fuel on board, the plane would have enough range to return to Yellowknife most nights.

Things proceeded pretty much as Chuck wanted. Early each morning he and his pilot for the day took off in a Cessna 185, which carries up to six passengers, but in this case was loaded with jerry cans of extra gas in back. Once they got in the air, Chuck told his man what heading to take, and they sailed off to open country. To him there seemed no romance or mystery in crossing the tree line; he was seeking samples, as usual. They turned north. The spruces diminished, rock outcrops muscled into empty spaces, the otherworldly panorama of endless lakes opened. They were over the Barrens. One more *kwet'i* looked down for a place to dig.

Few pilots asked what he was after, and when they did, he talked vaguely about gold. Usually he switched the conversation to an excited jabber of jokes, stories, and small talk. He pulled out a plastic bottle of vitamin C tablets, tipped back a few into his mouth, and, chewing on the tablets, shoved the bottle toward the pilot. "Vitamin C?" he queried brightly. The pilots usually smiled and passed. They found Chuck more entertaining than your average prospector. They liked to see an optimistic, energetic fellow. Occasionally he fell silent and scowling, seemingly off in his own world.

They went first to Baker Lake, an Inuit settlement near Hudson Bay. This was 800 miles beyond Blackwater Lake, more or less along the Keewatin ice divide, the farthest point. On following days they bounced around the tundra randomly, looking for a good place. Most of the land was a boggy mess. Dig a hole through the tough tundra plants, and there was usually mucky organic matter, rocks, and water immediately filling in from the sides. A foot or so down, below the active layer that melted each summer, was rock-hard permafrost. From the air Chuck could see well-drained eskerlike deposits where permafrost did not start for three feet or more, and plenty of lakes to land on; but the sampling sites and the landing sites were rarely near each other. By rule of thumb, pilots cruised a lake at eighty miles an hour, and if they saw twelve to twenty seconds' worth of rock-free water, it was long enough for landing and takeoff. Many were too cramped. They sometimes went 200 miles before stopping.

Since most places had no names, he noted only the coordinates and names of the nearest features where he took samples: Muskox Lake, Ghost Lake, Ajax Lake. His log showed a stop near Contwoyto Lake, then along the Coppermine River, along which blond-haired Samuel Hearne had followed 214 summers before. Chuck had never developed the habit of reading and so had never heard of Hearne or any other explorer. But like them he discovered that the upper reaches of the Coppermine barely passed as a river—more a series of lakes connected with confusing mazes of sluggish stream channels. He did, however, manage to spot a place along the Coppermine where the current in a relatively riverlike channel played itself out as it hit a lake and had dumped off a small delta of sand. This looked good.

The procedure here was the same as everywhere. The pilot dropped in with the lovely, soft landing that only float planes can make, cut the motor, and drifted toward the beach ahead of his wake. In the shallows, Chuck leapt off the pontoon in hip waders and charged like a one-man amphibious invasion. Tossing his knapsack onto land, he immediately set to digging sand from under a foot of water while the pilot stayed inside to avoid the gathering mosquito clouds. Chuck lifted the shovel straight out each time with water streaming off the edges, flicked rocks off with subtle tips of the wrist, and dumped the load into his sieve. He squeezed in some Sunlight dish detergent, then jigged the material in his water-filled bucket, rapidly reaching in and tossing more rocks over his shoulder.

Now, the pilots had all taken out gold prospectors before, and they knew gold prospectors usually kept the rocks and threw away the sand. This man was doing the opposite. Some speculated he was prospecting not for the hard-rock deposits that Yellowknife lived on, but for old-fashioned placer flakes—in which case he was crazy, since no one could hand-pan enough gold to pay the flying bill. They also noted with amusement the dish detergent; normal men did not launder dirt. Within a half-hour Chuck waded back lugging his little bag of sand and told the pilot where to head next.

A few nights they camped on lakesides, but generally they made it back to Yellowknife in light that lasted until around 11 P.M. There Chuck pulled out his roll of $100 bills and peeled off seven, the established daily rate. Then he left, looking dead tired. No one could blame him. It was hard work out there in the sun and bugs.

Chuck quickly gained a nickname behind his back: the Sandman. No one could understand what this strange little man was doing with his sand. When he left the docks, the pilots pulled out their forty-ounce bottle of Bacardi rum, some cans of Pepsi, and glasses. "Who's taking out the Sandman tomorrow?" someone would ask. Then they would all laugh. Occasionally after hours the fliers saw him down at the Range, revived and laughing it up with the locals. Somehow Chuck never got a bottle broken in his face. Chuck liked the Range, and it liked him; it was his kind of place.

On one of his last days, he had the pilot fly over the source of the Coppermine: the curving Lac de Gras, sprawled over the map like a forty-mile caribou rack. There was no place to dig along its rocky, windblown shores, and directly north was a wild maze of smaller lakes and streams, all unlandable. Then they came to something quite different: the huge central esker, running east to west. As they passed over, Chuck could see a place where it curled dramatically into a perfect S-curve. On its south flank was a small but landable lake nestled into the curve, and in the cove, a beautiful beach of light sand. Even from the air he saw that the strand line, where modest waves lapped up, was marked by a contrasting stripe of dark sand. The esker had concentrated sand once; and here the waves had performed a second cycle, washing away light fractions and leaving just dark heavies. What more could he ask? They came down and drifted practically to the beach. He splashed the last few feet through the shallows to the spot that had so enchanted the young GSC man Bob Folinsbee in 1947, when Chuck was one year old.

It took him only minutes to dig and sieve the purplish-black sand—an unusually easy stop. He bagged it and marked it with a serial number—G71. He liked this spot, and took a rare moment to climb up the esker's dunelike side. On top, the esker was remarkably broad and smooth, like a road or an airport runway. A cool, steady wind kept the bugs down. The flanks were laced with deep braided trails, where generations of caribou had followed exactly in each others' steps. Wolf droppings white with minced-up caribou bones and hair lay alongside. The immensity of the tundra rolled off to the horizon on all sides, the sky, the rocks, and the plants painted in every color of the rainbow. It was as if he were walking on water and had just pitched up to the top of a great ocean wave. He galloped down the side and splashed back to the plane with sample G71.

From August 17 to 27, 1985, the Sandman dug ninety-six holes and flew over 5,000 miles around the Barren Lands. Except for his pilots, he never saw another human, nor signs of one. On the 28th, he departed for Kelowna.

Stew was stunned and hurt. He could not understand how Chuck could leave without him. Chuck sounded sheepish; he said he thought Stew had changed his mind. When Chuck described his stops, Stew became even more distressed. Except for the esker, it sounded like a helter-skelter dash. Stew wanted to know why so much flying, so little planning. Chuck insisted he had done his best and had been unaware of Stew's plan. Amid the confusion, one thing was clear: All that long-distance flying had cost thousands more than they planned. Stew now had to ask: Where did all that money come from?

Chuck owned up: Dave Mackenzie. He had gone to the wealthy flier again and told him he needed cash for what he vaguely described as "broad-scale diamond exploration in the north." Mackenzie gave him $67,000 in return for a share. In effect, Chuck had brought in a new partner.

Stew slapped his head, moaned, and pounded on the table. This was horrible. Mackenzie did not know exactly where Chuck went, but that hardly mattered. Stew saw that Mackenzie's investment potentially diluted the original investors' percentage—to say nothing of his and Chuck's. He felt everyone had a right to know if someone else came in. Then Chuck mentioned the further twist: So that Mackenzie could reap certain legal tax advantages and have an investment in something tangible, he had converted much of his investment into shares of Dia Met stock. That meant that Dia Met was in, too.

Stew went berserk. In his eyes, Chuck had betrayed him. Dia Met was a public company; it had no place in their private venture. It compromised secrecy; it was confusing; he was not even sure of the legalities. There was only one answer, he said: Give the money back. But Chuck could not give it back; Mackenzie had already filed the tax papers. Everybody would get in trouble if they backed out now. Chuck saw nothing wrong anyway. They had needed cash; he had taken it. Stew raged, but without effect. To him, this was the ultimate miscommunication. They had been like a boys' club with a secret fort in the woods; now outsiders were invited.

Shortly, affairs worsened: Chuck prevailed upon Stan Emerson and Bill Shemley, his original Blackwater investors, to convert their interest into Dia Met shares too. Then he had Mackenzie installed as Dia Met president, while retaining for himself the title of chairman of the board. To protect secrecy, the shareholders agreed to remain uninformed about geographic or technical details, but these moves set up two camps: Dia Met and Stew, even though Stew had a multitude of Dia Met shares himself. Stew wondered whether Chuck had planned a hostile takeover all along. Social visits to Kelowna dropped off. Stew kept his Dia Met shares but within months resigned his board seat. He tried explaining to Chuck that he feared the financial machinations would backfire and felt he had to protect himself from liability, but Chuck did not seem to comprehend. Stew did not try explaining at all to Marlene. He hardly ever spoke to her again. Only later did he wonder if his reaction made his old friends feel as abandoned and sad as he himself now felt.

Whatever was going on, they both recognized that they now had to finish this thing one way or another. When they managed to temporarily quit arguing, they resolved to bring in a senior partner—a big company like Superior to finance expenses and pay them badly needed salaries in return for yet another piece of the pie. They visited a half-dozen companies, vaguely described the project, and were turned down everywhere. In 1986 Chuck even called Hugo and invited him to become president of Dia Met, with the idea that somehow Hugo's muscle would get things going. But Hugo knew how Chuck drove people, and in any case did not know where any prospecting had led since he left Superior, because Chuck would not tell him. Chuck did not even want to pay a salary. Hugo politely declined.

Through 1986 and 1987 Chuck and Stew were frozen—out of money, ever more fearful of unseen competitors, and with the faint scent of distrust hanging between them in the air.

They also were stopped partly because Chuck seemed so slow at analyzing the samples. The maps of his Barrens travels lay locked in his chaotic office, available only to himself and Stew—actually, photocopies of maps with all identifiers and names of features blanked out, so if someone stole or glimpsed them accidentally, they would be useless.

Chuck only seemed to be slow; in reality he was insanely, quietly busy. He had a new tool: the formula of the G10 garnet.

It had been only a matter of time before it leaked. Gurney was not the only one who had it. Nikolai Sobolev, heir to the Siberian diamond program, had been working on garnets from the start. He had figured it out around the same time. De Beers, temporarily sidetracked by the disastrous R-Factor, recovered and by 1982 had also figured out the real formula. They all kept it secret from one another, but by 1984 enough scientists for other outfits were working along similar lines that word circulated; Sobolev planned to publish. Gurney was horrified; if Sobolev published first, Sobolev would get the credit. So Gurney himself rushed out a paper in an obscure Australian publication, giving the basics of the low-calcium/high-chromium G10. It left out proprietary subtleties in interpretation and all the secrets of diamond-indicating ilmenites and chromites; but it sufficed to give Chuck and anyone else who cared to read it a start at their own analyses. Chuck realized Hugo had been lying to him all along, but he managed to remain angry only briefly.

Chuck got hold of a high-priced scanning electron microscope—a sort of poor man's microprobe, slower and less precise. Some of the money came, again, from Dave Mackenzie. The rest was from a GSC scientific grant—more or less a government subsidy—courtesy of Walter Nassichuk, a highly placed GSC man and, as chance would have it, a childhood friend of Stew's from Powell River. Nassichuk had taken an interest in Chuck's lab. He did not know exactly what was going on now, but he thought the lab's work might eventually help other Canadians refine their uses of heavy minerals.

Within a year after his Barrens trip, Chuck had learned how to use the electron microscope and was doing his first grain analyses in a windowless interior room at the lab. The walls were lined with sheet metal to keep out electrical interference, the door locked to keep out people. It was slow going because Gurney's publication was so sketchy and because Chuck did much of the picking, mounting, numbering, and cumbersome analyses himself to make sure no one else had the data. Thousands of tiny flecks were emerging from the processed samples, and he had to investigate every one.

To keep afloat, he hustled more Dia Met shares to anyone who would buy and continued poking unsuccessfully around the Jack Pipe and the Mark, a nearby cliff on which Mark had almost broken his neck several times. As for Mark himself: After quitting high school and becoming a dad, he had tried a string of menial jobs. But after life with Chuck, normal indoor workplaces felt claustrophobic and ridiculous. To support his new daughter and wife, Leslie, he grudgingly came back to Chuck and risked his life again at only slightly higher wages than before.

During this period Chuck also spent much energy spying, a pursuit at which he had grown ever more brazen and self-confident. De Beers continued to follow its southern indicator trails, and in June 1987 staked claims at a wheat farm outside Prince Albert, Saskatchewan, near the old penitentiary. Chuck went to investigate reports they were digging at a block of greenish rock. The road to the site, which was in a big field, had a gate, so he parked his truck, walked around, and headed for the pit. The only person around was a slight, shy De Beers warehouseman, left as guard because he had sprained a finger and could not work. The man walked over and politely asked Chuck if he had permission to enter. Chuck mumbled something, barged into the pit, and began shoveling De Beers's ore into a flour sack he had brought for the purpose.

"Could you maybe leave, and come back when you talk with the supervisor? I'm sure you'll be welcome if you just ask," said the injured employee. Chuck shrugged, moved to a new spot, and started digging another hole.

"I don't think you're supposed to be here. You have to leave," insisted the man. The yammering was beginning to annoy Chuck. He hated wimpy little guys. After he finished filling his sack, he pushed by, returned to his truck, and drove to the next town. As he entered the limits, a police car whooshed by in the opposite direction with lights going, obviously responding to a call from the pit. That was a close call, thought Chuck. Then the cruiser screeched into a U-turn and came after him. The wimp had phoned in his license-plate number.

The police made him follow behind them to the station. On the way Chuck stuffed kimberlite into various pockets of his vest to confuse them, and a deck of cards in case he had to go to jail. At the station, they took away his wallet and made him empty his pockets. Various phone calls were made to lawyers and the De Beers hierarchy. After some consultation, it was decided not to press charges; the De Beers lawyers said it would be too much hassle to prove the rocks Chuck had on him were the ones taken from the pit. And as De Beers already knew but Chuck did not, the rocks were barren. They offered no useful information.

Among the files at Diapros headquarters in Toronto was a report from unnamed sources that spoke of a mysterious gold prospector who had haunted Yellowknife a couple of summers previous. They called him "the Sandman." Joe Brunet, now in charge, suspected the Sandman was not looking for gold, as Yellowknifers assumed. And from the physical description, he had a good idea who the Sandman was. Next to the Sandman file, the latest Fipke sighting at the penitentiary pit was duly noted.

Chuck tried finding out what Jennings was doing but could glean little. There was a reason for this: Jennings was lost. His crews had picked up the indicator trail east of the Mackenzie, but then they bogged down in Glacial Lake McConnell. Jennings was behind in his remedial reading on Canadian geology and was unaware it was even a mapped feature. Had he known, he probably would have skipped right over it and kept going. Instead he found an expansive high spot, the Horn Plateau, marooned amid the old lakebed, and decided the minerals must have come from there. He sent drillers to the Horn. They had been there for three years now without finding a thing.

While Jennings floundered, Chuck finished analyzing the tundra samples. He was always too secretive for anyone to pin him down on when, but it must have been no later than May 1988, for that was when a great flurry of activity took place at Dia Met.

Just beyond the tree line around the Coppermine, the indicators persisted as before, scattered ten or twenty to a sample. East of Lac de Gras, they disappeared completely—nothing, zero. In between, in sample G71, from the band of purplish-black sand on the esker beach north of Lac de Gras, was something unbelievable: 100 ilmenites, 2,376 chrome diopsides, and 7,205 grains of pyrope garnet. The pyropes ran to lavender, light pink, reddish-purple, and a curious bluish-gray purple. Almost all of the latter were G10s— about 1,000. Other samples from fifteen miles off or more totaled hundreds of G10s—including the one from the Coppermine itself. Some were so big that they were easily visible to the naked eye. He was sure he had found the headwaters of the indicator train.

It was a dizzying discovery, and a dangerous one. The minerals on the curving beach were only a sign; he doubted if the pipe lay directly underneath. Millennia of pounding by glaciers, meltwater rivers, then wind and waves had all combined to concentrate the minerals here, but likely they had arrived from some unpredictable spot—or spots. He was close, but if he tipped his hand now, De Beers, Jennings, and anyone else with big money would pile in, stake for miles around, and shut him out in an eyeblink. He had to go back and find the spot—very, very quietly.

Thus the esker became Chuck's crushing, awful secret. He was afraid to tell anyone—including Stew, who remained in the dark for many months. Chuck reasoned that Stew was busy anyway. Stew was in Manitoba, Idaho, New Mexico, looking for gold. He was consulting at copper properties in Nevada and California. Chuck did call and tell him the general truth: They had something big and needed to jump in with a big crew. Stew, lacking maps and details in front of him, fretted about the cost and, once again, the complications of all these partners. Chuck tore at his remaining hair. In his eyes, Stew did not want to do this—but he did not want to let anyone else do it, either. They argued back and forth. Finally Stew acquiesced; he said Dia Met could raise more funds for the venture. However, by the rules of the game, he and Chuck now also had to put in some money to maintain their own large shares.

By this time not only did airline captain Eric Cartmell and Stew's other original investors not want to put up more money; they wanted out completely. This meant Stew was faced with the additional burden of buying them out, if he wanted to maintain a balance of power between himself and Chuck. What had begun as a simple outdoor adventure was becoming an accounting nightmare. He was unwilling to back out now, though, and went to the bank. With a home mortgage eating his and Marilyn's incomes, they scraped nearly every dollar from their accounts. Stew sold stocks he owned, and they took out a $7,500 loan against Marilyn's used Volvo. Marilyn did not know who she was angrier at: the bank officer who had the nerve to ask how much she spent each month on clothes or Stew, who raided her Air Canada paychecks as soon as she got them and turned them over to Chuck. "You better straighten this out," she told her husband sternly. "We haven't got a nickel left." They raised $24,000—enough for now.

The Fipkes were struggling, too. One more time, Chuck put on his rumpled blue suit and went to see the Vancouver financiers. He convinced one high-risk mining-capital firm, an outfit called First Exploration, to buy $125,000 worth of Dia Met stock—in their world, a piddling amount.

Stew's buyout of his investors was tricky, because he did not want to be accused of knowing more than they did. Thus, Chuck's silence about the particulars suited him in a way. He did not probe too hard. Then, in a fit of extra carefulness, he announced to Chuck that he could not go back into the field to help until a fair price was arbitrated, because that might give him too much information. Chuck was now left pretty much in control of everything.

He knew he could not possibly finish the job alone, but he had to figure out who could be trusted. At this point he was afraid to take along even the regular samplers. He hit upon a partial solution by chance.

Earlier that year Chuck had flown to New Zealand to consult for a company using heavy minerals to locate platinum. There, in the hotel of the small town of Riverton, he had met a weather-beaten man in his fifties, Ray Dawson. Dawson was built broad and towered six feet, eight inches. He was a freelance deer hunter, gold panner, and beach-comber of valuable abalone shells—the kind of tough, self-reliant outdoorsman Chuck liked. They befriended each other and spent every night the rest of Chuck's trip drinking and carousing. Chuck invited Dawson to drop by if he ever came to Canada. Lo and behold, in the summer of 1988, just as Chuck was planning a return to the Barrens, who should show up on his doorstep but Dawson, footloose and ready for anything. He crashed on their couch for weeks. To Marlene's disgust, soon the two men were making nightly expeditions to the Willow to view what Dawson called the "peelers" on stage. Chuck always made sure they arrived early enough to get the first-row "snifters' seats," named for a strategic advantage perhaps better not discussed here.

Chuck formed a plan. He figured Dawson would not mind being kidnapped for an adventure in the north; and afterward, Dawson could fade back to the New Zealand seashell beaches, where there were no nosy Canadian competitors for him to brag to. One day out of the blue he invited Dawson to come with him and Mark for what he claimed would be a few days' gold prospecting in Alaska. Chuck said little to Mark about the destination, but Mark was used to that. Dawson was game. They packed camping equipment, piled into a car, and drove to the town of Smithers, British Columbia. At a small airport there, they were greeted by a tall, gawky pilot with scraggly blond hair and a young, pimply face—a sort of overgrown kid. This was Dave Mackenzie, Chuck's chief investor and, now, confidant.

Mackenzie had kept putting more money into Dia Met and hanging around the lab out of curiosity. Finally Chuck had broken down. He told him of the esker sample—its existence, though not its exact details or location. Chuck needed another helper, and Mackenzie appeared ideal. He now had a powerful little Piper Super Cub plane, which would come in handy; as a shareholder, he would work for free; and, obviously, he would keep his mouth shut. Chuck convinced Mackenzie that he could land the wheeled Cub on smooth esker tops, so they could get at the many places inaccessible by float plane. They would need a float pilot, too, but they could get one in Yellowknife. Mackenzie got the Cub fitted with oversize, cushiony "tundra tires," designed for back-country landings, and practiced coming down in pastures and fields. Just before they went up, Stew heard of the plan and called Mackenzie to pass on some information: Those tires were used mainly as a last resort, not for daily chores. Of a half-dozen fliers he knew who had used them on near-identical planes, every one eventually crashed. No one had been killed—yet. Mackenzie stayed the course; he badly wanted to be included.

They flew north over the British Columbia mountains with Dawson, Mark, and the gear crammed in back while Chuck and Mackenzie sat up front, arguing over a map. Dawson gathered they were lost over the forested wilderness. Mark, now sporting a beard, rolled his eyes over at Dawson silently, like a foxhole-dwelling dogface who has seen it all. Finally Mackenzie found a locator beacon emanating from another small-town airport, and they landed to stay overnight. The next day, Mackenzie made an unexpected turn, and they were over the flat, swampy forests east of the mountains—not the way to Alaska at all. They came over a huge lake and landed. They were in Yellowknife.

Chuck disappeared, then returned with an aging float pilot named Bob Jensen, companion of the local female undertaker and a man with a particular reputation in town for silence. The slow-moving, meticulous Jensen never asked about anything, or even spoke much at all. After loading supplies, Chuck said to everyone, as if spontaneously: "Hey. Why don't we go to the Arctic for a while?" Mark just rolled his eyes again. A few hours later they were in the middle of the tundra.

Smaller lakes were still frozen blue, but the big lake they landed on had only a few ice floes clicking against the rocks in bays. Otherwise it was totally silent except for mosquitos buzzing. As their headquarters, Chuck had picked the shore of Desteffany Lake, part of the loose drainage of the Coppermine River, about thirty miles southwest of the garnetiferous esker beach. The Desteffany site, hidden between two rocky bluffs, was sheltered from wind and had a nearby esker for Mackenzie to land on that was not much worse than a regular gravel landing strip. But mainly, Chuck picked it because it was well away from the spot that really interested him. Anyone chancing to fly over would think this was their prospect. To strengthen the illusion, he had them stake a small legal claim around it. They set up two small tents and dug a hole in the permafrost for a refrigerator. Chuck set up in one tent with a big roll of maps and began issuing orders.

Mark was assigned to travel in a Zodiac, a collapsible boat of orange fabric powered with a small outboard motor. This would get him to beaches, sandbars, and creeks flowing into the biggest lakes along the trunk esker. Bob Jensen would pick up Mark with the float plane at each day's end and would fly the boat to the next lake when needed. Dawson was to be dropped off by Jensen to take samples on foot. Chuck would continue on with Jensen to various stops for his own reconnaissance. Mackenzie would take the Super Cub and land on eskers where no one else could go.

At first Mark found the Zodiac nice: He could speed over the surface ahead of bugs in a cool breeze until he got to shore. On the other hand, the boat offered a fast death from freezing or drowning if he fell out.

One of Mark's main targets was a big, twisty water body called Yamba Lake, along whose southern shore the great esker ran. Yamba is eight miles across at some points, and in the middle sudden winds can whip up big waves. He kept to the edges. However, one day he was running late and feared Chuck's wrath if he missed meeting the plane on time. Toward the end of the day he cut across a big arm. Out in the middle the wind came up, and within minutes choppy waves grew to four-foot swells. Before he knew it, water was breaking over the rail and rushing in. Frantically he lashed a tarp over the bow to keep from swamping. Simultaneously he had to keep the motor running and moving the craft with the bow into the wind so he would not capsize. He managed it for a while, but the light craft was not made for such conditions. Presently the bow bounced up on a wave, lifted above the waterline, and the wind slipped underneath. In one sickening second he could feel the boat being lifted on its end like a piece of cardboard.

Mark threw himself onto the bow to force it down. The Zodiac wavered at an angle, ready to flip backward. He tried to make himself heavier, but it stayed suspended. Then the wind paused, and the boat slammed back into the water. He had to leap backward to keep from falling out and control the outboard. He stuck his leg over the side like a rudder to steer.

Mark Fipke had become very good at not being killed. He got the Zodiac to shore and met the plane on time. Later that night, around the Coleman stove, he told the story of what had happened. Chuck had little comment, and later showed him the maps of where to go the next day.

Of course by now, bear stories were a regular part of campfire lore. The barren ground grizzly, a scrawny, underfed subspecies, was different. Like its shy cousins in the trees, it eats berries and roots and usually flees humans; but amid the thinned-out resources of the tundra, it can be carnivorous, devouring passing caribou and digging up ground squirrels that dwell by the thousands in sandy eskers. People did not seem to be out of the question. One day while hiking at the edge of a bog, Mark saw something floundering in the water, and started; big aquatic creatures do not exist here. It was a downed caribou, its back end raked by some big claw. The animal quivered in agony, tried to rise, then plopped back into the water to watch him with great dark eyes. Mark tiptoed away; the assailant could be hiding anywhere. Later he was hiking over the trunk esker, and as his head crested the top, he saw a grizzly with its back to him. It was moving in a low, careful crouch, like a cat stalking a mouse, toward a grazing caribou. Mark quietly backed down the esker and retreated. Chuck was predictably angry when Mark came back without the sample. He called him a coward.

Chuck was now literally willing to face down bears when he thought he could get away with it. He had Bob Jensen fly him to the garnetiferous beach—he was still the only one who knew the site—but just as he reached land, a grizzly came around the corner and sat down directly where he wanted to dig a new sample, 300 feet downshore. Jensen quietly called from the plane for Chuck to get back in. Chuck instead yelled and waved his arms in an attempt to dislodge the animal. It would not budge, so Chuck stood up as tall as he could and roared like a bear. The bear responded, as bears often do, by standing on its hind legs and sniffing the air energetically to see what the unfamiliar creature was. It put its huge paws back down in the sand and watched Chuck steadily. Chuck settled for a place near the plane, close to sample G71. He dug even faster than usual, keeping one eye on the unmoving animal.

Meanwhile Mackenzie was having his own troubles. Many eskers were less forgiving than the one near camp. They looked smooth while he was inspecting them from the air, but once he skidded down, they were often full of treacherous dips and rises, and waterworn boulders the size of basketballs sticking halfway out of the sand. Some were sprinkled with automobile-size boulders between which the plane's wings barely fit. It was quite upsetting to land and then glance rapidly side to side to see those big rocks whizzing by so close, while also keeping an eye ahead for potential disasters. He did not regret it, though; he was having a genuine adventure, complete with caribou herds milling in swales along the eskers and the possibility they might spot a diamond at any time. Each day he accepted Chuck's maps, risked the landings, and dug.

One day, in addition to Mackenzie's usual duties, Chuck told him to drop Dawson on an esker bisected by the Coppermine. Mackenzie was leery that the big man's extra weight might affect his landing, but he obeyed. He found what looked like a good spot and touched down without any problem. Then with too little warning, a big, collapsed ground-squirrel excavation loomed up. The wheels dropped into the depression and hit the opposite lip hard. The plane shot ten feet back into the air and relanded hard, jolting to a halt.

They got out uninjured, but a cable holding the wheels straight had snapped. They improvised a repair out of some wire taken off a pack frame, but Dawson refused to get back in. Mackenzie agreed to leave Dawson behind, with the plan that he would fly back to camp and have the float plane pick Dawson up at a landable lake they saw on the map, some fifteen miles off.

Dawson was nearly to the rendezvous at 11 P.M. when he heard the buzz of the float plane behind him. It came into sight and circled, and he could see someone waving. He waved back. The plane circled again and suddenly something plummeted out. Dawson sprinted over to see what it was: a sleeping bag. Inside was a can of corn that had exploded on impact, another map, and three sample bags with a note in Chuck's handwriting. It explained that the lake ahead was too rocky for landing. He needed to walk twenty miles to another one. They would meet tomorrow night. And on the way, since he was going there anyway, could he please pick up three samples and carry them along? The map contained three Xs along a meandering line. Dawson shook his fist and cursed as the plane disappeared over the horizon.

Dawson was doubly enraged because the usual routine was bad enough. The quota on foot was three stops a day, over twelve to fifteen miles. Equipment weighed 25 pounds, and each sample weighed 25, so he was lugging 100 pounds by the end of the day. It could take twelve to fifteen hours, depending on terrain and how many detours you made around ponds and swamps. Esker fragments were great for hiking, but they could not be depended upon to go where you wanted. Some spots marked for sampling turned out to be solid rock. At least this offered a reprieve, because there was nothing to carry from that spot.

However, when Dawson got back to camp, he saw Chuck was now collecting rocks, too. Chuck came back at night with his pack full of them, and after supper pulled them out to squint intensely through the magnifier he kept around his neck. He never seemed to find any that satisfied him, though; he kept scowling and throwing them away. He did not tell Dawson—who still thought this was a gold hunt—that he was looking for kimberlite float. Chuck was sure they should see visible signs by now. He kept seeing kimberlite-like rocks, but on closer inspection, none of them ever were.

Dawson was by now thoroughly sick of this trip, so to add confusion and get a small bit of revenge, he began gathering random rocks from near the tents and sticking them into Chuck's pack while Chuck was not looking. He watched with great pleasure as Chuck settled down after supper to empty his pack of rocks. Each time he got to one of Dawson's, his face clouded with a confused look that said: Why did I pick this one up?

The trip lasted close to eight weeks, with no days off. North of Lac de Gras, they took hundreds of samples from nearly 600 square miles. As it got toward September, the twenty-four-hour daylight disappeared, days shortened, and a heavy chill entered the air. Even Chuck had to admit that the returns would decline from here on. On September 1 they packed up, leaving the decoy claim posts and a cache of food and fuel under the rocks. A couple of hours later they were on the ground in Yellowknife.

After the silence and endless space of the tundra, Yellowknife seemed like a strange metropolis of the future. The straight lines of buildings looked like doorways to another dimension. Faces and figures of so many human beings crammed into a few acres seemed like bedlam. The sight of an innocent pedestrian coming down the sidewalk aroused fear of imminent collision.

That night they hit the Gold Range. A recent big fight had left holes in the ceiling, and a wall mirror was now reduced to nothing but bits of adhesive sticking to the tile. The usual mélange was yelling and flirting and threatening to punch one another, while ladies bearing bundles of single, wildly expensive red roses imported from the south circulated, trying to sell them to males who looked about to score. An Inuit rock band played "Mamas Don't Let Your Babies Grow Up to Be Cowboys" more deafeningly than anyone had remembered was possible. Beer flowed. They were back.

CHAPTER 16

The Barren Lands

Dawson disappeared back to New Zealand to hunt seashells, and Chuck kept the lab open twenty-four hours a day. He arrived at 10 each morning and stayed far past midnight, with a usual 10 P.M. supper break. The chemical analyses were now speeded by the arrival of Rory Moore, a young protégé of John Gurney. The formulas of the other diamond-indicating minerals were emerging in published papers now, so Gurney agreed to have Moore show Chuck the ropes on them all. Heavy pockets of indicators popped up all over the maps above Lac de Gras. All suggested diamondiferous pipes.

However, the ice had left a terrific mess. The best signs of ice movements were usually bare rock outcrops, polished smooth and scribed with parallel striations made by stones the ice had dragged. Normally such striations are straightforward, but out here they were mind-boggling. Striations in different locales showed the ice lumbering west, southwest, or northwest. A few pointed due north or even curled around and headed back northeast. Just above the garnet-heavy esker beach was a big field of drumlins, hills of glacial till, whose narrow, teardrop-shaped forms pointed west again. Everything had been stirred like a gumbo. There was no way to tell in what order the moves had occurred. The esker itself showed only how debris had been collected during final melting—not exactly where it came from.

The sheer quantity of information scared Chuck as much as the first discovery on the esker beach. That had been a single loud howl arising from the dark; now voices of unseen, buried kimberlites were coming at him from all around. He kept his maps in a separate windowless room next to the electron microscope, where he could lay them out. As he tallied grain counts for each sample, he put a small color-coded circle on the map with the numbers—red for garnets, green for chrome diopsides, dark blue for chromites, light blue for ilmenites—and overlaid maps of possible ice movements. Sometimes Marlene came late at night, and they did the marking together. But mostly, he was left with his original information: The indicator train stopped dead at the eastern end of Lac de Gras and the epicenter seemed to be the curving esker beach above it. On the north, indicators petered out somewhat about twenty-five miles above the lake, but he was not sure what that meant. West and south were much vaguer; strong signs persisted out as far as he had gone.

He feared that if he went back for more data, his enemies would finally detect him. He wondered if he should start staking now, outward from the esker beach. But they would need hundreds, maybe thousands, of square miles to cover the possibilities. That revived the perennial plague: money. Just the staking itself would cost more than they had; plus, to claim government land, which this was, you must pay 10¢ an acre to register it, then spend at least $2 per acre per year prospecting it. And of course once it was registered, everyone would know where it was.

Once again it was Wayne to the rescue. Wayne had just quit his longtime job managing the Citadel Theater in Edmonton, moved back to Kelowna, and was casting about for something to do—perhaps poetry- or novel-writing. He had saved some money, which allowed him to be impractical, at least for a while. Then Chuck started hinting that something big was up in the north, requiring a large amount of capital. Wayne had never been there, but he had a romantic vision. Being of a literary bent, he had read that his name, if spelled "Wain," referred to the seven bright stars of the Great Bear constellation of northern stars. Being also of a somewhat mystical bent, he also noted that when the moon *waned*, mariners navigated by the North Star, at the end of the handle of the seven-star Little Dipper constellation. Now here was Chuck talking about the north. At this crossroads in his life, it all seemed like more than a coincidence. Perhaps his destiny lay north. Wayne hatched an idea: He would raise money for Chuck, if Chuck would give him a job. Wayne wanted to be president of Dia Met.

Chuck laughed openly. His little brother—actually, grown a few inches taller than himself—had no experience. Also, given their childhood hostilities, he did not think it a good idea to be in business together; old rivalries might break out. Wayne, however, kept talking, a talent they shared. He offered to put most of his savings into Dia Met—$40,000. He offered to work for pitiful wages and to reinvest them in Dia Met. His main compensation, he said, could be stock options; if Dia Met did well, he would do well. Wayne's friend Mackenzie was tired of being president anyway. Chuck shrugged, and in January 1989 Wayne got the job.

He turned out to be brilliant. Endowed with the same manic hustle as Chuck, Wayne grew himself a black goatee to match Chuck's, hammered together a tiny office next to Chuck's out of two-by-fours and wallboard, got a phone installed, put on a three-piece suit and started working everyone he knew, was related to or had heard of, as long as there was a chance they could be persuaded to part with their money. Chuck still did not totally trust Wayne, so Wayne had only the vaguest idea of where the project was; and since Wayne had no geology training, he found it hard to say why diamonds might be found there. But he promised one thing: Everyone would get rich fast. His zeal quickly grew to religious intensity. Temporarily, the treasure hunt moved from the rocks of the tundra to the bank accounts of Kelowna.

Wayne's obvious first target was family. Their parents had no savings, so he ignored them. Their kid sister Carol had gotten married and moved away. She did not have much either—she was a haircutter—but Wayne harassed her with phone calls every few days. Upon discovering that Carol owned a fax machine, he clogged it with missives describing the riches she would soon possess if only she would invest in Dia Met. Carol resisted quite a while, but finally the sheer calling volume and excitement in Wayne's voice subdued her. Over nervous interjections from her husband, Carol put up several thousand dollars. Wayne immediately started working on her for more.

The youngest Fipke brother, Neil, was tougher to crack. Chuck and Wayne had never been too close to Neil, who now was a forester, but around this time Neil visited Kelowna, and Chuck playfully pulled out a big quartz crystal he had found near Lac de Gras. He truthfully told Neil it came from "the north." Wayne was standing right there when Chuck casually said, "Hey, look at this!" and scratched a windowpane with it. Chuck never actually told Neil the thing was a diamond and did not mention that diamond is not the only substance able to scratch glass, but Wayne saw how wide Neil's eyes grew. Within a week he had Neil taking out a $10,000 bank loan on the apparent premise that diamonds had already been found. Wayne worked the in-laws, too. His mother-in-law and his wife's brother bought tens of thousands of shares.

Then came friends and acquaintances. One of the first and best was an older Greek immigrant, Pavlos Zaphiris, who owned Talos Restaurant, where Chuck ritually ate twice a week during his 10 P.M. break. Often Marlene joined him at this hour, tucking the younger kids half asleep into lambskins Zaphiris had scattered on benches. Zaphiris loved the Fipkes; he saw them as a nice young family led by a nutty professor who knew nothing beyond geology—not art, not literature, not current events, nothing. When Wayne came around for his contribution, the restaurateur did not hesitate. "Your brother, he's very smart in his job, isn't he?" he said. Wayne enthusiastically agreed, professing to understand as little as Zaphiris—only that Chuck's impenetrable smartness meant Chuck must know things they did not. "OK," said Zaphiris. "If I lose this money, I lose honest. I know he's not a thief." Zaphiris bought shares whenever asked, eventually upward of 100,000.

The fact was, Wayne had correctly assessed their greatest fund-raising asset: Chuck's image as idiot savant. He took pains to paint his brother as a mad genius, so great at one thing he was necessarily incompetent at all else. He harped on Chuck's forgetfulness and made sure prospects toured the lab's strange sights. It was a boon that Chuck mumbled to himself and did not like wearing shoes, often padding around the parking lot trailing untied laces, in mismatched socks without shoes, or simply barefoot. The more obtuse and help-less Chuck appeared, the more honest people assumed him to be. Workers in the fishing and upholstery shops by the lab were used to seeing Chuck shoeless, so Wayne took their money easily. A nearby gunsmith bought shares, as did workers at auto-electrical and truck-spring shops. Wayne even got to the man who delivered coffee to the lab. Wayne was the other half of the equation: the reasonable man in a suit who could consolidate the proper image for his sibling and assure small investors that at least one person in the family spoke in whole sentences. Chuck, no slouch at raising funds himself, could not help but admire Wayne's acumen.

Wayne went on to research everyone in Kelowna with money—doctors, dentists, lawyers, store owners. When he discovered three brothers who were all orthodontists and partners, he invented his "brotherhood concept." Chuck and he were brothers, too, he said, and had always stuck together, just like the brother orthodontists. Family is everything, wouldn't you agree? suggested Wayne—espe-cially when it comes to a big, secret diamond mine in the mysterious north that will soon make everyone in this whole town rich. The orthodontists pooled $50,000. Backyard barbecues with the neigh-bors became a selling opportunity. It never occurred to Wayne that if they lost their money, he would be on their shit list forever, and he would still be living next door.

By May they had several hundred thousand dollars, but Chuck was nervous. He began backpedaling and saying they should go up only for more exploration. He was unsure about the staking; he did not know if he had the right ground. And for the first time in a while, Stew was on his side. Stew was still sitting on the sidelines because the buyout of his partners had hit legal snags. However, based on his limited information, he feared Chuck had not gone far enough up-ice. He could not understand why no pipes had been spotted. Stew suggested they wait until he could rejoin, and then just the two of them could go, like in the old days. They could spot the pipes from the air or with magnetic instruments. Chuck, who could still be easily intimidated by Stew, wondered if he was right.

When Wayne and Mackenzie heard this, they wanted Stew's blood. As far as they were concerned, Stew Blusson was a fool. He had done nothing for years. His only function now was to complain, snivel, and get in the way. Mackenzie was sweating; if someone else got there first, he would lose every cent. Wayne, at first intrigued by Stew, began hating him. He felt Stew acted like he was smarter than Chuck and treated Chuck like an inferior little brother. As for Chuck, Wayne felt his brother was suddenly being far too scientific and cautious. His and Chuck's desks were only ten feet from each other, separated by the thin wall Wayne had built. While they both sat on their separate sides, Wayne dialed Chuck's fax machine and sent a printed threat to quit if Chuck did not stake.

Chuck waffled right into June, then wavered to their side. He told Wayne to write up a budget and present it to Stew as a fait accompli. He resolved to camp on the esker beach and stake out from there, grabbing as much ground as they could before anyone caught on.

C huck's fear of followers was growing all the time. To foil listening devices real or imagined at the lab, he started taking Mackenzie and Wayne for walks outside to talk. When it occurred to him that someone with binoculars might read their lips, he made them shield their mouths with their hands.

He decided they needed a corporate disguise, and he had one ready-made. To lower his taxes, a year before he had legally created a dummy entity through which to channel certain funds to or from Dia Met and C. F. Minerals. This was Norm's Manufacturing and Geoservices Ltd. Chairman of the board, president, and sole shareholder was Norm Oftedal, the middle-aged lab janitor, who lived with his mother. Among other menial tasks, Norm repaired samplers' sieves in a claustrophobic loft above the drying oven, sweating in summer like a coal shoveler in a ship hold. After Norm's Manufacturing was incorporated, Marlene would occasionally call Norm down to sign Norm's Manufacturing checks, invoices, and his corporate tax return. Bashful and perhaps not the lab's most brilliant personage, Norm received no compensation as CEO; but it broke up his day, and he did possess a very impressive-looking signature.

Suddenly Norm began working overtime. He was organizing an ambitious Arctic expedition. Norm chartered planes, stockpiled supplies, bought maps, and revamped ailing vehicles. Norm savvily contracted the making of thousands of pointed four-foot wooden stakes to a Kelowna workshop run by disabled people because it was economical and because buying stakes in spy-ridden Yellowknife would be suicidal. Norm designated Wayne as head staker; the ex–theater man would see the north. Norm himself would be far too busy to go.

Chuck sent Mackenzie and Wayne up to scout. They discreetly checked at the Yellowknife mining recorder and Bob Jensen's hangar to see whether anyone had been up to Lac de Gras. All was quiet. Then they flew to the previous year's camp, and Wayne got his first look at the tundra. Bears had demolished their cache—even aluminum propane tanks were caved in with tooth marks—but the decoy stakes were untouched.

In late June the Fipke brothers borrowed their father's ancient, rusted five-ton dump truck and loaded it with all-terrain vehicles and the stakes, packed in orange garbage bags to keep out prying eyes. Chuck rounded up his faithful old samplers, and they piled into Chuck's now-antique black Olds and Chevy Blazer for what Chuck said was a gold-staking expedition. There were ten people, including Mark, Mackenzie, and a couple of inexperienced Kelowna teenagers whose fathers Chuck knew. They convoyed at high speeds for two and a half days, and when the last few hundred miles of road turned unpaved, they kept the windshield wipers going to fend off a constant barrage of dust and flying gravel.

They reached the Old Town float docks early in the morning of July 1, 1989. By a great stroke of luck, the night before all Yellowknife had performed an organized pub crawl in honor of Canada Day. Attendance at all twenty-four bars had been required for the award of a commemorative T-shirt, so the few citizens visible were lying in various positions of repose on the streets, unaware of new arrivals. The rest appeared to be sleeping it off at home. The laconic

Bob Jensen knew they were coming, though. He had a huge World War II–era twin-engine Beech float plane waiting. They transferred the first batch of garbage bags and were off by 7 A.M.

The samplers were dumped into the treeless land at the edge of the curving esker beach. None had ever seen such a landscape. As the big Beech taxied off for another load, Brent looked around with more than vague unease. The lack of trees made him feel naked. If he ever got lost out here, there was no shelter, nothing to make a fire, no way a person would find any other person in all the unmarked emptiness. This spot here did not even have a name. They pitched their tents a few steps above the band of dark sand and gave the place a name: Norm's Camp.

Mackenzie was not scheduled to fly in for a few days, but Chuck arrived with the next load some hours later and ensconced himself in an army-surplus tent in the center. He laid out piles of maps on a plywood sheet, and a newly purchased base-camp radio. Everyone got a walkie-talkie. Having had little time to plan at home, he began platting land for miles north of Lac de Gras in contiguous rectangles to be staked. Directly, the rest set out on foot in all directions, staggering under prepackaged bundles of sixteen stakes each. Wayne put in the first stake a few hundred yards east down the esker.

Government rules were specific. A maximum claim was 2,582.5 acres, its lines oriented as close as possible to the four points of the compass. Wooden stakes were to be a minimum inch and half square, and numbered. You started at the northeast corner and went clockwise, firmly pounding in a post every 1,500 feet. If a lake was in the way, you walked around and marked each shore. The corner post of each claim got a stapled-on official aluminum tag, obtained at the mining recorder, with name, date, and time. To lay another claim next door, you started where you left off and did it all over again, except that it was contiguous, so you already had one side staked. It was legal to disguise claims with phony names, as they had in the past, and you also did not have to reveal what you were looking for; but once you started, you had to register claims within sixty days. When Wayne planted his first post, the clock started ticking.

The all-terrain vehicles arrived, which Chuck thought would speed things immensely, but it quickly became clear they were useless. The tundra was too boggy, bumpy, and blocked with boulders or ponds. The first vehicle pitched directly off a rock into the lake at the end of the esker beach. Another sank in muck a bit farther on.

Wayne was not used to outdoor activity, but bravely led the charge. He was alarmed the first hour when three large white wolves popped their heads over a rise and peered with ears erect. Fumbling with his radio, he called to an experienced sampler, visible in the distance. The radio crackled with a laugh. "Wolves don't eat humans. They're just curious." Wayne was not so sure. The wolves followed his every step. When he stopped, they stopped. When he moved, they moved, eying him all the time. Sometimes they dropped behind a rise, then reappeared unexpectedly farther on, sitting on their haunches, looking. After a while he tried to ignore them.

In following days the wolves did not return, and he felt incredibly lonely without them. For hours at a time there was nothing but empty land. When the wind started up, it seemed to blow away even his own thoughts. Occasionally a lone caribou appeared by surprise, then galloped off. A ptarmigan flew out of the stunted vegetation with a heart-stopping burst of sound into the bright sunlight. Unseen loons called from ponds among the rocks.

A few miles away, Dynamite Dan saw nothing move all day until he came over a rise and saw a living hallucination. At first he thought it was a bear cub with its back to him. He froze; where was mama bear? Then the creature turned to confront him. It was the size of a small boy, had long, matted hair, saucer-size paws and a dark face that was a combination of mountain lion and dog. In all, it added up to something eerily human. It bared its fangs and spat out two earth-shaking snarls, then loped clumsily over a rise. It was a wolverine.

Between strange encounters they walked and walked. They were supposed to make the stake lines straight by navigating with compasses, but the needles often swung 10 degrees to one side, then to the other, and refused to settle. In some zones, they swung as much as 50 degrees. There were detours around water everywhere. Vegetated ground that looked level from afar was actually made of protruding tussocks of peat and plants, underlain by bog. You could either hop clumsily from tussock to tussock or sink into the matted goop around them; either way you lost. The boulder fields were the worst. They hopped from one precariously balanced stone to the next, one rock sometimes rolling slightly and clicking against another. Some days it rained briefly, and lichens on the boulders turned greasy. This led to some unforgiving, painful slips. Getting a stake to stand upright anywhere was tough. Often they just shoved it into a crack between rocks.

Within the first week, they were well beyond easy walking distance of camp, and a second pilot working for Jensen picked up the job of dropping off workers with a smaller float plane. A day later, Mackenzie screeched down onto the top of the great esker with his Cub. As promised, it had a relatively smooth runway. Mackenzie's job became stake resupply. At midday he flew along each line until he found the man working it, and radioed down to see how many stakes the staker needed. Then he circled back low, opened the window with one hand, and jettisoned a packet. To him this was great fun, much safer than landing in chancy places to take samples. He was glad that part was over.

Chuck decided he was not done with sampling, though. After issuing orders for the day, he had himself dropped off to stake, scout, and dig all at the same time. Mark was sent in the Zodiac to dig as well. During these trips Chuck always kept an eye out for odd rocks and rock formations, but he never spotted anything. Then one evening Wayne whispered in his ear that he had something urgent to show him. They walked over a rise where no one could see. Wayne glanced around, then reached into his pocket, nearly shaking with excitement. Five clear, peanut-size crystals lay in his hand.

"Look."

Chuck bent over and peered at them. He looked deeply into his brother's face.

"Wayne, that's quartz," he said.

Chuck broke out laughing. Once he got going, he could not stop. With disappointment, Wayne looked down at the crystals in his palm.

"Are you sure?" he asked.

Chuck just kept guffawing. The upside was that now he knew Wayne probably would not rip him off if they actually did find something.

Three plagues came upon them: biting bugs, biting bugs, and biting bugs. Northern summers were always bad, but this was the worst anyone had seen. The temperature ascended uncharacteristically into the nineties and stayed there. The usual wind stopped dead. Humidity rose. Perfect bug weather.

First were mosquitoes. A person returning to camp could be spotted a mile away before his actual figure was visible, by a certain spectral darkness blemishing the horizon. It was a cloud of female mosquitoes, converging on a point food source. They needed a blood meal to lay eggs and make more mosquitoes. At night, mosquitoes got into the tents. Lighting a smoke-producing coil helped, but only a little. Several mornings the men awoke to a patter of heavy rain on the tents. Actually, it was only mosquitoes, trying to get in.

A week later came blackflies. Smaller but much meaner than houseflies, the northern variety specialize in biting four or five times in quick succession, leaving a blistery, linear red wound that festers and itches painfully for three to six weeks. At their height, they entered mouths and nostrils in such quantities that the men sometimes gagged. The more panicky sometimes felt suffocation coming on. The blackflies were impossible to wave off, dodge, or outrun.

Finally came deerflies. These were big enough to take on a bumblebee, and generally flew at about the speed of a professionally pitched baseball. They operated by crashing into the target, bouncing off hard, and tearing off a morsel of flesh as they came away. Constructed to penetrate the tough hides of caribou, they easily bit through clothing. It was like getting hit with a whip.

Most men wore bug jackets and bug repellent, but it was never enough, so underneath they also wore heavy wool pants and shirts, with cuffs duct-taped shut. They sweated horribly as a result. Dynamite Dan, always ready for anything, wore two bug jackets, heavy clothing, gallons of repellent, wraparound sunglasses, and his biggest hunting knife, but he still got bit. They had toilet disinfectant for their crude outhouse—actually, outhole—and when Dan somehow discovered it temporarily relieved itching, he doused himself liberally with that, too. Even Chuck was suffering. In his broiling tent, black-flies swarmed his eyes and bit him so much, his maps were smeared with his own blood. "Geez, they're bad this year, hey?" he would comment loudly over the buzzing to whoever came by. He almost broke down and put on the bug jacket, but not quite.

The first casualty was an adolescent named Todd, whom Chuck had hired as a favor to Todd's father. One day while walking the tun-dra amid the bugs, round-the-clock sun and solitude, Todd simply lost it. He dropped his bundle of stakes and pack, wandered off his com-pass line, and vanished. The pilots searched in systematic grids for hours. The float pilot finally spotted him stumbling around and retrieved him at a lake. Todd had pulled the hood off his bug jacket so the insects could get it over with. His head and neck were a continu-ous welt, and by evening they had swelled nearly double. Todd went out on the next plane to the hospital in Yellowknife and did not return.

The next casualty was Arnie Baslaugh, a seasoned sampler whom Chuck had employed on and off for years and thus was not expected to crack. Baslaugh ended up similarly hoodless and wandering. Back at camp, he told Chuck he was going mad and begged to be taken out. Chuck took pity and complied.

After three weeks Wayne was perhaps the worst mess of all. He had come with brand-new hiking boots that did not fit. Soon both feet were one large purple and white blister, except for a narrow strip in the middles, protected by his arches. He acquired a terrible, unending case of diarrhea, even though every lake was clean enough to drink from and no one else got it; maybe it was the food. This, combined with massive sweating, quickly sapped him of twenty pounds—not to mention what the bugs did to him every time he had to go to the bathroom. Like everyone else, his face was fissured and nearly black from sunburn and dirt.

With the usual black humor that the crew mustered for such situ-ations, the samplers took out bets with Mark on when his uncle would crack. The pool grew quite large, but the big moment never came. Not only did Wayne not fold; he rarely even limped, and, to the crew's surprise, never complained. As boss staker, he cracked the whip when anyone else lagged. His last name was not Fipke for nothing.

As August wore on, Chuck was becoming frantic. With the failure of the ATVs and loss of two workers, staking was not proceeding fast enough. He still was not even sure they had the right place. He lengthened the twelve-hour day to eighteen. To glean more information, he now ordered the strongest stakers to also take samples. This was almost too much for anyone to bear; at least when they were only staking, the load grew lighter as the day progressed. Now they were given shovels, and the weight of disposed-of stakes was steadily replaced by twenty-pound sample bags. Late at night, he made everyone number the next day's stakes and sample sacks with Magic Marker. "When you're not eating or sleeping, you're working for me!" he snarled when he saw someone merely resting. The sun started its yearly journey downward, but this did not stop Chuck. In the morning twilight he rose before everyone, threw a pound of coffee directly into a big pot of water, and began kicking sleeping bags. They all hated him.

Even the usual escape route seemed cut off: Chuck was chief party animal in this zoo, but he decreed Norm's Camp alcohol-free. For good reason: Norm's was ripe for out-of-control drinking, and anyone drunk out here could be deadly to themselves and the others. There were too many things that could go wrong, and no way to right them this far out. Out of deference, everyone took care not to become too visibly smashed on the many bottles of booze Dave Mackenzie had smuggled inside a duffel bag, wrapped in clothing so they would not clink together. Aside from enjoying a little nip himself now and again, Mackenzie figured it would keep the nonshareholders from quitting. That and Chuck's longtime promise to the sufferers: "Hey guys, when we make it big, we make it big all together, are you with me? We're in this together, hey?"

Toward the end of summer they had staked close to 385,000 acres—600 square miles—in row after row of 2,582.5-acre rectangles. Chuck still did not know if it was enough, especially south, where they had not reached the shores of Lac de Gras itself. That country was too stony to get a float plane into, and so remained poorly covered by mineral samples. At this late date, he decided he needed at least one more sample from down where the water bleeds the west side of Lac de Gras and becomes the Coppermine River. He ordered Mackenzie to go fetch it.

Mackenzie protested; he did not want to do any more such landings, given what had happened last year. Chuck lost his temper and accused Mackenzie of shirking his duty. He told Mackenzie this was a key sample. Mackenzie relented; he had too much invested to say no. He was also afraid to cross Chuck in Chuck's overexcited state.

Early the next morning he flew the twenty-five miles to the headwaters of the Coppermine and searched for a place with the right texture and drainage features. Pickings were slim, but eventually he spotted a short but apparently landable esker segment. He headed down. A northwest wind had been blowing all day, and it swept the esker sideways. That was OK; he could adjust for a crosswind. As he came down, he saw the wind needle shift—probably a momentary aberration. The ground came up fast, and he realized he should have paid more attention to where he was landing. It was pretty rocky. Just as the tires touched, he realized the needle shift was real; instead of a crosswind, he now had a tailwind, which was driving him hard from behind. This increased his groundspeed, which meant that he was now going too fast to stop in the allotted space. A few seconds ahead was a cliff. He panicked. He could have taken off again, but in that split second instead he slammed the brakes onto both wheels. One tire hit a rock. The nose pitched forward and caught the ground. The plane rose entirely on end and flipped over. Dave Mackenzie, chief investor, Dia Met president emeritus, and overgrown kid, had attempted his last sample.

It was evening when Jensen happened over the area in the smaller float plane and spotted Mackenzie's familiar craft. It took him a moment to realize it was belly-up. Unable to land anywhere near, he radioed camp, where Chuck was sitting at his plywood sheet.

"Does anyone know first aid? Does anyone know first aid?" cried Chuck in a high voice as he ran out of the tent. Everyone looked at him with mouths open. No one had brought much more than Band-Aids.

The float plane was useless for getting close, but they waited for Jensen to return and made preparations to land as close as they could, then walk to the crash site. Jensen was back within ten minutes. He started refueling and checking over his plane for another takeoff as everyone looked on in dread.

The deep curve of the esker beach lay washed in declining light. Five minutes before they were ready to go, a tall, skinny figure rounded the bend a few hundred feet away. It came walking along the sand. It was Mackenzie.

"There he is!" they all yelled at once, and sprinted. "Are you OK? Are you hurt?" Everyone was crowding and touching him gently to make sure he was real.

"No, no, I'm fine, just a little tired from walking," said Mackenzie in his best offhand voice. "It's OK." He tried to wave them off. His only visible wound was a big cut on his scalp.

Before flipping over, he had had the presence of mind to shut off the gas and electrical systems to prevent fire. Then, he found himself hanging upside-down in his seatbelt, banged hard in the head. He unclipped himself, dropped down, and opened the door to crawl out. He cursed himself. He had fucked up. His plane was injured, but his pride was destroyed. His glorious flying adventure was over. The only good thing was that he was still alive and he knew where he was. He pulled out a map and compass, took a reading, and began walking toward Norm's Camp. He felt more depressed with every step. He was not in the mood to detour around knee-deep lakes, so he walked straight through, soaking his feet repeatedly. He stumbled over boulders and clawed his way straight over hills. Twenty-five miles later, here he was.

Chuck was pawing Mackenzie with everyone else. When they were sure he was not a ghost, Chuck spoke up.

"Did you get the sample?"

Everyone shut up. Mackenzie stared at him, wide-eyed.

"Where's the sample? You didn't bring it with you?"

Mackenzie wanted to believe Chuck was kidding. He looked into Chuck's face and saw plainly that he was not.

"No, I didn't get the sample," he replied.

"You, you left it behind somewhere?"

"I crashed my plane, Chuck. I didn't take the sample. I don't have it. I just walked twenty-five miles."

"Dave, you, you, you, go to all that trouble, you crash your bloody plane, you walk twenty-five miles, and you don't . . . you . . . *you come back without the fucking sample?*"

In that moment Mackenzie did not know whether to commit suicide or murder.

That night Mackenzie descended into despair. He felt ashamed and useless. Especially when he overheard Chuck in the next tent ranting to Wayne.

"He's useless! Send him back! Now we don't even have his plane! Send him back!" shouted Chuck.

Wayne remonstrated. He reminded Chuck that Mackenzie had risked his life for them. They were all friends, and had been for a long time. They had to stick together. Only wolverinelike growls emanated in response. Wayne came over to see Mackenzie a while later.

"Chuck doesn't care whether I live or die," said Mackenzie, near tears. "I should get out of here."

"We're here, we're together, we're doing what we came to do," said Wayne. "It's not over."

"What's my role?" asked Mackenzie.

"You're part of the team," said Wayne. "You're strong. You just gotta play a different position. You're strong."

Wayne kept saying the mantra: "You're strong. You're strong."

Next day, Mackenzie was playing a different position, carrying his quota of sixteen stakes on foot.

Then came the fire. Jensen saw it first, consuming Mark's tent. He grabbed the fire extinguisher from his plane and went tearing over. Flames and thick black smoke were billowing out the front. Jensen got it partly out, and several other men came running with water. They screamed for Mark—who just at that moment came running, too. They all gasped in relief; he was not inside being burned alive. Then someone remembered what was inside: Mark's shotgun, loaded for bear, and boxes of spare shells. Everyone scattered to dive into low spots and behind rocks some distance off. The tent rekindled and they let it cook down like a marshmallow.

When it was safe to come out, it was determined that Mark had left a mosquito coil burning and must have knocked it over when he came out—a careless thing to do. In the melted pile lay the remains of all his clothes, packs, his gun, a plastic lump that used to be a portable radio, the shotgun shells (which fortunately did not go off), and a scorched, empty whiskey bottle.

Chuck turned purple when he saw the bottle. He blamed the fire on Mark's drinking and stupidity. Mark told him it wasn't so, but refused to reveal the source of the booze. Chuck began shouting about how Mark never did anything right, how worthless he had always been, how he always goofed off.

Mark stood and took it, then said he had to go back to Yellowknife to reoutfit. Chuck told him he wasn't going anywhere.

"Then I quit," yelled Mark.

"Quit! You can walk back to Yellowknife. You're not going out on any plane *I* pay for," said Chuck.

Wayne stepped between them to keep things from going any further.

Mark was not allowed to quit. Nor was he allowed to go to Yellowknife for clothes; Chuck was afraid he might not come back. The samplers felt deeply sorry for Mark. After things died down, they mumbled among themselves and took up a collection of spare clothing, a lot of it several sizes too big. It was all they had to offer.

It was nearly September. Early snow flurries nipped them a couple of times, and nights returned. The insects abated. In the dark after work, they all dropped into their own personal comas, their dreams accompanied by choruses of high wolf howls from somewhere on the other side of the esker. It was almost time to leave.

Wayne got up in the middle of the night to pee, then lay on his back in the cold moss and looked at the night sky. Ursa Minor, the Little Dipper, was out there, the reliable old North Star shining at the end of its handle. On the tundra it was infinitely brighter than any southerner ever saw it.

Despite the awfulness, Wayne had come to love this place. They had never seen any other humans, or signs of humans, the whole time. On the tundra everything was crystal-clear, quiet, unbroken, like a waking dream. But then, he thought: We are suffering so much. Maybe we are crazy to be out here. Maybe Chuck is wrong.

He half-dozed, his lids lowering over his eyes. He prayed: Dear God. If this is real, what we're doing out here, show me something. Show me a sign.

In an instant he opened his eyes wide. A gigantic meteor shot through the open sky, then blazed its brightest within the cup of the Dipper. As he sat up, the meteor vanished at the cup's bottom. He gasped. That's it, he thought. The sign. The North Star, the Dipper. The meteor dropped in the cup. It must be our cup. It's going to be filled. There are seven stars in the Little Dipper—four in the cup, three in the handle . . . and seven stars in the Bear constellation. The Wain. Seven stars. That can mean only one thing: seven mines. There will be seven diamond mines, right here, where we are now. He shivered and went back to bed.

After they flew to Yellowknife, then home, everyone got a week off—that was all. Chuck decided they were not done for the season. Instead he sent Mark, Dynamite Dan, Brent, and a couple of others back to build a capital for their new empire and prepare to expand it even more before winter. Brent was given a bookful of Norm's Manufacturing checks, and in Yellowknife he used them to buy plywood, chipboard, nails, and two-by-sixes. Jensen flew them back in, and they began work on a six-berth bunkhouse and small cookshack along the esker beach.

The weather was at first merely bracing, but soon the wind was blowing so hard that Mark and Brent's aluminum-frame tent was picked up while they working and sent downshore, twisted into a piece of modern art. They slept each night in the mangled remains, but Mark was having some kind of recurrent nightmare. "Get out of here! Get out of here!" he screamed at someone, then began snoring again. Mark never woke from this dream, and Brent never knew what it was. But every time it happened, Brent bolted up in terror from his own sleep, fingers on his shotgun stock because he thought a bear was in the tent. With the caribou gone, the only life around was a single black raven that flew directly over camp at the same time each morning. Then it returned the other way at the same time each afternoon, as if commuting to and from its workplace—probably some carcass.

After a couple of weeks Chuck and Mackenzie arrived, and Chuck announced that everyone was to do more staking and sampling. Off they went, with a chorus of complaints. By now it was freezing, and they were forced to wear heavy mitts and coats. Work proceeded at a crawl because the top of the ground was hardening. Chuck often spent only a few hours in camp, then was off again; he was now registering the claims back in Yellowknife, but trying to do it piecemeal so as not to attract too much attention. He borrowed the names of Norm, the stakers, and various Dia Met shareholders to disguise the claims. At this point the samplers came up with a new nickname for him: Captain Chaos. He was worse than usual. Every time he came back, he upset their own finely arranged routines with demands that they drop everything to sample some new, distant place he had just thought of. Unable to keep focused, he would spend five or six hours trying to figure out problems that should have taken an hour. Mark had his own nickname for Chuck: Der Führer.

While Chuck was gone on one of his registration runs, they built a fine outhouse for Norm's Camp, complete with a crescent moon sculpted into the door. They felt it needed still more ventilation, and since everyone now owned a firearm of some sort, decided to have some jollity. They lined up on the esker at a range of seventy-five yards to ventilate it with lead. Five or ten minutes of potshots did the job, their aim sometimes disrupted by bursts of laughter. Then someone piped up jokingly: "Hey! I think Chuck's in there taking a crap!" Instantly a spontaneous thunder opened up. A fusillade from Mackenzie's .30-.30 rifle, a heavy .30-.06, two shotguns, a .455 Magnum, and Mark's semiautomatic assault weapon blew jagged holes in the walls. Chips of the outhouse went sailing in all directions. When the magazines were exhausted, those carrying extra sidearms unholstered them and kept shooting while the rest struggled to reload. When the smoke cleared, the nice new outhouse resembled a colander that had been beaten with a crowbar.

It was October, and the wind howled without stopping. No one could leave Norm's Camp. Clouds of fog, sleet, and snow were blowing through. Bob Jensen, gone to town with his big twin-engine Beech for supplies, could not come back, and unused stakes sat stacked outside the bunkhouse door. For ten days Chuck and everyone else slept, played cribbage, read the same magazines over and over, and watched the food, coffee, and fuel for the kerosene heater dwindle.

One morning they had to push hard on the door to make it open; snow had drifted in front. Next morning it would not open at all. They took turns drop-kicking it, and it burst open. The wind had been blowing so hard it was hurling sheets of spray from the lake onto the cabin, freezing the door shut. The next morning the edge of the lake itself was frozen a half-inch thick. Chuck had done it again, and pushed things too far.

A few days later the weather cleared, and they heard the heavy drone of the Beech coming. By now only the middle of the lake was open, and most of the shoreline was frozen four inches thick; landing would not be easy. They cheered when they watched Jensen splash in among the floes, then plow close to shore like an icebreaker. They walked onto the near-shore ice and unloaded a cargo of fuel drums with only one person falling through and getting soaked to the waist.

At this point, Chuck tried persuading everyone to wait for the weather to improve, then return to work—though he himself was going out on this plane. "You'll be OK," he kept insisting. "The season is over, Chuck," they all shouted in unison. Next morning they filled the bunkhouse with bundles of stakes and packed up.

Overnight the big plane's wings had become covered in six-foot icicles, and the floats had frozen into the lake. With spring nine months off, the samplers started a betting pool on whether the plane would have to stay here till July and a helicopter would have to evacuate them. Jensen did not appreciate this, and he made them form a bucket brigade to ferry hot water from the bunkhouse to melt the icicles. He fired up the engines, then had Mark and Brent kneel on the floats with axes to chop at the ice in front. When they had loosened enough ice, he powered up and edged the floats' curved fronts onto the frozen surface; then the whole huge plane. He traveled slowly a quarter-mile on the ice to a remnant patch of open water along the beach and dropped the plane back in. They hastily ferried their belongings over. Chuck told Brent to go back and padlock the bunkhouse door.

"Huh? Someone's gonna come here and steal the stakes?" said Brent.

"You never know. Just do it," said Chuck.

In the last bit of unfrozen lake, Jensen opened the throttle. They bumped a few times, skimmed the surface, and were in the air.

As Chuck filed the claim maps, he feared competitors would notice and rush in to outflank them. Now that it was winter, he did not think they would try, but he dreaded the spring. They had little money left now, so he felt doubly defenseless. Wayne had already remortgaged his house to buy more Dia Met shares, and even Mackenzie was tapped out for the moment.

In April the Barrens are still wintry. Ice and snow cover the land, and grizzlies sleep protected by snowdrifts. But by local standards it is spring; days lengthen with bright sunshine, and temperatures may rise to 10 degrees below zero. Winds moderate. In April 1990, as orchards in Kelowna were blooming, Chuck informed Mark and Mackenzie that they were going back to get as much new ground as they could pay for. Wayne was to stay behind to raise more cash.

They arrived in a chartered helicopter on April 8, 1990. Norm's Camp lay in the lee of the esker, where snow drifts the deepest, so only the peaks of the hut roofs were showing when they got there. Chuck had chartered a helicopter for this last-ditch effort because he thought it would be the fastest way to stake. Including fuel, it was $1,000 an hour. Mackenzie was left to dig out and heat up the place, while Chuck and Mark took off. Chuck sat up front to navigate while Mark occupied the back with stakes. The pilot flew one claim's width outside their last perimeter. Then every 1,500 feet they touched down briefly, and Mark popped the door to lean into a blast of arctic air and drive a numbered post into the snow. When he shut the door, he yelled "Go!" and they were off again at full speed to the next spot. The pilot was mystified; staking was not usually done this way; no one ever even flew out here in cold weather unless they were on their way to somewhere else.

With snow obscuring lakeshores and only high spots raked bare by wind, sometimes they were not quite sure where they were, but the pilot did some careful dead reckoning and always found his way back to the winding great esker, generally snowless on top. They proceeded rapidly, concentrating on the southern end, where Chuck had been so unsure how far out to stake. Toward the end of April 13, they knew they had one more day; the Dia Met treasury was gone.

They were at the southeast corner when they reached the region's single named land feature: Le Pointe de Misére, or Misery Point. It had been 101 years and six months since King Beaulieu had stood in the snow here with bloody icicles draping his beard, cursing Warburton Pike. Chuck of course had never heard of Pike. He was navigating alongside the esker by Misery Point when he noticed a vaguely ovular lake a quarter-mile away, within a swath they had just claimed. "That's not right," he muttered. The east and west shores were hard to make out under snow, but the north and south ones were clear: They were exposed vertical cliffs directly opposite each other, dropping straight to the ice. He asked the pilot to pull in close and circle at 100 feet. The lake seemed to cover about forty acres. The cliffs were cut from a single unit of biotite schist, a gray metamorphic rock foliated into wavy bands of dark minerals alternating with lighter ones of mica—typical here, but not usually dissected in such neat textbook fashion. It was as if a hole had been punched from above—or below. He tried not to grow too excited, but marked it on his map. They resumed staking and headed back to Norm's Camp at dark.

Chuck gobbled the supper Mackenzie had made. He had all his maps showing sample results and afterward spread them on the wooden bunkhouse picnic table. It so happened that they had taken a sample just southeast of the odd lake on Misery Point. The sample was barren of indicators. A sample had also been taken on the other side of the lake, a quarter-mile northwest, near the trunk esker. This one held a few kimberlitic chromites. A third sample, from a creek that flowed from the lake itself, had more chromites. Chuck checked for ice direction; as far as he could tell, the anomalous samples lay down-ice. This all certainly fit a hypothesis that the lake was the top of a pipe.

That night Chuck could not sleep. Every time he began dropping off, the image of the ovular lake between two dark, windswept cliffs filled his head. Finally it all faded to white, and he dreamed. In the morning he told everyone they were skipping the last bit of staking. They would not get much more anyway.

Intending to go straight to Yellowknife afterward, he had the pilot fly twenty-four miles to the ovular lake's southeast side. They landed on the ice and got out in heavy mukluks and parkas. The temperature was far below zero, and wind was driving sharp snow crystals at them.

Chuck told Mackenzie and Mark they were now going to dig a sample from the lakeshore. They thought he had finally cracked. He had always told them never to dig samples from frozen ground; it was useless, and also more or less impossible. Nevertheless they followed him as he stumped over the frozen lake to a low spot along the shore. Mark was already yelling when they got there.

"You'll never get a sample! We're wasting our time! What's wrong with you?"

Snow had drifted deeply down the sloping edge of the lake, so only rounded outlines of snow-covered boulders could be seen. With some guesswork, Chuck picked what he thought must be the water's edge, where waves might work up a sandy deposit among the rocks. Not having planned on this, they had only a small collapsible shovel and a couple of geologists' hammers.

The snow was over five feet deep and hard-packed. They took turns digging. Chuck was practically in over his head when they bottomed out in the snow and saw dark ice with small boulders sticking out—the frozen shallows along the lake's edge. "Good," said Chuck. He began pounding with the blunt end of his hammer at the ice around the boulders, then prying each one out with the pick-end. Each time he dislodged a rock, he tossed it over the hole's lip, along with ice. He was soon exhausted and climbed out to let the other two take a turn. The pilot stood, looking on.

After an hour and a half, they seemed almost through the ice and near the sand. Mark was in the hole. Chuck told him to get out and jumped back in himself. As they peered down, he cracked through the last inch or two with his hammer. There were more rocks. Thinking the sand must lie under them, he clawed at them and pried them out one by one, dusting out pulverized ice with his mitt. Chuck pounded and pounded. Below was a terrible sight: more rocks. He had misjudged; there was no beach here—just stones. They had dug the hole for nothing. "That's no good," he muttered. "Loon shit," he hollered up. "It's loon shit!"

Chuck climbed out and looked around. Mark pointed to the distant esker, on the other side of the lake. On a high mound of till near it, about 800 feet from shore, the wind had bared a high spot. "Let's go over there. At least there isn't any snow," said Mark. Mackenzie nodded in agreement, but Chuck ignored them and walked to a spot thirty feet from their hole. "Start digging," he ordered. Mark protested, but Chuck glared. Then it was Mackenzie's turn. At the bottom, Chuck jumped in to claw at the ice. Presently, there it was again: loon shit. Chuck's face took on a rare look of discouragement. Mark was yelling down upon his head.

"This is useless! You fucking idiot! What are we doing here? It's too cold! Are you crazy? You want to kill us?"

On Chuck's orders, they dug yet another hole. After a total of five hours, they stood amid a crazed semicircle of excavations. With Mark cursing about how they were going to die, the pilot pointed to the time; it was going to get dark if they stayed much longer. Despite all the heavy clothes, Mackenzie felt his feet, hands, and nose numbing. Since frostbite was coming, it did not seem appropriate, but Mackenzie started laughing. "Are we really doing this?" he said.

Finally Chuck listened to Mark. The windblown bare spot across the lake was, in fact, down-ice. It might contain minerals from the lake. They trudged up to it and, now on high ground, were totally exposed to the wind. Unsorted till peeped out between sharp wind-carved snow ridgelets. It was not ideal, but better than nothing. They knelt in a tight circle, pulled out the two hammers they had among them, and took turns whacking away, shielding their eyes from flying pieces with their mitts. The hammers bounced off, sending painful vibrations up the handles to their freezing hand bones.

Chuck carefully swept each loosened bit of till into a plastic sample bag with his mitt. Below the initial layer, it loosened a little. It was a well-drained spot that did not harbor too much moisture, so gradually it grew less difficult. In an hour—no one was sure how they did it—they had three bags almost filled with unsieved material.

By this time Mark, still in the flush of youth, was the only one going strong and taking hard, effective swings. They watched sidewise as he felled one particularly hard blow. Then Mark stopped and stared. He removed a mitt and reached into the hole. Between thumb and forefinger he picked up a bright green object attached to a bit of frozen clay.

"Hey, look at this." He held it up. "Pretty nice, huh?"

Chuck pulled off both mitts and leaned over.

"Oh, hey, look at that," he said, as if he were a parent admiring a child's drawing. "It's a chrome diopside."

Mark dropped it into his hand. It was bigger than a cooked pea, formed so perfectly it could be mounted in a pendant. Any chrome diopside that big meant they were practically on top of a pipe. Chuck turned the thing calmly over in his hand and dropped it into the last sample bag.

Events moved forward rapidly from here. Within days Chuck analyzed the sample of till and found it loaded with all sorts of indicator minerals. The same week, Stew completed the torturous buyout of his partners. And Hugo rejoined the team.

Hugo had landed a job at Broken Hill Proprietary (BHP), the big mining company Chuck had visited in Australia. BHP was a world power. It mined copper, manganese, iron, and coal, with operations in fifty-some countries from the Pacific islands to Canada. They gave Hugo a big office in San Francisco and an impressive title: Manager of Exploration, North and Central America. His main job: seeking out joint-venture deals with prospectors. Chuck and Stew had always promised to come back to him if they could. They were all immediately on the phone to each other.

Chuck and Mackenzie pushed hard for a deal; Dia Met hovered near bankruptcy. Also, rumors were floating in Yellowknife: It was said De Beers was snooping around town and that Jennings had a twenty-five-man crew staking for something out near Hudson Bay. Stew, however, was in no hurry. Now that he was back, he had resumed his daydream of two simple prospectors in the wilderness. He was already mapping out what he and Chuck would do next, on their own. Chuck hollered at him: The claims were too big. They needed geophysicists, fleets of aircraft, millions of dollars. Stew reluctantly agreed to negotiate.

In May they showed Hugo the collected maps and data. He was awed at the sweep of the travels. Stew had his own eye-opener: Up to now, he had known nothing of the spectacular sample on the esker beach. The big chrome diopside near the ovular lake was impressive, but the thing Hugo liked most was a fresh report from John Gurney. Chuck had sent Gurney raw data from mineral analyses throughout the claims, and Gurney saw many G10s had only 2 percent calcium—fantastically low. This, he said, increased diamond potential even more. The ilmenite chemistries, too, were outstanding, he said. He couriered an independent report: "The dataset represents the best for diamond potential that we have seen anywhere in the world and we have no doubt that highly diamondiferous kimberlite is the source of the heavy minerals."

With this as leverage, Chuck and Stew ganged up and bullied Hugo mercilessly. Hugo wanted this deal, but they played good cop/bad cop on him. Chuck would agree to a particular contract clause; but then he said he knew Stew would not go for it. And vice versa. In this way they raised the stakes week by week. Hugo shouted that they were robbers. Chuck and Stew hung together, working better together than they had for years, while Hugo surrendered to every new demand. Behind the scenes Chuck and Stew argued bitterly over everything: Chuck's secrecy, Stew's absence, Dia Met's still-disputed role, what everyone would get if they found something, whether to take on yet another partner in BHP. In this way the swift arctic summer wasted.

Finally on August 31 they struck a deal. Chuck and Stew agreed to each receive a 10 percent personal stake of any possible mine. Dia Met shareholders got 29, further bolstering Chuck's and Stew's portions, since both owned many shares. BHP got the remaining, controlling interest—51 percent. But only after it fronted a staggering $500 million for exploration and development. Staking was now to continue, largely on BHP's money, in a new four-mile swath around the existing claims, to create a "buffer zone" against competitors. BHP would send in specialized crews to track down pipes.

For the moment, Chuck remained in charge of operations. Of course everything would collapse if BHP decided there was nothing there and pulled out.

Days after the signing, Hugo, Stew, and Chuck were standing at the lip of the lake on Misery Point with a small entourage of BHP executives. The now-melted waters of the oval lake—at this date almost ready to refreeze—were cobalt blue. Tundra plants had already turned brilliant autumn oranges and reds.

Hammer in hand, Hugo ran to the southeast shoreside where they had dug in the snow. He upset rocks and ripped into gravel, half expecting to find exposed kimberlite. There was nothing, so they headed to the northwest side. The hole they had chipped in the till was still visible. They inspected it briefly, then walked toward the big esker. At bedrock outcrops Stew picked out striations that confirmed the ice had come this way. There were round, yard-wide frost boils scattered around, where till had been pushed out by freezing. In them they saw something they had seen only once before: big indicator minerals sparkling on the surfaces, like on Tom McCandless's desert anthills. Here, the minerals had literally erupted from the ground. They fell to their hands and knees and filled their pockets with pea-size chrome diopsides. Smaller, glittering pyrope grains lit the tops of surrounding mosses like Christmas-tree ornaments. The mineral swath was exactly the width of the oval lake, and ran straight at the esker.

They hiked onto the ridge and followed for a ways in the direction of the distant Norm's Camp. Even here they could pick out occasional minerals by eye. It was for Stew the biggest day of his life; everything, billions of years of geology, was neatly collected and laid out within walking distance like a museum exhibit.

When they turned back to the lake, Hugo shouted, "I'll give $5,000 to anyone who finds a diamond in one of those frost boils." Instantly everyone was down on their stomachs, picking. Nobody found one, but no one worried. That night at Norm's Camp, Hugo and Stew were too giddy to stay in the bunkhouse. They insisted on cramming themselves together into a tiny pup tent on the beach and talked all night. Chuck was partying with the executives in the bunkhouse.

Chuck instantly had a big crew spirited up to dig more samples and continue staking. Within days, they were hindered by thunderstorms blowing rain horizontal to the ground and, at the end of September, the usual blizzards. One worker walked ten feet from the bunkhouse and almost did not find his way back. The 1990 season was over. They waited for a clearing in the weather and fled once again.

In May 1991 a small BHP geophysical crew returned to investigate the crater lake, which looked promising, to say the least. By this time it had a code name: Point Lake. This was partly after Misery Point, but mainly because there was a much larger Point Lake on the map west of the claims, and they hoped this would confuse outsiders.

Spring came on suddenly, and the surface of Point Lake turned to treacherous slush. The geophysicists waited until 2 A.M. each night, when it refroze, then crept out gingerly and worked by moonlight. Soundings showed it was 150 feet in the middle—unusually deep. They traversed the ice with special instruments to send electromagnetic waves down and measure the strength of the returns; these showed a formation of something highly conductive underneath, vaguely carrot-shaped. A plane with similar instruments was sent overhead, and they lit up with a bull's-eye magnetic anomaly. The head geophysicist, an intense young man with green eyes named Ray Ashley, counseled caution; not all such anomalies were kimberlites.

This made Chuck nervous on more than one count. This was still their only possible pipe. They did not want to pin all their hopes on it; if they drilled and came up with no kimberlite, or kimberlite and no diamonds, BHP could easily pull the plug on the whole thing. He put off drilling and went looking for other targets.

During all this time, reaction from the competition was strangely lacking.

Jennings and De Beers were close behind—or perhaps ahead. After Jennings's search had sunk into Glacial Lake McConnell, he quit Selection Trust and caught up on his reading. It was then that he saw his mistake; he should have headed east, into the tundra. He took a new job as exploration chief of another mining giant, International

Corona Resources, run by the flamboyant Toronto mining promoter Murray (The Pez) Pezim. The Pez's thing was gold, but as soon as Jennings got to Corona, he redirected money from a couple of his budgets to look secretly for diamonds. He never got permission.

From June to September 1990, while Chuck and Stew were wrangling with Hugo, Jennings sent a freshly minted but highly competent geologist, Leni Keogh, in a float plane to sample the tundra. On her own, Keogh decided she liked Lac de Gras because of the fine, sandy beaches along the esker system. She took quite a few samples. The land was so huge that she never noticed any of Chuck's stakes. In early September, Keogh landed just outside the claim lines, across the Lac de Gras–Lac du Sauvage narrows from Misery Point. There she found another curving esker beach, much like the one at Norm's Camp. And on the strand line was an identical band of countless ilmenites and pyrope garnets.

"I don't believe you," said Jennings, when she called him from a satellite phone at Baker Lake. "That's not possible. I've certainly never seen such a thing." She told him to come see for himself. He caught a plane up. In front of Jennings, Keogh spilled out the bag of sand onto a table. "Get right back out there," he ordered. Then the same thunderstorms and blizzards that had driven out Chuck's crews drove off Keogh.

The field season was over, but a week or so later, a revealing nine-sentence story came out in the weekly *Northern Miner*. BHP and Dia Met would have liked to have kept their deal confidential, but Dia Met was a public company operating under Canadian law, and this meant it was obliged to release at least the essentials of any event that could affect its stock price. The story only said that the two had gone into partnership and vaguely mentioned large "Canadian claim blocks." Most people paid little attention to it, and the stock price was little affected. But Jennings figured it all out instantly. He knew Hugo had moved to BHP, and he knew Chuck ran Dia Met. He knew in which direction Chuck and Stew were headed when last spotted. He sent Keogh to the Yellowknife mining recorder, and she spotted the registered claim blocks quietly expanding month by month—nearly to the edge of the Misery Point narrows. All were listed under obviously false names. Norm's Manufacturing my arse, thought Jennings. His adrenaline rose when their own sample from the esker came back. In addition to the indicator minerals, it contained two 0.1-carat diamonds.

Jennings ordered every available government map and photo of the region, pored over them, and gloated. From everything he could see, Chuck had misjudged. There was not just the big esker; there were swarms of eskers, converging on it and feeding it from the east, northeast, and southeast. In his judgment the source was not on the BHP-Dia Met claims but farther up-ice. In July 1991, with the BHP crews already out in the field again, Jennings began writing long memos and pounding on doors at Corona to demand a massive staking campaign to outflank them on the east.

De Beers was closing in, too. Barry Hawthorne's rejected case to prospect the Barrens rebuilt as more detailed GSC maps of the Barrens came out, showing clearly the outlines of the ancient Slave craton and the movements of glaciers off it. Analyses of indicator minerals were becoming more refined each year, too. Lab scientists could now calculate the conditions under which they formed. These were seen to be abnormally cool, abnormally compressed—the kind of equation that could exist only within a theoretical lithospheric root— a cool, deep keel of solid rock poking into the earth's soft underbelly. The most likely place for one was under a craton, and academic geophysicists were now equipped to confirm whether such cratonic roots existed. A growing global network of seismographs had been measuring departures and arrivals of waves from earthquakes and, occasionally, nuclear bomb tests. By the 1980s computer links were up, and the academics started adding up data into underground tomographic maps, like doctors looking for tumors. In 1987 University of Illinois geologist Stephen Grand made the first one of North and Central America. It showed several roots, including the most dramatic, 180 miles deep or more, under the Slave.

In 1989, three months after the Fipkes started staking, Massachusetts Institute of Technology geologist Samuel Bowring published the 3.96-billion-year dating of the Acasta gneisses, a new record for the world's oldest known rocks. Found with the help of the GSC, they lay some 100 miles northwest of Norm's Camp. Also by this time, geologist Herwaert Helmstaedt of Queen's University in Ontario was studying much younger volcanic dikes lacing the Slave—formations that could have wrecked diamond deposits if they reached down deep enough. But, he asserted, the dikes were "diamond-friendly"—that is, the magma had merely cracked the surface, not the diamond-producing regions. In fact, the dikes might have provided potential weaknesses for kimberlites to emerge through, he announced to an international conference. The De Beers men did not have to go any farther but their library in Johannesburg to see that all the elements were there. It was time to go back north.

De Beers's Canadian subsidiary, now renamed Monopros, had lost track of Chuck—that is, until the *Northern Miner* article about the BHP-Dia Met deal came out. Joe Brunet saw it too and made the same deductions. He personally went to the Yellowknife mining recorder, where he found the same giant claim blocks spreading out under the same phony names. Loose mouths down at the Gold Range confirmed that a harried-looking man named Chuck had been rushing in and out a lot lately, looking confused and carrying big rolls of maps.

As soon as the ice melted in spring 1991, Brunet was out with a couple of assistants, secretly taking samples just beyond the claim lines. Toward the end of summer results began filtering back from South Africa, and he practically did backflips. Unlike Jennings, who had decided Chuck had not traveled far enough east, Monopros decided Chuck had traveled *too* far. Their samples detected a huge swath of indicator minerals to the *west*. That, combined with seismic data pinpointing the westerly region as the center of the lithospheric root, convinced them. They laid plans to outflank BHP-Dia Met on the other side.

Not only did all this happen behind Chuck's back; there was someone in his own ranks unwittingly helping attract the enemy. His own brother.

After Wayne had his meteoric vision on the tundra, he returned to Kelowna to fulfill his destiny. This, he perceived, was the sale of ever more Dia Met shares at ever higher prices, so that he could finally make some much-needed money. Here he was president, and he was poor. He set out to publicize their work, running straight up against Chuck and Hugo's fanatical efforts to hide it.

Wayne gave himself a crash course on diamonds, devouring translations of Sobolev's incomprehensible Russian tomes and Gurney's latest charts. Then he molded his new "investor awareness program" on the premise that they had on their hands a pretty much proven "world class discovery." He pumped out press releases, brochures, easel displays, a videotape, and a growing stream of letters and phone calls to stockbrokers and investors. He bought more stock for himself, and if he heard someone was planning to sell theirs, he called to upbraid them; too many sales would make the price fall. He could not reveal the location of the claims, but any other ploy seemed fair game.

Among the items Wayne produced was a sixteen-page color report that listed all of Chuck's life achievements, starting with his record as a Boy Scout and on to his worldwide prospecting experience. Then it wandered into dangerous territory. It talked about claims in the north, including "80 geophysical anomalies and distinctly separate indicator mineral trails," with an actual chart of some of Chuck's own geochemical analyses. One page contained a map with mysterious arrows emanating from Hudson Bay into the surrounding tundra. Wayne even sketched his "Artist's rendering" of the imaginary mine, adapted from pictures of Kimberley, South Africa. Sprinkled throughout were head shots of Chuck, Mackenzie, and Wayne, as well as Wayne's heavily made-up teen daughter Leila. In each picture Leila sported a large borrowed engagement ring on her middle finger in various Playboyesque poses. In the closing photo Leila licked the bejeweled finger and eyed the viewer with a certain tingle in her eyes. "Diamonds are forever," whispered the caption. "Leila's bottom line 'yes' commitment will be based on the potential a man shows to deliver on his promise."

Chuck had at first admired Wayne's artistic side, but saw only destruction here. Wayne patiently explained to Chuck that people would not buy stock unless he "educated" them. To this end, he started manufacturing a daily press release about their progress. He finished it in longhand around noon and sent it to a secretary for typing. Chuck, after doing his rounds at the lab, intercepted the release each afternoon promptly at 3 and stormed with it in hand to Wayne's office at 3:02. Together they screamed like twelve-year-olds, usually till about 5:30.

Few of the press releases made it past Chuck, though somehow the Leila Report got printed up. Few readers believed Wayne anyway. He soon developed a reputation among mining-stock analysts. "Bullshitter," "carnival huckster," "bold-assed promoter," and "inflammatory son of a bitch" were some of the milder opinions expressed. It never bothered Wayne because he knew they were wrong. He worked conventions and mineral shows with a traveling display. Finally, at the yearly Toronto Prospectors and Developers convention, he met a small knot of men who were wandering about, though without the usual nametags. They listened raptly, asked questions, and each respectfully took a brochure. They were all Monopros geologists.

Hugo was also there, walking around. He missed the hated Monopros men, but on the convention floor spotted Mackenzie, whom Wayne had recruited to help with publicity. Sensing that something unsavory was up, Hugo walked up to Mackenzie and stood chest to chest with him. They were about the same height, but Hugo was a lot wider. "Oh, say, what's *this*?" asked Hugo casually, and wrestled a telltale bulge out of Mackenzie's sport-jacket pocket. It was the Leila Report.

One day shortly after, Wayne arrived at his tiny office to find that all the data sheets and maps he normally had access to had been removed, to what Hugo termed "a secure area." A board meeting was called, and Hugo told Wayne in no uncertain terms to shut up about their activities. Hugo then set up his own part-time office in the lab to keep an eye on things. Wayne had expected Chuck to come to his defense during this apparent coup, but instead Chuck told Wayne to shut up, too.

At this point some of the earlier, biggest institutional investors started selling out. They were apparently disappointed with the lack of results and surfeit of hype. This included First Exploration, a Vancouver mutual fund that at one time had owned a full 30 percent of Dia Met's stock. Many of First Exploration's shares went to the same few buyers for pennies: Mackenzie, Wayne, Norm's Manufacturing, and, supposedly, a certain Kelowna businessman named "Mr. X," whom Chuck and Wayne had advertised would absorb whatever came on the market. Chuck, C. F. Minerals, and Norm's Manufacturing had been amassing major blocks all along; the lab did all the operation's processing and analysis work, and it had to be paid somehow.

At the end of summer 1991 everything approached critical mass. BHP-Dia Met had taken thousands of samples and expanded the claims to a virtual kingdom—1,500 square miles, in a rectangular mass reaching almost to the shores of Lac de Gras itself. This included 510,000 acres of "buffer zone" land staked around the original core claims. Most of Misery Point was taken, though the southern end was left open, as well as the territory across the narrows. The only thing missing was a proven pipe. High BHP executives started wondering what Dummett was doing in this empty quarter. Soon it would be winter again. Money was flowing out with nothing to show. Hugo sensed he had better pull the rabbit out of the hat now, or see the corporate guillotine sliding toward his eyes.

Hugo informed Chuck he was suspending the staking and was ordering a drill flown in to Point Lake. It had to be not just a kimberlite; it had to be a diamondiferous kimberlite, or they were finished here. Hugo showed no signs of flinching, but in this moment of truth Chuck chickened out. He protested, argued, and whimpered to somehow put it off, so they could look for other drill targets. Hugo said no.

The drill rig was flown in pieces by Twin Otter on September 1, and by the 5th a hired crew from Winnipeg had it reassembled and going. Sited 150 feet from the shore, it was headed at a 45-degree angle toward the middle of the lakebed. The roaring machine drove a hollow, industrial-diamond-tipped bit at the end of a $2^3/_4$-inch-wide pipe—in search of gold, the drillers were told. Once each section of pipe sank to its end, the drillers pulled up another cylindrical core of rock from inside, screwed on more pipe and kept going.

The drill wheezed through hard metamorphic rock, mostly metagraywacke. The drillers laid each largely identical core alongside its mates in shallow, coffin-like wooden boxes. This went on around the clock in twelve-hour shifts for four days. Perhaps because Chuck could not stand the suspense, he stayed away; supervision was passed to Kelowna geologist Ed Schiller, a minor member of the Dia Met board in whose name many claims had been staked. Like most Dia Met shareholders, Schiller knew far less than Chuck about the prospect.

At 7 A.M. on September 9, the drill foreman came to Norm's Camp to tell Schiller they had entered a section of weird, soft stuff at a depth of 455 feet. Schiller went flying out there and saw several cores of it already lying in the boxes. It was kimberlite.

Per Hugo's instructions, they had communications codes. Schiller ran for the camp's radiotelephone to call Chuck. After a few tense pleasantries, Chuck inquired: "So, Ed, how's the fishing?"

"We just caught the biggest fuckin' fish you ever saw in your life," replied Schiller.

"It's about time!" shouted Chuck.

Several days later a float plane landed with Chuck, Hugo, Hugo's boss, Mackenzie, and John Gurney. By chance Gurney had been attending a symposium in Saskatchewan when the news came, and Hugo spontaneously invited him. The cores were laid out inside a plywood shack, and everyone stuck their faces to the greenish rock with magnifiers. Gurney saw right away that it was "juicy"—nicely loaded with dark garnets and other pretty minerals that to his eye looked excellent. No diamonds were visible, but they almost never are. Hugo's boss asked, "What is this, is this good news?"

"Yeah, absolutely. I'm sure this is good news," said Gurney, speaking as the expert.

"What if you're wrong?" said the executive.

"I'm not going to be wrong," said Gurney.

When the drill was shut down at 920 feet in late September, it was still bringing up kimberlite. Chuck had the dozen core boxes nailed shut and sealed with wax to prevent tampering. He feared someone might hijack them on the road, so he sent the only man whom he both trusted and who knew Yellowknife intimately: Stan Emerson, the ex–gold prospector who ten years before had put up some of the original money. Emerson drove a heavy diesel truck up. When the Twin Otter came in from Norm's Camp, he helped unload the core boxes directly into the truck at dockside. Then he hid out for the night at the cabin of an old friend. Starting in early morning, Emerson drove forty-eight hours to Kelowna without stopping, taking just a little nip of warm beer from an open bottle every once in a while. He had discovered years ago that this would keep a man from going to sleep; so he believed, anyway. When he got to the lab, he and Chuck pulled the cores into a garage bay and bolted the door. The only people allowed in were Chuck and Gurney's man, Rory Moore, now hired on for the duration.

There was some question whether to analyze the cores. Hugo needed results, but Stew lobbied to store them unexamined for a few months. This was not as crazy as it sounded; the Misery Point drill hole was still perilously close to the edge of their claims, and Stew, like everyone else, wanted to keep staking. If they analyzed the cores and they did contain diamonds, Dia Met's involvement would come back to haunt them, for they would be legally obliged to announce the results. This, Stew predicted, would spark the staking rush that so far had failed to materialize. He aggressively pushed this view on Hugo and Chuck but never got an answer. Chuck made himself unavailable for comment, and after a while Hugo stopped returning Stew's calls. Hugo and Chuck could not wait to see what was in those cores, and figured they could keep it hushed, whatever it was.

Just after the drilling, Stew and Marilyn took a vacation in Mexico. While they were gone, Chuck selected several small intervals of the core totaling fifty-nine kilograms and shoved them into a heavy revolving drum to break them up. When they were reduced to powder, he had the powder run through the lab like any routine heavy-mineral sample, telling the workers nothing. Secretly, he lurked in the background, eyeing every step. At the end Chuck had a technician dump the heavies into a container of high-potency acid, which would dissolve most anything but diamond. When he came back, he saw the acid had left a little pile of light-colored sandy residue. He seized it and locked himself in his office to pour a batch onto a petri dish. When he looked under the microscope, he nearly had a heart attack.

Among some whitish grains of zircon, apatite, and a few light-green olivines that had survived was a gigantic bluish-white octahedral diamond. He jerked his head away. It was not actually gigantic—it was only 1.1 millimeters, just a big grain. It had only looked big under the microscope. But he pulled off his glasses and looked around the lens directly at the dish. He realized it was big enough to pick out with the naked eye. Within minutes he found sixteen of similar size. After three unbroken hours at the microscope, he had eighty-one tiny stones.

The grading of a diamond pipe from such a small core is impossible. But the presence of any diamonds, including microdiamonds, could be considered very encouraging. In secrecy he flew to San Francisco.

It happened that Jerry Ellis, chairman of BHP, was visiting, so Chuck got to show Ellis the diamonds. Ellis was skeptical when he saw how small they were, but Hugo swooned. From everything Hugo knew of the latest arcane studies regarding the grading of kimberlite cores, eighty-one stones of any size in just fifty-nine kilograms was spectacular. Lately, theories had been developing that certain quantities of microdiamonds from a small sample could predict certain quantities of macrodiamonds in a larger sample. As long as the small ones were of good quality, there was a chance they were fractured or melted off big brothers somewhere. These little diamonds were nice. And where there was one pipe, there had to be more. Hugo now wanted to resume staking—grab everything, the whole central Barren Lands if they had to. They would screw De Beers, Jennings, and all their enemies forever. He had only four words for Chuck: "You can't tell anyone."

Around this time, Dia Met stock began moving upward imperceptibly. It had started with the drilling. All kinds of companies drilled holes all the time, but it seems the pilots or perhaps the drillers sensed a certain extra excitement at Norm's Camp. Some probably heard Gurney or Chuck discussing the kimberlite. There were now too many people around to keep every conversation private. Norm's had a full-time cook, a young woman named Diane Peterson, who swallowed the gold story, but when she came back to Yellowknife, it seemed like she was the only one. "So. Did they find any diamonds out there?" asked a local businessman. Some in town began buying shares of Dia Met, currently trading at about 50¢. The price edged up.

Chuck's samplers and lowly labworkers had always known this was no gold project. Each Christmas they had been getting a hundred or so Dia Met shares as bonuses—a pathetic holiday perk that would not even buy a turkey—but they sensed a change now, a certain edginess about Chuck. They knew him; something was up. Some actually *bought* shares. Norm—the genuine, janitorial Norm—must have gotten a hot tip somewhere because he began taking up a major position relative to his meager pay. Dynamite Dan had helped with the drill hole, so he knew more than most. One day Chuck visited the remaining core, rolling up the garage door partway to admit sunlight. While he walked away momentarily, Dan stuck in his head to admire the sparkly indicator minerals. Then Chuck returned.

"Did you see that core? No eyes are supposed to see that core!" screamed Chuck.

"Sorry Chuck," shrugged Dan.

"Get out. You're fired. Never set foot in this lab again!" yelled Chuck. Dan had never seen him so upset. Dan instantly mortgaged his small house and used the entire proceeds to buy Dia Met, now at 68¢.

To the now-figurehead president Wayne, Chuck told nothing. But Wayne knew Chuck better than anyone. Even before the core was processed, he borrowed anywhere he could to buy more shares. His most loyal investors did the same, proselytizing for Dia Met at cocktail parties with renewed vigor. By the time the diamonds secretly appeared, the stock was hitting $1.

Hugo, as a BHP employee, was forbidden to own any junior partner's stock, and he watched with dismay. He was certain a leak had taken place. He kept quiet, but a few days later he got a call from his boss's boss, a man just below Jerry Ellis. The executive had been consulting with company lawyers. From the stock price, it looked like they had the appearance, even if not the reality, of insider trading. They had to, he said, report the core to the stock authorities. Hugo argued and tried wriggling out, but there was no recourse. The executive ordered him to announce the results immediately. Hugo called Chuck, and reluctantly they agreed on the wording. No more extra staking had been done along Misery Point, or anywhere.

On the morning of November 7, 1991, Dia Met and BHP faxed a press release to selected media: "The following information has been released to the Vancouver Stock Exchange, Canada: The BHP-Dia Met diamond exploration joint venture, in Canada's Northwest Territories, announces that corehole PL 91-1 at Point Lake intersected kimberlite from 455 feet to the end of the hole at 920 feet. A 59 kg sample of the kimberlite yielded 81 small diamonds, all measuring less than 2 mm in diameter. Some of the diamonds are gem quality."

PART IV

CHAPTER 17

The Great Staking Rush

The morning the press release scrolled off the fax at Monopros's office in Thunder Bay, Ontario, the geologists huddled around their map table. They had been preparing for a showdown—but not quite this soon. Now that diamonds had been found and everybody knew, they foresaw a rush from companies and every small-timer with visions of wealth. "Holy shit, here we go," said Todd MacKinlay, a young Monopros geologist who knew how especially rabid Yellowknifers could be.

Two weeks before, they had called their main contact in Yellowknife and told him to gear up for an unspecified staking expedition. This was Brian Weir, the contractor who had helped bring the cement mixers and Dene workers to Blackwater Lake. There had been no hurry two weeks ago; but hours after the Point Lake release, Monopros faxed Weir maps showing 1 million acres, mostly in a block west of BHP-Dia Met, and told him to go.

They wanted a discreet surrogate, and Weir was a good choice. As one of the main gossips who hung around at the diner next to the Gold Range, he knew everything going on in town but did not tip his own hand. Within a few days he quietly chartered aircraft and rounded up stakers. He told them they were working for him—not who he was working for. He disguised a huge order for aviation fuel by buying it through a friend's small airline. He snuck up the back stairs to the mining recorder and bought all the claim-corner tags they would let him have—1,000 of 2,000 on hand—and snuck back down. At Johnson's Building Supplies in Old Town, he bought them out of four-foot pickets, orange flagging tape, staplers, and staples. "If you tell anyone about this, you will be toast," he promised Karl Lust, the mild-mannered German-immigrant owner. Lust delivered the supplies to the float docks after dark. On November 11 at 8:30 A.M., the hour of freezing arctic sunrise, Weir's first helicopter roared out of its berth near Ragged Ass Road.

C hris Jennings had been poised, too, but he was in big trouble again. In the months preceding, he had been writing one memo after another demanding Corona stake east of Dia Met-BHP. His bosses refused. Specializing in gold, they had no understanding of his data; besides, he had done this without authorization. They had no intention of further endangering lives and money out there. Jennings was ready to explode. Just before the Point Lake press release, he had an all-out screaming fight with his boss and was fired on the spot.

Jennings, now fifty-seven, still taut and in a boxer's fighting shape, flew immediately to London. There, he tried raising money for his own expedition but had no luck. On the day of the Point Lake announcement, he was back in Toronto, setting up an office in his home basement, trying to figure out what to do next. He felt depressed when he saw the BHP-Dia Met announcement. Then he received a call from a longtime northern prospector, Grenville Thomas.

Thomas was a beloved character around Yellowknife. He had come from his native Wales in 1964 after doing time in the Welsh coal mines, and as a miner and prospector fast developed a reputation for honesty and good instincts. He worked and drank hard, telling endless stories at night, always rising first in the morning, and never getting a hangover. His bright little daughter Eira had camped with him summers from the age of six. He taught her to prospect by looking for "sparkly things" in the rocks while friends kept an eye out to make sure she was not eaten by bears. The Thomases inspired fondness, but also sorrow: In the high woods where he usually worked, the perspicacious Gren Thomas repeatedly found thorium, beryllium, tantalum, and other valuable deposits, but somehow the ground always slipped from him and the profits went to others. Now living in Vancouver, he ran Aber Resources, a tiny, ailing exploration company much like Dia Met. Eira was completely unlike her counterpart, Mark Fipke. Now about the same age as Mark, she had just gotten a geology degree, worked harmoniously with her dad, and was eager to follow him.

Gren Thomas happened to be in Toronto the day the press release came out, and he called a meeting of friends who had similar tiny companies. He was one of the few to take real note of the release; many readers of the *Northern Miner* assumed it was a scam. At best, eighty-one microdiamonds did not sound very impressive. Thomas, who knew nothing about diamonds, certainly did not know what it all meant, but he had a feeling. He remembered meeting Jennings once at a professional conference, and Thomas realized that Jennings was one of the few people in Canada who knew this arcane specialty. With the meeting in progress, Thomas called Jennings at home and asked if he might come over and talk to them.

Jennings was overjoyed. He sped across town in his car and launched into the whole long tale of the secret diamond search. They listened, rapt, then electrified. Within five days, Aber and the others had a syndicate with $250,000 in the combined treasury and Jennings as their main consultant. A day later they headed to Yellowknife and bought a great deal of heavy clothing. A day after that, Jennings and Thomas were setting up camp east of Dia Met-BHP in the middle of a snowstorm.

Less savvy outfits—basically, everyone else—were unaware the rush was already on.

To the west, Brian Weir flew in plywood platforms, heavy-duty tents, oil heaters, stakes, and stakers for the Monopros operation. They set up at Obstruction Rapid on the Coppermine, where in 1825 Franklin's men had crossed the river and started eating each other. After getting the shelters up, they located Dia Met's western-most posts, not far away, and started tying on with new claims.

Jennings and Thomas were some fifty miles off, on the east side of BHP-Dia Met. Since Monopros had already chartered most of the region's aircraft, the Aber syndicate at first had only a single helicop-ter. Within days they had two more ferried up from the south, plus more crew. No one could believe they were doing this in winter.

The temperature was 22 degrees below zero when Jennings arrived, not counting wind chill. The first day, wind blew down the pipe of their oil stove and filled the tent with soot. The place started freezing instantly as they coughed and scrambled to relight the stove. Soon gas-powered generators and electric heaters arrived to keep the helicopter engines unfrozen when idle. Initially this was most of the time, since there were only five daily hours of light, and it was usually storming. They waited out the weather by writing serial numbers on stakes. Jennings, who had never been here in cold weather, was excited. He wanted to go ice fishing, so they donned parkas, went out, and spent a while chipping a hole in a nearby lake. It was frozen to the bottom. Jennings saw green and white auras shimmering across the sky—his first glimpse of the northern lights. It all seemed harm-less, as long as they stayed warm. The only casualty, at least at this stage, was the cook. He developed diarrhea and kept running to the latrine, dug into a snowbank along a nearby esker. After several days he was evacuated to Yellowknife with serious frostbite on his penis.

When the weather cleared, Jennings wanted to head for Dia Met's northeast corner, where he believed the minerals on Keogh's beach came from. However, Monopros was too fast; its crew had already come over and grabbed the choice parcels there. The others wanted to go directly for the southeast corner at Misery Point. Point Lake's location had not been released, but things had a way of leaking; besides, it was near Keogh's beach. They headed for the tip of Misery Point and started staking there, then across frozen Lac de Gras itself, onto small islands, and finally back onto the mainland farther east across the narrows, where the trunk esker wound off into the distance.

The kind of high-speed helicopter staking Chuck and Mark had done the year before was almost unheard of; now Monopros and Aber did it, too. As days passed and light worsened, both worried other competitors might arrive. The staking did not seem to be going fast enough. And, as it happened, electronics makers had just invented a way to go even faster: small receivers that unscrambled signals from the U.S. military's Global Positioning System (GPS) satellites, which allowed travelers to locate themselves anywhere on earth's surface down to a few dozen yards. The availability of GPS for nonmilitary use was a huge event: It made erratic compasses, poor maps, and dead reckoning instantly obsolete. Once the stakers got hold of GPS, they felt ready to go anywhere, fast, even in poor visibility. They started skirting the rules. Instead of messing around with maps and stopping for stakers to fix pickets, they told the pilots to fly forty or fifty miles an hour, as low as five feet off the ground, without stopping at all. When the digital GPS readout showed a staking point, a front-seat "navigator" called "Go!" and a "bombardier," as he came to be called, cracked the back door to toss the picket out. If he got it perfect, it stuck upright in the snow, but that was rare.

This high-risk change in procedure scared older pilots. Many were grasping the control sticks with whitened knuckles anyway, not sure they trusted the new instruments, and some feared a stake would fly back and catch the tail rotor, causing a crash. Many quit in the first few weeks. That left mainly younger, nervier fliers who liked new technology and had less sense of their own mortality. Some carved windows in the plastic bubbles so stakes could be efficiently jettisoned without opening the door. The stakers, mostly longtime Yellowknife roustabouts, viewed it as exhilarating sport, especially since they got a handsome $300 a day and all they could eat. Occasionally they saw the wake of whipped-up snow and ice fog left by a competing helicopter in the far distance, and this pumped them up even more. Soon the crews had it down so that each helicopter was doing 25,000 acres a day. They figured on doubling that when daylight lengthened.

The staking steadily picked up speed until about December 20. Then there was a mutiny. Crews on both sides wanted to go home for Christmas and New Year's. The bosses tried to keep things going, but the pilots ganged up and said they were going home whether anyone liked it or not. A cease-fire was called, and everyone headed out for southern holiday revelry. On January 2, 1992, they were back.

During the break, hired workers attended parties and family gatherings in Yellowknife, Vancouver, and other cities, spreading word of their adventures. The life span of the secret was clearly limited; even Monopros acknowledged this. By New Year's, the company released Brian Weir from his vow of silence, and he began telling relatives and friends, too.

By mid-January, small Vancouver mining companies were calling Weir and other Yellowknife contractors to see if they could get land staked or were assembling their own operations. Few knew anything of Lac de Gras, diamond prospecting, or northern logistics. They were just following the old dictum: Whenever some lucky fellow hits it rich, stake nearby. If there is not more of whatever he found, you can fall back on Mark Twain's observation: The real secret to mining success is not in discovering a mine but in convincing *someone else* you have found one, then selling the land to him.

Weir and others quickly discovered they could stake anything and sell it sight unseen, as long as it was vaguely near Lac de Gras. Soon calls from small southern juniors were coming in back to back. On January 13, Weir picked up a call from Vancouver mining promoter Duane Poliquin, whom he had never heard of. Poliquin politely inquired about ground for several small companies he represented. Weir, busier than he had ever been in his life, surprised even himself by blurting, "I can't even think about it unless you wire me $50,000 by tomorrow morning." They both paused. Then they both laughed at the absurdity of what he had just said. Poliquin responded: "Here I am, talking to a voice on the phone, to a person I don't even know. And you're gonna go and stake me some ground on no geologic premise, and I don't even know where it is. And you want a $50,000 bank transfer in the morning?"

"Yup," said Weir.

The $50,000 was there the next morning.

By early February Aber and its partners had 2,400 square miles. That was it for them at the moment; their $250,000 was gone. Monopros would never run out of money. Fielding a much larger force, they now had 4,200 square miles and were still going.

With most of the territory adjoining BHP-Dia Met now taken, latecomers swarmed about the periphery like flies around a newly manured alfalfa field. They tied on to the Monopros and Aber claim lines. Claims started building outward in continuous swaths of land starting ten miles, then twenty miles, then thirty miles, from Lac de Gras. The others were mostly obscure or hastily assembled outfits: Horseshoe, Dentonia, Northwind, Totem, Etruscan, New Dolly Varden, Slave Diamond Syndicate, Kalahari Resources, Heard Syndicate, Pure Gold, Gerle Gold, Thermal Exploration, Inukshuk, Kettle River Resources—the latter known among friends as "Ma and Pa Kettle" because it was run by a husband-wife team. Major mining powers like Ashton Mining, Cominco, Lytton Minerals, and Kennecott Canada moved somewhat slower, but one by one they jumped in, too, often by making deals to option the ground of the faster-moving juniors.

Weeks after Weir staked ground for Duane Poliquin, Poliquin resold an interest to Kennecott for a 1,200 percent profit—and that was only a one-third interest. The Aber syndicate made an even sweeter deal for the ground they had so smartly seized near Misery Point. Kennecott agreed to fund all the exploration there, while the syndicate got a huge share of whatever they found. Jennings got a 1 percent royalty for his help—a sizable sum if they struck something. The big fish were already eating the little fish, and the little fish were happy.

Jennings knew a bonanza when he saw one. In March he stopped working for the Aber syndicate, bought the shell of a defunct public company, and turned it into his own diamond-lands conglomerate, SouthernEra Resources. Soon SouthernEra had hundreds of small claims peppering the region. Jennings quickly made profitable deals to finance, resell, or explore them among an ever more bewildering web of partnerships involving as many as seventeen different outfits at once. He also continued developing various other properties in conjunction with Aber, which now had money again and was expanding its holdings. Stock in SouthernEra, Aber, and the many other companies involved multiplied fast.

The Dia Met-BHP principals were aghast at the monster they had created. The day after the Point Lake press release, Stew and Marilyn had been flying back from their vacation in Mexico, unaware anything had happened; Stew did not even know the core had been processed. In Los Angeles he picked up an in-plane copy of the Toronto *Globe and Mail* and, by habit, flipped to the stock tables. Overnight Dia Met had doubled, to $2. Though few credited Dia Met with a real strike, stock speculators were ready to buy anything they thought might turn a quick profit before nose-diving, so Dia Met suddenly became popular. Stew was so mystified, he wanted to call Chuck from the plane, but it was too expensive. He waited until they landed in Vancouver.

"Chuck. Stew here. What the hell is going on?"

"I had to do it, they made me do it," babbled Chuck.

"Had to do *what*?"

"Well, you know, see, this is the thing, Stew, hey. They made me tell, they said we were gonna get in trouble if we didn't tell, hey? Well, maybe I should send you a fax."

Stew was furious: He had been left out again. Chuck said Stew knew little because Stew contributed little. They fought, but as usual got nowhere.

Within days it was clear their borders were besieged, their chance at the extra staking lost. A brighter side soon emerged, though. Even before news of the Monopros and Aber staking leaked, more speculators flocked to Dia Met stock. Within three weeks it reached $7. Chuck, Stew, and the few others with a lot of stock were suddenly becoming quite wealthy, at least on paper. It did not help internal relations, though. The better things looked, the more Chuck resented the fact that Stew was doing almost as well as himself. Chuck was coming to view himself as the sole discoverer; Stew, not only as a do-nothing, but as a leech.

Wayne was happy to see the stock go up finally, but it was a mixed victory. With his role as president now reduced to almost nothing, buried sibling hatreds between him and Chuck sizzled. Trapped together at the lab, they fought almost continuously over petty and not-so-petty matters, including stock options. It all culminated one night when Wayne and Mackenzie joined Chuck for his usual 10 P.M. dinner at Talos Restaurant. Over drinks the brothers had a squabble about office furniture; then a more serious fight about who had possession of certain warrants to buy 25,000 now-skyrocketing Dia Met shares. Mackenzie joined in; he wanted to know about those warrants, too.

Wayne tried defusing things by suggesting their family saga was getting less like *Dallas* and more like *Treasure of the Sierra Madre*, the old Humphrey Bogart movie about three gold seekers in the Mexican desert derailed by greed, suspicion, and attempted murder. Wayne analyzed the movie and said Stew could play Bogart—the mean, crazy spoiler who gets killed in the end by bandits. He thought this jab at Stew would mollify Chuck, but the non-moviegoing Chuck got the wrong idea. He assumed Humphrey Bogart must play the hero. It touched a nerve. They were finished with dinner and on their way out the front door when Wayne laughingly mentioned *Treasure of the Sierra Madre* again. Chuck, leading the way, wheeled and karate-kicked at Wayne's crotch. At the last instant he had a second thought and pulled back, but did make some kind of contact with his brother's scrotum. Wayne doubled over and groaned deeply. Mackenzie comforted Wayne while Chuck stalked off, yelling that everyone was betraying him now that things were finally going right.

Shortly Wayne turned over the presidency to Jim Eccott, a Kelowna lumber dealer and minor shareholder to whom Chuck had apparently switched loyalties. Wayne was not quite done, though. Newly idle and sensing that people would soon want to read about their adventures, he dashed out a novel, *The Star of Canada*. About two brothers' long search for a diamond mine, it involved much skullduggery and many thrilling wilderness adventures, including a high-risk scuba dive to a diamondiferous arctic lake bottom. The protagonist was "Jack," a suspiciously Chucklike character married to a suspiciously Marlenelike wife. In the end, the brothers find the diamonds, but with a twist: Jack gets a younger woman pregnant and disgraces the whole family.

To say the least, Chuck and Marlene did not appreciate Wayne's literary effort. Chuck screamed at Wayne on the phone, then retained a lawyer to threaten action if he published the book. Wayne put *The Star of Canada* on a shelf and retired to the sidelines, from where he issued occasional vociferous criticisms of the way Dia Met business was being handled, alternating with predictions to stock analysts that Dia Met would hit $900 a share.

Since more staking looked hopeless, BHP-Dia Met concentrated on what they already had. As soon as the ice was frozen hard enough, Hugo ordered a larger drill with a $10^{3}/_{4}$-inch bit airlifted to the frozen surface of Point Lake. In February and March, it made thirty-seven big, deep holes through the ice and into the bottom—an intermediate test of the pipe's grade. Chuck landed in a ski plane to supervise the initial work, then deputized Rory Moore to continue.

Some 160 tons of kimberlite were transported to Yellowknife, then via a ten-truck convoy 2,000 miles to a familiar place: the failed Sloan Pipe in Colorado. Not only was Hugo's old processing plant still there, but Chuck had been smart enough to snap it up at a fire-sale price from another now-defunct mining company that had bought it from Superior. When the convoy arrived, Mark figured out how to reactivate the machinery. Moore and Chuck sorted through the resulting concentrate on a broad tabletop, using knives to push the dregs through a small hole in the middle.

There were more diamonds—bigger ones. The total from the 160 tons was 101 carats—very respectable. A quarter of the stones appeared to be of gem quality or close to it, and a few weighed over 3 carats each. Chuck jumped on every new stone and peered at it with great excitement through his magnifier. The diamonds were then sent off for a long round of appraisals in London and Antwerp to see whether Point Lake was worth an underground shaft for a final sample.

Meanwhile news of the 101 carats was published. This was in May 1992. At this point mining-stock analysts, previously ranging in attitude from cynical to skeptical due to Wayne's wild prognostications, were swept in. Most knew nothing about diamonds, but they quickly gathered information and speculations from experts. Many analysts were soon comparing Point Lake's ore grade to the best South African mines. One extrapolation held that the pipe's first 700 feet must hold some 46 million carats. Others were wary of extrapolations and, more significantly, questioned whether any mine could meet the huge expenses of operating on the tundra—or exist at all, given the conditions and remoteness. Optimists outnumbered pessimists, though, and Dia Met doubled again, to $14. Chuck, Wayne, Stew, Mackenzie, and fifteen or twenty relatives and friends in and around Kelowna were now officially multimillionaires. The effects on their lives were explosive; the effects on the Barrens, more so.

At the first signs of snowmelt, great numbers of animals began moving back into the tundra. By now dozens of companies were joining the rush. They staked outward in all directions from Lac de Gras, so that a huge, solid rectangle grew like a stain, preceded by speckles of smaller claims laid out by pioneers farther out. On the south end, one of Brian Weir's crews hit the first high outliers of boreal forest, and saw migrating caribou strung out in the snow, headed north. One bombardier aimed carefully and managed to bean one animal in the butt with a picket. No harm was done, but the caribou ran like hell. Packs of whitish wolves could be seen trailing behind, looking curiously up at the aircraft. Within days the wildlife movement became a flood. Migratory geese, ducks, cranes, and redpolls appeared in such masses that pilots had to keep constant watch to avoid tangles.

As staking penetrated the trees farther, a competitor of Weir's, Yellowknife contractor Lou Covello, made it to Snare Lakes, a settlement where a surviving handful of nonurban Dene had hunting and fishing headquarters. The Indians were angry to see strangers swoop in and stake all around them. Covello landed at Snare Lakes and tried explaining that the stakes only marked out two-year subsurface rights. He added what he believed to be the truth: There was probably nothing here, and the claims would expire when the time was up. Even the young Dene who spoke English showed no comprehension. They went around pulling up stakes, but stakes multiplied too fast for them to keep up.

When the worst cold weather was over, the real boom started. Even Hugo sent out last-ditch teams to take over outlying land. Many other outfits obviously did not know what they were doing. As melting proceeded, it turned out many stakes had been dropped onto frozen lakes. They floated in slush, then water. One particularly brash outfit, Canamera Geological, camped on ice, apparently unaware breakup could happen so suddenly—especially with powerful oil heaters in the tents. One Canamera pilot was in the field when he got a panicky call from the camp engineer: "The tents are sinking! The tents are sinking!" The pilot returned in time to see the crew scampering off a tilting ice floe.

Elsewhere, crews worked so fast in the growing light that some accidentally laid competing claims to the same ground. For the most part, such run-ins were settled amicably; no one wanted a lawsuit. Hugo got involved in one such affair after his men staked a parcel, then spotted overlapping posts set out by Weir's crews. It was not clear who had gotten there first. Hugo was extremely suspicious of Weir, but back in Yellowknife Weir invited him for breakfast, and Hugo consented. Breakfast consisted of three big glasses each of boiling water and 100-proof rum, served in Weir's living room. When the atmosphere was sufficiently friendly, Hugo started in.

"Well, we have a bit of mess here, don't we?"

"It's a lose-lose situation," said Weir. "You know if we go to court, it will last forever. Really, it's just acreage isn't it?"

Hugo agreed. They pulled out their maps and compared. The claims overlapped by 150,000 acres. Within minutes they settled it with a pencil line down the middle. Weir took the western 75,000 acres, Hugo the eastern 75,000.

"A pleasure doing business with you," said Hugo as they shook hands at the door.

The rush got stranger when Stew jumped back in separately. He could not figure out what was hurting him more: being locked out of business at BHP-Dia Met or the whole business itself. He hated to see hordes running around pristine wilderness driving in stakes, leaving litter, chopping blazes in stunted little trees that probably had taken 500 years to grow. Yet he regretted having done so little, in the end stages, to get it all started. He knew Chuck regarded himself as the discoverer; and so, to make his own discovery, Stew started yet another company to stake new claims—in effect competing with himself and his partners. He dubbed it Archon Minerals and invited relatives and friends to invest. Since Lac de Gras was so crowded, he laid out his own ground far west, nearer Blackwater Lake, where no one had yet reached. He kept it small and simple, as he always liked—

one float plane and a few trusted buddies—but his old methods suffered a major change. Chuck had been right about one thing: Stew was afraid his number was up. Unwilling to take the risk anymore, he sent field crews in without him, and when he himself visited, had a hired pilot at the controls.

Ed Schiller, the Dia Met director who had supervised the original Point Lake drilling, also split to found his own company, Tanqueray Resources ("Tonic Over Ice" was the motto on one report). Tanqueray picked up some ground Brian Weir had staked in among the expanding De Beers claims. Afterward, there was muttering around Dia Met that "Fast Eddie," as some nicknamed Schiller, had taken unfair advantage of his insider status to compete with his supposed friends. However, in Schiller's opinion, it was not he who had taken advantage. He claimed Chuck had taken advantage of him and the other minor Dia Met shareholders by not telling them of the fabulous indicator-mineral reports and other signs early on. Schiller claimed that if he knew what Chuck knew at the time of the drilling, he would have sold his house, car, and dog to buy more Dia Met stock.

Diamond stocks became a rage; diamonds themselves were beside the point. Many small companies ran up their prices by claiming minuscule parcels attached to the magic words, "Lac de Gras." In reality the real estate might be a three-week walk, but Canadian mining investors, long deprived of the excitement of a big strike, rushed to buy anything. Many company principals sold out their own shares at fantastic profits because they knew they might stay hot for only a few days or weeks. Money raised on initial public offerings practically evaporated; investors later suspected that some company officials had no intention of exploring claims and just devoured cash by sending themselves up for fast, expensive arctic adventures. Most companies got away without legal problems; everyone acknowledged these were high-risk ventures. It should be mentioned that one entrant was Bre-X Minerals Ltd. This was the same Bre-X that later became famous for setting the world's dollar-amount record for mining fraud in 1997, when its salted "gold" lode in Indonesia cost U.S. and Canadian investors $6 billion. Executives moved to the Caymans and Bahamas, while chief geologist Michael de Guzman reportedly plunged from a high-flying helicopter into the Indonesian jungle. Some people say the corpse found was not his.

The most infected diamond investors were Yellowknifers; nearly everyone had friends or relatives in the field or had just returned from the field themselves, so everyone thought they had the inside story. Yellowknife went temporarily insane, buying every new diamond issue, sometimes by mortgaging homes. Instead of calling south to ask advice from big stockbrokers like Richardson Greenshields, working-class Yellowknifers were receiving calls from Richardson Greenshields asking *their* advice. Yellowknifers had plenty of cash for investing; the rush meant local businesses were booming, and even unskilled workers commanded high wages. And farther south one man made money faster than almost anyone: the retired GSC man Bob Folinsbee, who had spotted Chuck's G10s on the esker forty-five years before. He had dumped his savings into Dia Met early. When he sold out, he made more money in one minute than he had made in his whole fifty-year career as a geologist.

The extent of the rush would not be tallied for several years, for that is how long it took mining companies to claim the expanses they coveted, even with modern technology. In total, it may now be said that some 260 corporations from Canada, Britain, Australia, South Africa, and other nations claimed about 100,000 square miles of the Barren Lands. The claimed land exceeded the combined areas of Hungary, Portugal, and Ireland, with room for several other small countries to fit, running in an almost continuous block around Lac de Gras down to the trees and up to the Arctic coast, plus many outliers east and west. At the coast, an outfit called Caledonia Resources staked from Bloody Fall down to the mouth of the Coppermine River. From here they kept staking into the river's underwater delta in the Northwest Passage, on the theory that diamonds could have washed down from Lac de Gras and settled on the bottom of the northern ocean.

CHAPTER 18

The Corridor of Hope

As the rush continued into spring 1992, Chuck uttered a half-joking boast to a *Northern Miner* editor. It came to be known as Fipke's Curse. "We got all the good ground, hey?" he said. "Nobody else is going to find anything."

Many took the Curse as a hostile challenge. As spring came on, both Norm's Camp and opposing camps grew dramatically. Twin Otters hauled in sprawling caches of yellow aviation-fuel drums and equipment. Soon an estimated forty-five helicopters and 200 fixed-wing aircraft were operating around Lac de Gras, all going in so many directions in so little space that Transport Canada issued an alert for everyone to be extra careful. Hundreds of geologists and hastily trained samplers set to digging thousands of little holes in the tundra.

On the BHP-Dia Met claims Chuck now ran a multimillion-dollar operation, though BHP called the shots. Ray Ashley, the geophysicist, took a lead role, with Hugo making the bigger decisions. Senior samplers were Dynamite Dan—Chuck had rehired him—and Brent. Both still worked directly for Chuck, who charged a hefty fee for his management services.

With still no other pipe but Point Lake visible, all contenders set out to fly geophysical instruments over their new dominions; word had spread that Point Lake gave off a definite signature. Helicopters and planes had instruments hung underneath in "birds"—thirty-foot torpedolike apparatuses full of gear to measure underground conductivity and remnant magnetism. Each craft flew repeated grid lines, each line just a few hundred feet apart—hundreds of thousands of miles, back and forth. Everyone knew it would be tricky; kimberlites were small, and those of different regions could look different, sometimes showing as highs, or lows, or indistinguishable, depending on the kimberlite and surrounding country rock. In the Barrens it was worse, because confusing anomalies also emanated from the numerous metal deposits, nonkimberlitic volcanic dikes, lake-bottom sediments, and even ground struck by lightning. BHP's readings over Point Lake became of intense interest: Competitors figured that if they knew the readings, they would know what to look for.

Hugo knew what they were thinking almost before they thought it, and he planned to stop it. He was not about to help them. There was no law prohibiting anyone from flying over, but BHP controlled the ground. Hugo put Ray Ashley in charge of protecting it. They discarded a joking idea of surface-to-air missiles, then an only half-joking one of barrage balloons to block low-flying aircraft, like in Nazi-besieged London. Finally they settled on a giant electric wire surrounding the lake. Theoretically they could loop it around, switch on the current, and watch aerial snoopers' sensitive instruments go crazy from the field it created. Within days Ashley had a long half-inch insulated cable unspooled around the shores, attached to a gasoline generator in a plywood shack. It worked: They flew over with their own instruments, and the normal peaks and valleys of electromagnetic readings went flat. A revolving crew was appointed to stay around the clock in a cabin near the shack.

Sure enough, enemy aircraft showed up while there was still heavy ice on Point Lake—two on the first day. When the crew heard the drone of engines, they ran out, fired up the generator, and ducked; the airplanes came in low and fast. When the craft passed, the crew howled with laughter; they could imagine the looks on faces up there as technicians tried to figure out what had gone wrong. The enemy obviously figured it out fast, for the same aircraft came over repeatedly, sometimes skipping a day or two in apparent hope of catching them snoozing. Spy helicopters began showing up, too. One came down and hovered so close to Ashley and his crew they could see someone in the front seat videotaping them. It was a Monopros man. They screamed obscenities, stuck up their middle fingers, and dropped their pants to show their bare asses to Johannesburg.

One most persistent plane came from Geoterrex, an Ontario geophysical contractor hired by the more minor companies. The captain, Alan Capyk, overflew Point Lake on five straight days in early June, but they beat him to the generator every time. He decided on a ruse. On a Saturday, he and his crew returned to Yellowknife, hit the Gold Range, and let it be known they were giving up and flying home. They trusted some BHP spy in the room would radio this news back to Point Lake, and they were right. When Ashley heard Geoterrex was pulling out, he let down his guard. Also, even on the tundra most folks tend to sleep extra hard and late on Sunday mornings.

At 5 A.M. Sunday Capyk zoomed in at 134 miles an hour over Point Lake. The generator was off and no one was visible on the ground. As the plane came over, beautiful peaks and valleys spiked on its instruments. They needed more data, so Capyk banked sharply and came back the other way. As he was overhead getting more readings, he saw the door to the guard-cabin door fly open and Ray Ashley stumble across the melting snow barefoot, in a pair of white longjohns. Two other figures pushed past him, pulling on pants and jackets. Capyk laughed hard, turned, and made a third pass.

The guards had reached the generator shack and were yanking the flywheel cord, but nothing was happening. "Shit!" screamed Ashley. "Out of gas!" While they fumbled with a gerry can, Capyk came over a fourth time. Exhaust belched from the generator stack and the instruments went blank. With every possible reading they could wish, Geoterrex flew back to Yellowknife, and their clients, in triumph.

In retrospect, the Battle of Point Lake may not have been a turning point in the war. Within weeks BHP's own airborne geophysics teams spotted dozens of targets much like it. Most were under lakes. Across the borders in other domains, other airborne teams were doing the same, with or without the spy data. The targets were obvious. There appeared to be scores, maybe hundreds, of pipes.

BHP, Monopros, Aber, and others soon had drills sucking on the tundra like mosquitoes. Since many geophysical anomalies were under lakes, barges were brought in for the best-looking; the rest were saved for winter ice. Espionage, counterespionage, and dirty tricks multiplied apace.

Hugo hired a southern outfit, J. T. Thomas, to drill, but was angered when he went out to inspect. The company had enclosed the drills with plywood painted bright yellow-and-blue with the J. T. Thomas maple leaf logo. Spy helicopters could spot them miles off, and some were circling to see what they were doing. On Hugo's orders, a BHP worker chosen for artistic talent showed up at the drill pads with paint in tundra browns and greens, and repainted the sheds in camouflage patterns. A few days later, big army-surplus artillery-battery camouflage nets arrived to be thrown over fuel-drum caches and the drills.

Spy craft also hounded the samplers. Brent and Dynamite Dan were so spooked to see aircraft looking for them in the middle of nowhere that they hid among the rocks whenever they heard one coming unless they knew it was friendly.

The Point Lake loop was dismantled and Ashley's ground crew moved on to refine outlines of targets spotted by air crews. Extra drilled-up kimberlite from the pipe was left in huge cloth sacks in the care of an Inuk from Coppermine named—by coincidence, probably—John Franklin. Someone was watching. No sooner had Ashley left than a helicopter landed and four unknown men emerged to ply Franklin with aggressive questions, peer at equipment, and generally act like they owned the place in the absence of other white people. Franklin, also known by his traditional name, Kaodloak, did not argue. He called to his wife, Mercy, for his rifle. The men left in a hurry. Later that summer Mercy was alone one night when flash-lights silently came over a ridge in the dark. Consumed by an ances-tral fear of raiders, she fled. Come morning, each kimberlite sack had a neat slit, some of its ore gone. A segment of core from one drill site also vanished. A half-mile from one BHP camp a helicopter landed in broad daylight, and an enemy sampler got out to dig a hole. This was too much for BHP's geologist in charge, who called up his own helicopter and pursued the other craft for miles. He caught up, flew parallel, and radioed furiously for them to land. The occupants were all students hired for the summer by Monopros. They claimed, perhaps honestly, that they thought they were on their own ground. The geologist threatened a citizen's arrest, but settled for watching them dump out the sample.

Things reached a gratuitous pitch of nastiness when it came to young Eira Thomas, who was put in charge of Aber's dirt-bagging operation by her father. Eira was camped in a pup tent with her dog Thor, a huge husky-German shepherd who was supposed to protect her from bears, but instead whined piteously and tried worming into her sleeping bag whenever one wandered by. One day a single wolf trotted into camp and Thor trotted after it—the end, feared Eira, since wolves sometimes lure naive domestic canines around the bend, where the whole pack waits to tear them to pieces. The next day she was greatly relieved to get a radio call: Thor was safe at a BHP drill pad twenty miles away, identified by his tag. She offered to pick him up, but BHP refused; they wanted no Aberites near their drills. Neither would they deliver the dog; they claimed to be too busy. Instead they exiled Thor on the next supply plane to Yellowknife, where he spent a week in the dog pound before Eira could rescue him. After that the Thomases referred to their cowardly pet as "Thor the Spy Dog" and gave him special treats.

The summer of 1992 brought fast, spectacular results. BHP drilled nine new targets—and hit nine new pipes. All were within walking distance of Norm's Camp, all under lakes, and all diamondiferous. The best several were in a tight cluster about eight miles from Norm's, about a third of the way to Misery Point along the trunk esker. Aber also drilled ten holes and hit ten diamondiferous pipes. What De Beers had no one knew, but rumor had it that they and several other companies had something.

Each new BHP pipe was assigned a number and, to confuse radio-traffic eavesdroppers, a code name. The cluster of good ones were dubbed Fox, Panda, Koala, Grizzly, and Leslie. Among the poorer ones was Tri-Hump—for what a male and female geologist who worked for BHP did there one day after being left alone for a few hours, as wags at Norm's had it, but more likely for three prominent rises of nearby glacial debris.

Leslie, named for Mark's wife, got its name because Mark helped find it. While out scouting, he had stopped for lunch on a boulder, then saw a certain odd texture to it. He put aside his lunch to crack open its weathered, lichen-covered surface with a hammer. Mark had done it again; it was intact kimberlite float. He followed a trail of float boulders, some big as autos, straight to a small lake, coincidentally also targeted by geophysical readings. Lugging back a forty-pound piece of kimberlite, he temporarily pleased Chuck. But this did not last long. Soon Chuck was hollering at Mark as usual, and Mark hollered back. Some fights took place in front of the whole camp and ended in Mark's firing or quitting once again, but the two always came back to each other. Neither seemed to know what else to do.

Chuck pushed everyone as hard as ever. He sent BHP staffers to do impossible tasks and worked them impossible hours. However, unlike Brent, Mark, and the others who still worked directly for Chuck, the highly trained BHP men were not used to this. It engendered bemusement among some, as he entertained them daily with his odd combinations of the meticulous and the mercurial. One day they glanced over from Norm's and saw him at the edge of a pond in his jockey shorts, minus his glasses. Apparently he had seen a rock or some other thing that so excited him, he had turned his head too fast, and his glasses flew off into deep water. They watched in suspense as Chuck dove straight in and disappeared. Seconds later his head bobbed out, wearing the glasses. Just then a spy helicopter came over a rise and hovered. The dripping Fipke mounted a rock and held his clothes to his chest, jutting out his goatee silently in defiance under the wash.

When a white female wolf started wandering into camp, Chuck could not resist trying to make friends. He named her "Snowball" and took to setting out buckets of pork-chop scraps and other goodies. Soon Snowball was there nightly, letting Chuck photograph her in various poses. She came so close at his coaxing—"Here girl, here girl, hey, that's a nice girl"—she snatched food practically from his hand. Such tameness was strange—and considered bad form by wildlife officials. Wild animals habituated to humans often become nuisances and wind up getting shot. However, Snowball showed no aggressive tendencies.

Awave of misfortunes now descended. South of BHP, a syndicate made up of Jennings's SouthernEra, Aber, and a few other smaller outfits drilled a kimberlite core that contained five small diamonds, including a couple of yellowish "canaries"—considered fancy, and potentially very valuable. SouthernEra, the main partner, sent out a press release on a Friday morning, and all the companies' stocks soared.

Jennings was in Toronto feeling jubilant, but over the weekend he developed an uneasy feeling. He had not yet personally seen the diamonds. He had them couriered over, and when he and his right-hand man, Lee Barker, examined them, they grew alarmed. Low-value synthetic stones also could be yellow, and these had all the markings of synthetics: dodecahedral shapes; pebbly surfaces; sooty things in the middle, possibly nickel-iron powder used as nucleation points in the factory. What the hell was going on? They were not salted. The other possibility: They had fallen off the drillers' diamond-tipped bit, a not-unheard-of occurrence. They called their field camp and had the bit flown out. When it arrived in Toronto they pried out some stones for comparison. They exactly matched. The next day Jennings was forced to send out a new press release: They had screwed up. The stock of all companies involved plummeted. Gren Thomas went to the hospital with chest pains. Jennings went out of town, leaving Lee Barker to answer phone calls from angry investors.

By the end of August the BHP encampment also was swept up in the joy that comes before a fall. Ashley and the others were working eighteen hours a day—and loving it. Though Chuck drove them hard, each day brought a new discovery. It was virgin land, with nearly every question remaining to be answered—paradise, for a scientist. When they were not out probing with instruments, they were working up new computer programs and ordering hardware to collate their ever vaster quantities of data into ever fancier color maps. Targets were everywhere—so many that the drillers were continuously on the move from one to the next. They had not missed hitting kimberlite once.

By August 20, 1992, things were winding down for the summer. Some crews were leaving. There was only one regular pilot left at BHP, a friendly forty-ish man named Howard Damaron. On August 24, Damaron was slinging drill rods from the so-called Willy Nilly Pipe to the next site. This involved hovering over the drill pad and lowering a steel line forty or fifty feet down from the craft's belly. Men below grabbed the end and hitched it to a bundle of pipe. Then Damaron rose slowly and headed over the horizon to the new place with the pipe on the line, and another crew unhooked the load as he hovered again. Generally, one winched the line back up and headed off for another load. Damaron seemed a little nervous despite his solid experience—he was not used to keeping his bearings in such wide-open spaces—but he was a nice fellow, and everyone quickly adopted him as part of the family.

In late afternoon Ashley and the drillers were at Willy Nilly when they heard the whop-whop-whop of Howard returning for another load. A moment later the helicopter appeared a half-mile off, on the far edge of a rise. Then the sound stopped. That was what they all remembered later—the sudden silence. They fixed their eyes on the machine, which remained in the air. One door flew off. Then the top blades. Then the tail boom. It was like a series of soundless explosions. Now the only thing left floating was the bubble—the part with Howard inside. The bubble fell, fell, fell, in slow motion, and disappeared behind the hill.

They all looked at each other, then bolted straight through a shallow lake, over the rough ground, and up the rise. When they reached the top they could not quite figure out what they were looking at. On the ground was a scattering of thumbnail-size pieces of plastic and metal. The biggest objects were: Howard's seat cushion, untouched; his headphones, untouched; and his jacket, without a rip and containing the cigarettes he always kept in his breast pocket. There was only a slight bend in one cigarette. The only thing missing was Howard. Could it be that a miracle had taken place, and he had landed somewhere amid the soft tussocks, merely stunned?

Frantically they ran to and fro. No sign of Howard was seen anywhere. Then they found what appeared to be the biggest piece of wreckage: the engine, three feet by three feet. Someone saw Howard's watch next to it. The watch was attached to something. It took them a minute to realize it was a small, detached piece of Howard. There was only one place the rest could be: under the engine. No miracle had occurred.

Several drillers sat and wept in sorrow, while the rest of the men stared off into the distance. They, the other drill crew, plus a half-dozen lone samplers awaiting Howard's pickup, were all stranded, with no way to even call for help. Chuck—back in Kelowna at the moment—was so afraid of eavesdroppers, he had forbidden drillers to have radios. Howard had been the one with the radio.

At Norm's Camp it was hours before anyone realized something was wrong. It was cold, dark, and near midnight when Aber received a radio call from them about a missing helicopter. BHP had been reluctant to ask for help, but now they were desperate. Lee Barker, in charge at the moment, did not hesitate. He threw food and coffee into Aber's helicopter and, following directions on Howard's probable route, found Willy Nilly at 3:30 A.M. The crew had retreated there to make a fire with leftover diesel poured into a depression on top of a boulder. They looked to be in shock, so Barker volunteered to send them out first and stay behind. Ashley refused; the Willy Nilly cores still lay in their boxes. "We can't leave you alone with our core," he said blankly. Barker stared, turned purplish and sputtered: "We're not here to do that. How can you think that at a time like this?" Ashley stammered incoherently. Suddenly he felt deeply ashamed. He did not know how he could think that at a time like this.

The BHP men were evacuated first in batches, but Barker was not left alone with the core at any time. By afternoon an RCMP officer and a federal aviation official had flown in to investigate. As far as they could figure, Howard had saved time by leaving the other drill pad with his empty sling line still hanging down—not the safest procedure, but common in these days when everyone wanted things moved at top speed. It was surmised that when Howard was near Willy Nilly, he had ascended to get a visual on it, then descended a little too fast while moving forward. The line swinging from his belly must have arced back into the tail rotor. He never had a chance.

The BHP men did not want to watch what came next, so it fell to Barker to help the RCMP officer and the aviation official turn over the heavy helicopter engine. Under it was Howard Damaron's remains—fifty pounds of torso, mostly burned meat and bone. Barker took a sampler's shovel, and as respectfully and gently as he could, shoveled the fallen pilot into a black plastic bag. A bit of Howard's pants was left, and out of the pocket fell a charred loonie— a Canadian dollar coin with a loon on the front. Barker picked up the loonie in his fingers and put it also into the black plastic bag.

Ashley fell into a deep depression; this no longer seemed like such a pure, joyful adventure. He and his crew hammered a bronze memorial plaque onto a boulder at the crash site. Then, like ancient Dene marchers, they left the dead behind.

Winter was usually time for retreat, but the staking rush had proved it possible to operate, as long as you kept warm and did not get lost. After a brief pause to let the lakes freeze solid, Chuck and Hugo ordered everyone back out. The four most promising pipes— Leslie, Fox, Koala, and Panda—were to get intermediate drill tests. Work was aided by extension of the winter road from the old Tundra Gold Mine, to the south. A huge fleet of heavy equipment was brought up to plow a path over frozen lakes and smooth a few spots on the frozen land so trucks could pass. Within weeks the temporary path arrived at Norm's. Chuck took advantage to have a scrounged-up vacation trailer hauled in, which he and Mark swathed completely with insulation and plastic for themselves. Visitors could not figure out how father and son avoided suffocating together inside.

Traveling and working were still dangerous. On clear days, ground geophysicists bundled up and went by aircraft or snowmobile to delineate pipes, but often fog and snow rolled in around them. It always cleared in time for them to be rescued, but everyone feared one day it would not. The drillers worked in heated plywood shacks, but they had to go outside to get fuel or assemble equipment. Exposed skin would freeze instantly, but it was practically impossible to attach bolts or do similar tasks with heavy mitts. Rubber and steel grew brittle and snapped in the cold. Several times drillers stranded by whiteouts were saved only by Kaodloak; GPS was not exacting enough to find something as small as a drill shack. Kaodloak climbed on a snowmobile, hooked up a sled carrying needed supplies, and told a convoy of white men to follow his taillight. He went to the spot every time. When they asked how he did it, he did not seem to understand the question. "I just go straight," he replied.

Mark and Ashley dreamed up a solution to the travel dilemma: a portable camp. It was simple, really: They ordered a big tracked vehicle like the ones that groom ski slopes, and a train of big sleds with heavy-duty tents on top—enough for a small lab, sleeping quarters, kitchen, generators, fuel, even a shower. Bristling with antennae for GPS, radios, and scientific instruments, it cost $1 million. It was hauled up the winter road, and soon they were moving up to thirty-five miles a day across the frozen surface, climbing out on targets at their leisure with handheld instruments.

With the big drills now turning around the clock over Koala, Panda, Leslie, and Fox, southern diamond-stock analysts were getting increasingly worked up. Some were already predicting BHP would open one or more mines by 1998.

Before the winter road collapsed, hundreds of tons of ore were trucked to Colorado. Then in June, with ice breakup pending, the intermediate grades were published. They were stunning. All four pipes looked like mines. One held over a carat per ton, with nearly a third of the stones judged gem quality. BHP cautioned that further appraisal was required, then true bulk samples from underground shafts. Dia Met hit $67 a share. Chuck and Stew were no longer just multimillionaires; their combined worth jumped past $1 billion.

In 1993 others rode the heightening wave. Jennings and SouthernEra recovered from the canary-diamond debacle. Jennings was still one of the country's main experts on diamond prospecting, and the only one willing, nay, anxious, to share his expertise. He emerged as lead promoter and oracle for the whole rush. Jennings's energy now was unbelievable. Besides running 5 million acres of joint ventures, he was constantly on the road barnstorming for Northwest Territories diamonds like a political candidate. He gave slide shows to hotel-ballroom audiences, interviews to the *Financial Times*, and five-hour private dinners to select stock analysts, during which he loaded them with grading statistics and prospecting science. With his help, analysts' reports grew into booklets, complete with graphs, charts, and obscure academic references. Jennings presented the strike as a neck-and-neck race. He predicted a minimum of five mines among the competitors; Lac de Gras was "the best bet in the world," he said. The stock pickers responded. One Jennings tutee, an analyst from California, was spotted in the locker room of the San Francisco 49ers discussing how young football players might multiply their newfound income with diamond stocks.

Jennings himself was involved in the season's hottest prospect, the Tli Kwi Cho Pipe. Hit in March 1993, it was owned by Southern-Era, Aber, and five other companies. Kennecott, the senior partner, controlled the pipe and did the actual work. The pipe—actually two bulging lobes joined to each other like Siamese twins—had showed up on airborne geophysics, the electromagnetic patterns standing out especially well—"like dog's balls," in the technicians' lingo. It was named Tli Kwi Cho, a Dene phrase, because one partner, George Stewart (the "Pa" of Ma and Pa Kettle), decided they needed a spicy moniker to boost public interest and queried Joe Rabesca, chief of the Dogrib, how to say "dog's balls" in his dialect. "*Tli kwi cho*," said Rabesca. It was sometimes more delicately translated in press releases as "sack for bullets" or "bag of jewels." The truth was, Rabesca might have felt offense at being asked to supply an off-color name for a *kwet'i* prospect. Unknown to anyone but the Dogrib, the syndicate had actually named its pipe "dog kennel."

During 1993 Kennecott drilled three dozen small holes in Tli Kwi Cho. The more they drilled, the better it looked; it had high diamond counts, up to 50 percent gem quality. Analysts compared it to Point Lake, then to South Africa and Russia, where similar multiple-lobed pipes were worth tens of billions. With each new announcement Tli Kwi Cho threatened to dwarf BHP—a bitter draught for Hugo, since it happened to lie on the 75,000 acres he had traded to Brian Weir during their rum breakfast the previous year. The Tli Kwi Cho partners were so confident, they announced they were skipping intermediate drilling and going straight in with a spiral-shaped underground shaft for 5,000 tons.

Ed Schiller, still in charge of the competing Tanqueray Resources, whipped anticipation even further by pointing out something no one had noticed: All the promising pipes so far formed a northwest-southeast line, coinciding with the so-called Mackenzie dike swarm. This eruption had cracked the bedrock through BHP and adjoining lands 1.27 billion years before, theoretically creating a single large weakness through which the cluster had emerged. Schiller predicted all the good finds would be made along this narrow trend. He dubbed it the "Corridor of Hope."

The Corridor quickly became a hit, a sort of geologic advertising jingle. Soon every analyst's report buzzed about the Corridor of Hope. Companies whose land lay along it obtained special cachet, and their prices went up. Coincidentally, these included Tanqueray, which had just found a promising four-pipe group along the Corridor. Shortly Schiller outdid everyone by signing up directly with the most impressive joint partner so far: De Beers Consolidated Mines.

The Corridor of Hope idea might have had some scientific merit, but Chuck viewed it as so much voodoo. To him, Schiller and Jennings were sideshow ballyhooers riding his coattails. By the end of the wild summer of 1993, BHP had found twenty-six pipes. Outsiders did not know it, but many were nowhere near the Corridor. More diamond counts were coming in all the time from drill holes, and BHP made plans for its own underground shafts into several pipes in its central group—the final phase before deciding whether to mine.

Meanwhile, Chuck seemed to be buying into his own voodoo. Southern newspapers, magazines, and TV crews were pouring in to cover the story. And whatever publicity Jennings and Schiller generated, however many people Chuck had worked with in the past, and despite the hundreds of scientists and billions in capital now fueling the strike, it was clear the show needed a star, not an ensemble cast. Only months before, Chuck had been an obscure, strange little man, probably thought by women not very attractive, definitely pitied by men for his hopeless quest. Now he was a hero.

The *Northern Miner* was first to hail "the hard work, single-minded dedication and the remarkable persistence of geologist Charles Fipke." The paper gave the first abbreviated, public version of the story, starting with Blackwater Lake, briefly mentioning Hugo, Wayne, Stew, and Gurney. Then in 1992 the *Northern Miner* made Chuck its "Mining Man of the Year"—the first award he had ever received. Mainstream reporters rushed in for more detail, including the *New York Times* and *Los Angeles Times*. The *Wall Street Journal* ran the same fine, flattering pen-and-ink drawing of Chuck repeatedly. Even *GQ*, the high-toned American men's magazine, sent a reporter up to meet Chuck, though for some reason the story never ran.

With each new article, the Fipke mythology grew. Chuck did his best to oblige listeners by telling the story over and over, but he made a fidgety, sometimes incomprehensible interviewee. He left out all sorts of details he could not remember and good-naturedly invented a few harmless ones to fill in blanks. The reporters had no time for subtleties anyway. The general impression was that Chuck had single-handedly camped for a decade in the Barrens, being chased by bears and wolves. Some believed it was all done on foot; even some prominent geologists came to believe that Chuck had carried his samples out hundreds of miles in a rucksack. With each retelling, other players faded and Chuck got bigger. His eccentricities, of course, only made him a better character. The Sunday *Financial Post Magazine* accompanied him to the tundra for some supposed prospecting but instead Chuck got distracted by a nice blueberry patch and spent a good twenty minutes on all fours shoving blueberries into his mouth, then a mushroom the size of a soup can. "Delicious, best mushroom I ever ate," he pronounced while the reporter and photographer watched for signs of poisoning. Chuck was invited to speak at a big scientific meeting in Toronto, but at the microphone he stuttered so badly, people winced in pain.

As the strike widened, organizations gave Chuck more awards with ever grander citations. Don Brown, senior mineral adviser to the Canadian government, proclaimed Chuck "the man who finally opened the north." The *Canadian Mining Journal* declared that "his achievement . . . has literally sustained exploration in this country." One government-lawyer-turned-author began work on a biography, *Fire into Ice*—the tale of how one man's flaming passion was forged into shining jewels in the crucible of the Arctic. Chuck was at first flattered, but he soon got nervous about what the author, Vernon Frolick, was going to say. He told Frolick he wanted veto power over the material—and got it. Movie producers inquired about feature films based on Chuck's life, but Chuck said he was too busy to talk to them.

Eventually the American actor Alan Alda, star of the TV show *M*A*S*H*, was sent up to narrate a segment for the TV series *Scientific American Frontiers*. In it, Chuck could be seen showing Alda how to use a prospector's pan in the tundra while Mackenzie whizzed by in his plane, and working alongside Marlene in the lab over the once top-secret maps. After airing, excerpts ended up as a continuous loop at a De Beers–funded diamond exhibition in the American Museum of Natural History in New York City.

It was not long before the fame and wealth began to backfire, first in small ways, then in big ones.

Chuck was no less vain than anyone else. He tried hiding his age from journalists, but this was easily gotten around by calling his secretary, Jennifer, who kept a photocopy of his driver's license so it could be replaced on the frequent occasions that he lost it. Jennifer, a friendly soul, just read off the birthdate: July 22, 1946. The reporters variously described Chuck as "balding," "stubby," or, in the case of the *Wall Street Journal*, "stumpy . . . with muscular forearms." No one meant any harm, but Chuck felt hurt. Especially when it turned out that Brent read the *Wall Street Journal*. After that, the samplers gleefully addressed Chuck as "Stumpy" whenever they felt like tormenting him.

They were madder than ever at him, and this time they stayed that way. They had done most of the stoop labor for pitiful day wages, always with the idea that they would share in the riches. Now the riches were here, and Chuck was not distributing one cent. Many who had bought Dia Met had sold out when it merely doubled or tripled, netting a few hundred dollars. Only a few made enough to buy a new car or boat. Derkson was dumbfounded; the only diamonds he had ever got were the nine minuscule ones on his wife's new engagement ring, bought with $2,600 earned from his last tundra trip. At this point Chuck would not even give him a full-time job with benefits. In disgust Derkson quit and went back to college to become a junior high school teacher. At Christmas, Brent, who could quote Chuck's promises going back ten years, received a commemorative Dia Met T-shirt for his efforts—and when he collected his next paycheck, discovered the cost had been deducted. Chuck could not understand why Brent then quit for a far better-paying job directly with BHP—fittingly, as manager of safety. Norm, who never got rich either, began slacking off at work in apparent silent protest. When things got slow, Chuck laid him off.

Dynamite Dan was the only one to do well, because of his full-body plunge into the stock market. Even so, he kept working until Chuck fired him permanently, after Dan pocketed some loose kimberlite cuttings of debatable worth lying near a drill. Dan said he had only wanted to use the indicator minerals in sculptures. He could not have kept at it much longer anyway: Only in his twenties, his spine was crippled from years of bending, digging, and sieving. He paid off the mortgage on his house and founded a New Age educational foundation in the States, which quickly drained his funds.

Jealousies were now spreading like mildew on all sides. Stew had almost as much money as Chuck, but not one-tenth the publicity. Stew initially liked it that way, but changed his mind when Mother, now up in her eighties, happened to see the Alan Alda program. Alda spoke of "one man's vision." At Edith Blusson's age, she had a hard time remembering everything, but she knew Stew had somehow been in on this discovery. She called. "I thought you had something to do with that big diamond business up north," she said. Stew assured her he did. "Well, there was nothing on TV about you," she replied. Stew was horrified; did his own mother doubt his role? She had never wanted him to be a geologist anyway; his success was tainted if he could not prove he had earned it. Marilyn saw this, grabbed up every news article, and seeing her husband mentioned almost nowhere, began believing Chuck was manipulating the press. "That little bugger," she muttered. "He's taking all the credit."

Presently the articles started going into too much detail about Chuck. Several described with relish the sneaky procedures he had used to keep everything secret, even from shareholders, before Point Lake. One group found this highly unentertaining: First Exploration, the mutual fund that had once owned 1,672,348 shares of Dia Met—until they agreed to sell it all back for 25¢ to 32¢ a share to Norm's Manufacturing or its designates, just before the strike. The designated buyers had supposedly included a "Mr. X," an anonymous company outsider willing to take the stock off First Exploration's hands. At the end of 1993 the fund filed a huge lawsuit against Chuck, Wayne, Mackenzie, C. F Minerals, Dia Met Minerals, and Norm's Manufacturing and Geoservices Ltd., alleging the defendants, all insiders, had "agreed to actively conceal and suppress their exploration results" so they and friends could buy back the stock themselves. If it was true, the results now were obvious. Well over $100 million was at stake, plus punitive damages. Everyone was dragged in for endless depositions, and lawyers on both sides started collecting huge sums.

Then Mackenzie filed a lawsuit. This was against Chuck, Marlene, and C. F. Minerals, seeking the same 25,000 Dia Met shares that Chuck and Wayne had been warring over when Chuck assaulted Wayne's crotch. Beyond this, Mackenzie was angry on a deeper level. He had provided so much money and risked his life so often, he believed Chuck would be nothing without him. Chuck, he said, had always been in it only for himself. When Chuck received the legal papers, he was stunned. He called Mackenzie.

"Aren't you rich enough?" Chuck asked.

"Chuck, ask yourself the same question," said Mackenzie, and hung up.

Wayne now reentered the fray, recruiting Mackenzie and eleven other major shareholders for a class action against both Chuck and Stew. This would allege the two had unfairly diluted down the other shareholders when they made the BHP deal because they had assigned themselves such huge personal interests in addition to their Dia Met shares. Among others, Ed Schiller and Bill Shemley were ready to join because, they said, Chuck had kept them so poorly informed early on. Some other Kelowna investors refused; they were happy to retire to their big, new houses along the lake—especially after Chuck hired a ferocious Toronto litigator to threaten them with a countersuit if they dared sue.

Stew found himself sued separately by Pioneer Metals, the company that had diverted his attention so much when Chuck went to the Barrens. Pioneer said Stew had been bound by his directorship of their company to give them an opportunity to invest in the project. Stew countersued, his lawyers saying pretty much what Chuck's lawyers said about Chuck's assailants: They were greedy failures who made the wrong decisions, and now wanted to pounce on the money in retrospect.

Then there was the worst lawsuit of all: Stew's action against Chuck. This hinged on a last-minute clause in the BHP deal that gave Stew rights to earn a major interest in the "buffer claims"—the 500,000-some acres BHP had staked around the core—if Stew spent a specified amount of money prospecting there himself. Stew had insisted on this at the last moment, and the clause was a bit murky. Now, still hurt that everyone had circumvented him, he wanted the extra ground. He went to a lawyer and tried getting a decree without a lawsuit, but the lawyer insisted that suing Chuck and Dia Met was the only route. A few days after the papers were served, Stew ran into Chuck coming out the door of a Vancouver stockbroker's office. The two circled each other warily, like surprised animals that could not decide whether to attack or run.

"Stew, why are you suing me?" Chuck finally asked in a sorrowful whine.

"I don't know, Chuck. It's just the procedure," said Stew lamely. They looked at each other but did not know what to say, so they both turned slowly in opposite directions and walked away.

About the only people who did not sue were Stew's original Blackwater Lake partners, who had long ago sold out to him. Two of the three were dead—stomach cancer and heart attack, respectively—and the third apparently believed Stew had acted in good faith.

The *Financial Post Magazine* reporter who accompanied Chuck for his mushroom-eating expedition helped inflict an extra-special piece of damage. She nosed into everything going on at Norm's Camp, including the fact that it held thirty-one men and one woman, an attractive young Scandinavian named Lela Petersen. Petersen, the *Post* reported, was Chuck's new "assistant." It described Chuck poring over maps while she "assisted" at midafternoon by shifting sleepily within a bunk inches from Chuck's drafting table. The magazine carried a picture of Lela sitting sideways in an open truck cab, about 90 percent Lycra, bare legs and blond hair, sporting untied hiking boots and what appeared to be a man's borrowed coat. Stew was among the thirty-one males in camp who could not figure out what she was doing there. They nicknamed her "Tinkerbell" and "Buttercup" and snickered behind her back. As soon as Chuck left, the BHP geologist in charge fired her.

Back in Kelowna, Marlene had bought a new house down by the lake, six blocks from their old one. It was a lovely, rambling affair with cedar roofs and many species of decorative trees. A story later circulated around town that on the Sunday morning the *Financial Post Magazine* came out, Chuck drove wildly around buying every copy so Marlene would not see it. He forgot they got home delivery. When he got to the new house, Marlene was waiting at the door with the paper, open to the story.

Thereafter the house was tense, and friends heard there were horrible fights going on inside. Perhaps by pure coincidence, the place sat almost on the site of the farmstead once owned by Percy Williamson and his wife. The Williamsons had long ago moved away, after their only son was killed while driving his Ferrari too fast somewhere outside Vancouver. The son was buried on a hillside high over Kelowna, and the farmstead split and built over.

At the same time Chuck's marriage appeared to be falling apart, he and Mark had their final falling out. Mark had been working almost nonstop up north for years. Finally he decided Chuck was just being too mean, and he quit for good. He hoped BHP would give him a managerial position. But BHP had enough people now, and he wound up without any job at all. When he came home, he announced that he and Leslie were getting divorced.

"Geez, how come you're splitting up? I thought you were getting along great," wheezed Chuck.

"Well, you know, Dad, in the last year you gave me less than two weeks off with my family," replied Mark. The implication was that it was all Chuck's fault. Chuck was crushed. He felt even worse when he learned Mark was remarrying—to an attractive young Dutchwoman named Marja, whom Chuck had met while on a trip to Europe and hired as a diamond sorter for the Colorado processing plant. Chuck said he did not think it was a good match and refused to attend the wedding. With financial help from Marlene, the newlyweds obtained a horse farm on the outskirts of town, and there hid out from Chuck and most everyone in the narrowing circle connected with him.

Jennings and others were hale and happy. Cold weather came again, and the calendar rolled toward 1994. The underground shaft into Tli Kwi Cho was rapidly spiraled down to 350 feet, and the ore was trucked over the winter road to a new processing plant Kennecott had built in Yellowknife. The many diamonds washed out of it were sent to the European appraisers. Tanqueray was quietly testing its own pipes in conjunction with De Beers, and Ed Schiller was riding high. Speculation mounted. Jennings, constantly quoted in the press, was enjoying the whole thing immensely. He continued touring around, giving speeches, making videos. He now wore a bright red sweatshirt reading "DIAMONDS NORTHWEST TERRITORIES '94—THE RACE IS ON" everywhere, even when he was in the tundra.

Finally the big day came: Kennecott got the appraisal results and was ready to announce them. Late one afternoon after the stock markets had closed, Jennings and Gren Thomas filed into a conference room at Kennecott's Vancouver headquarters, along with a half-dozen smaller partners. They seated themselves around a long table and coffee was served. Everyone was smiling and chatting amiably when John Stephenson, Kennecott's commanding executive, strode in through a private side door trailed by an assistant carrying a pile of ringed binders. Stephenson sat at the head of the table and everyone shut up.

"Right," snapped Stephenson. "I've got the results here, and I'm going to tell you straight off. The diamonds are not that good, and our feeling is that the values will not justify our continuing with this project."

There was silence for about thirty seconds. Jennings turned ashen and slumped in his seat. Thomas clutched vaguely at his chest. A few mumbles could be heard. "Well," said Jennings finally, stiffening back up. "There's a new one for the books." Stephenson let his assistant pass out the ringed binders with the particulars—too many industrials instead of gems, aggregate value of only $21 a carat, and so on and so forth—but everyone was ignoring him and running for the telephone.

The following day the companies' stocks crashed again, this time as much as 90 percent. The rest of the diamond stock market was incinerated in the shock wave that followed. Investors lost confidence, and a giant selloff took place. Some $700 million in values evaporated within days. Even Dia Met lost ground.

Shortly after, intermediate drill results came back from Tanqueray's claims on the Corridor of Hope. They were dramatically poorer than initial results—so bad that De Beers exercised its option to abandon the joint venture. With ruination running through the investment community, people muttered that someone must have salted the original cores. There was no evidence of this, and plenty of evidence to the contrary. In diamond exploration, results often are not what they seem at first, and the amateurs were learning the hard way. De Beers simply moved on. It made deals with four other juniors who seemed to have something and sent a steady stream of crews to its own ground.

Then the Point Lake Pipe fell through. After the initial tests, Hugo had sent in an augerlike drill nearly a yard wide. It brought up so many tiny diamonds that geologists occasionally saw them flash in the crumbly, fist-size chunks of kimberlite. Afterward, the European gem appraisers weighed in. Like the Tli Kwi Cho diamonds, these were insufficient. Those of any real size contained too many flaws; others were good, but too small. When they added up all the figures, it was clear the stones would never cover the fantastic expense of mining them. BHP had to abandon its flagship pipe. Dia Met fell further.

It was not quite over, though. BHP was now blasting big underground shafts from the shorelines of lakes toward Fox and Panda pipes, and had airlifted an entire disassembled South African diamond–processing plant right onto the claims. For the final push, next door to the plant a 180-person camp was going up, and partly on top of an adjoining esker, a 6,000-foot airstrip—long enough to handle Boeing 737 and Hercules aircraft, the world's largest air freighters. The material came from an esker dubbed Airstrip Esker. A cobweb of roads, also built with quarried esker material, was spreading to the test shafts. BHP had to make its decision quickly to cut losses in case things did not work out. They did one more thing to improve efficiency: They laid off Chuck.

The logistical and technological problems had grown too huge. The BHP people found Chuck too disorganized, too reckless with their lives and his own. As winter came on again, Chuck told one geophysicist to go check a distant spot and met with flat refusal; the man did not like the look of the weather. Chuck then had himself dropped to do the work himself—and got stuck in a whiteout. In visibility of five feet he tried hiking back miles using a magnetic compass. He was saved only when an aircraft spotted him through a hole in the mist. He never did completely figure out how to work a hand-held GPS receiver or how to answer the fancy new satellite telephone BHP gave him. Now that Mark was gone, he could not yell at anyone else to do it. The place needed corporate management.

Chuck was eased out. During the winter of 1993–1994, Hugo imported an experienced South African mine engineer, Jaap Zwaan, to get the shafts going. Hugo appointed Rory Moore as Chuck's "co-manager." Marlene, who had always organized for everyone, obligingly helped Moore assemble arctic clothing in Kelowna and got him a plane ticket for Yellowknife. Chuck's responsibilities were taken away one by one, until eventually he ran out of things to do. When Moore became sole manager, Chuck came home to Kelowna to run the lab, await the final results, and try to sort things out with Marlene. By summer 1994, he had not been north for months.

It was then that I met him.

L ike God's servant Job, Chuck had done well for a while, then hemorrhaged riches, love, and friendship. Also like Job, he saw a sudden revival in his fortunes.

There was no one dramatic discovery, no eureka moment. Rather it was an accumulation of information, a slow march. BHP was now running ore from the underground shafts around the clock. The cataract of data grew great, the evidence powerful: The pipes were good. Panda not only held plenty of diamonds; appraisers valued them at an average $112 a carat, reputedly enough to make an Arctic operation profitable. Stones from Koala were so fine that the appraisers themselves wanted to buy them. On Misery Point, BHP unearthed another pipe near the failed Point Lake: 93-J, or Misery. It showed a fabulous 3 carats per ton, including flawless, purest-white octahedra worth over $1,000 a carat. Geophysicists kept finding new, yet-untested pipes—now nearly 100. Skeptics still insisted the Barrens would drive BHP off no matter what through expense, danger, or the good likelihood that the government would refuse permits to disturb the wilderness. Yet privately, BHP geologists said the question was not whether to mine, but which pipes first. In May 1994, BHP announced a "feasibility study"—a semiofficial statement of intent. It made no secret that it was already drafting requests for environmental permits.

Dia Met stock reascended. Everyone started saying there would be a mine. By early July 1994, several major U.S. magazines asked me to cover this apparent climax of the strike. I visited Chuck, as described at the start of this book.

When we met, he seemed untouched by all that had swirled around him. He wore a pair of stained blue pants with a clumsily sewn rip in one thigh, and drove me around town in some scrap metal on wheels— an only slightly newer replacement for his old Chevy Blazer. While waiting, he seemed at loose ends, trying to figure out what to do next. Trying to take time off for a change, he was limping around in bare feet, having just gotten out of a cast after a ski accident. He volunteered that he was still in big trouble with Marlene, but they were working on it. He was thinking of building a treehouse for his youngest son, Ryan, whom he hardly knew because of the years of overwork.

After a few days we headed to Yellowknife to visit the claims. Stew was in town, organizing Archon's next moves. I met him too—separately. He was dressed in the same rags as Chuck, one toe peeping out a hole in a worn leather moccasin. He drove a faded 1979 Mustang with part of the steering wheel missing. He seemed unquiet, tense, a man looking for the next big thing.

After spending time with Stew, I rejoined Chuck and a small crew of BHP workers for the flight to the claims. An hour and a half later the awesome new complex of roads, sheds, and diggings loomed out of the emptiness. It was all allowed by environmental authorities under the rubric of "exploration," but it sure looked like mining to me.

Over the next days we visited far-flung samplers, surveyors, drillers, and geologists, using the only big toy Chuck had so far acquired: an A-Star helicopter, with its own pilot. Chuck alternated between silent brooding and trading wisecracks with everyone on site, laughing his characteristic loud laugh. At night he offered to have the A-Star lift me to any lake I liked, where I could fish for monstrous trout without competition. Clearly he had time and money for it all; everything ran without him. Some newer workers did not even know who he was.

Toward the end, we landed alongside a lake where a huge tunnel angled from shore. Workers had just broken through country rock into Fox Pipe. After a long, dark tractor ride down the shaft, we stepped off into the middle of the ancient eruption, 790 feet below the lake bed. Water droplets and indicator minerals shimmered on the walls. Sooty, raincoated miners took a few minutes to explain their procedures and pose for snapshots. One turned to Chuck.

"Are you a reporter, too?" he asked.

"Uh . . . no, no, I'm just with Dia Met," muttered Chuck, and walked off with a hammer to knock some rocks off the wall.

Afterward, we ascended to a cramped underground office to meet the shaft manager, Len DeMelt. DeMelt knew who Chuck was. DeMelt's grandfather was Ed DeMelt, one of the first Great Slave Lake gold prospectors, who had landed in summer 1918 with his bedroll and three dollars. The family had been looking for or working in mines ever since. None had ever hit it big, but they rooted for the other guys. Len now had a good-paying full-time job as long as this operation lasted. He shook Chuck's hand and slapped him on the back.

"I'm pretty pleased you found this place, Chuck," he said. "You know, people in Yellowknife used to think you were crazy."

"You know, sometimes *I* thought I was crazy," said Chuck. He laughed loud and long. "Sometimes I was *sure* I was crazy!"

CHAPTER 19

An Esker Runs Through It

On December 9, 1994, BHP applied for government permits to mine. This caused almost as much uproar as the diamonds themselves.

Others had now discovered the Barren Lands. Conservationists, some of whom previously could not locate the region on a map, suddenly saw it as the last wild frontier. The organizations ranged from a small group called Ecology North to the Sierra Club and the World Wildlife Fund. They objected not only to mines, but to the staggering scale of ongoing exploration itself. So little was known of the wildlife that no one could say what was threatened. Even before BHP sought its permits, they were lobbying government to halt everything so studies could be done.

The Dene rediscovered the Barrens. They worried about the migrating caribou, on which many still depended for meat when the animals came south. Older people especially subsisted on such "country foods," shot during community hunts. They worried about water pollution leaching from mine wastes; fishing was also still popular. They worried about looting of artifacts left by their ancestors.

Perhaps most of all, they worried they might be left out of the profits. They started building a case that they, not the government or corporations, owned Lac de Gras.

Because the Northwest Territories was not a province, people there got little direct say. The Territories were ruled more or less as a colony directly from Ottawa, through the Department of Indian Affairs and Northern Development. The name said something about the government's attitude. However, some territorial wildlife officials wanted to discover what was out there before it disappeared. Two biologists from the Territories' Department of Renewable Resources, Ray Case and Steve Matthews, got themselves some government funds and set up a research station amid the diamond claims the first summer I visited—three all-weather tents along big Daring Lake, in the southern lee of the trunk esker.

I visited repeatedly, and it could not have been a more spectacular study site. Just over the ridge from Daring lay Yamba Lake, where Mark Fipke had almost drowned, dotted with odd pingo-like islands. A few hundred yards west along the esker was a caribou crossing, like that above Misery Point. Here the esker sloped and broke, and the waters of Yamba rushed into Daring. Sometimes you could see thousands of caribou follow the esker top to the water's edge at the narrows. They hesitated, then plunged in to swim across in long lines, galumphing out on the far beach to continue on. As a result, the narrows was frequented by a halo of wolves, bears, and wolverines. Peregrine falcons and bald eagles flew nearby. When a person stood on the esker above the narrows and viewed the multitudes of colors in the skies, the lakes, the empty valleys and hills—then saw the caribou approach—that person knew he was in a very powerful place. On chilly late-summer nights when black clouds flowed like rapids over the moon and wolf howls ripped the air, the narrows looked like the entrance to hell. There was no sign of humanity. Only the abstraction of maps showed it to lie on Monopros land, just over the border from BHP.

Samuel Hearne's detailed notes on tundra birds and mammals provided good background reading here—in fact, almost the only background reading. Hardly any known person had been through

since. Biologists had recorded only caribou migrations with any reg-
ularity, and even this was sketchy. As far as anyone could tell, five separate herds crossed distinct regions each year. The one haunting diamond territory was known as the Bathurst herd. Roughly 350,000
strong, this group was believed to reach its northernmost point in
June near coastal Bathurst Inlet, a lifeless region bounded by high sea
cliffs and steep river gorges. Females gave birth there, then wandered
back to the trees by September, leading their calves. What they did
in between—and why they made the awful trip at all—were mysteries. Some speculated it kept them from overgrazing the southern
ranges; others, that it shook them of predators so they could reproduce in peace. The Dene had a saying: "No one knows the ways of
the wind or the caribou."

Wolves had their own devil's pact with the land. With caribou
their main food, they survived only by following, but could not
always do that. To reproduce, they were obliged to dig dens to protect cubs from eagles, wolverines, and bears and to stay put for up to
three months—right in the middle of the caribou's move north. This
could strand wolves far from their movable prey, so they were faced
with picking the least bad spot. In the early stages of the diamond
strike, one researcher discovered hundreds denning just past the tree
line. Some appeared to travel up to 100 miles to kill and gorge, then
return to vomit up meat for the others. But the trip could obviously
be too far; sometimes they starved to death. How many took their
chances at dens in the deeper tundra, no one knew.

The Daring Lake researchers quickly made startling discoveries.
For one, there were more animals out here in summer than anyone
had realized. For another, the eskers, made of well-drained, easily
minable sand and gravel and thus targeted for BHP's roads, airstrips,
and building pads, were a major habitat. Aerial photos showed they
covered only 1.5 percent of the surface, but most creatures seemed to
have some vital relation.

The first suggestion came from birds. Many species cross the
snowy Barrens in early spring to nest on the northern coast. Biologists studying these sites noticed geese and cranes arriving with
bills and tongues stained bright blue. Apparently the birds feasted
on frozen blueberries left over from the previous year, exposed on

windblown esker tops. It seemed possible that this single resource might make the difference that allowed some to survive and reproduce.

For ground creatures, eskers were home. At the outset of the Daring Lake operation Fritz Mueller, a biologist working with Matthews and Case, toured the region via helicopter for fifty miles around and spotted dozens of wolf dens, plus countless fox and ground-squirrel holes. Almost all were dug into eskers or eskerlike deposits. The reason was obvious: The rest of the land was rocks, permafrost, and water. Eskers were the only good place to dig. Wolves were pickiest, choosing den sites with expansive views so they could see what was coming, adjoined by dips and rises to shield their own presence. Further, they preferred sites with plants on top; the roots held up the sandy roofs of their human shoulder–width tunnels. They returned yearly to the same few good sites, developing whole complexes. Most were surrounded by big open-air cemeteries of caribou femurs, tibias, vertebrae, and sacra, which radiocarbon dating showed could be 700 years old. These were ancestral abodes.

Several miles from camp was one ravishingly beautiful wolf valley, hidden within a tangle of esker branches and low hills. From a sand mound, the wolves could spy the glimmer of Daring Lake down a long, boulder-strewn ravine. The den entrances were invisible even a few feet off; the only giveaway was a lush garden of purple-flowered fireweed spears rooted in the disturbed soil, fertilized by generations of wolf scat. Mueller quickly learned to look for purple fireweed and had himself landed some distance off, then walked over to investigate. It all looked unoccupied. As he was contemplating a belly-crawl in one hole to see what it looked like, a pure white female walked out literally under his nose. Seconds later, five tiny pups emerged and began swarming, tumbling and playing at his feet. The adult barked once, the little ones tumbled back in, and she trotted off to hunt. They had never seen humans. That would make them easy to exterminate, thought Mueller.

Mueller showed me the place through binoculars, and one summer day I returned alone. Again, it seemed empty. I also wanted a look inside. After waiting a while and seeing nothing, I put my face to the entrance and respectfully asked if anyone was home. No response. I swallowed my doubts and threw myself in head-first. I was startled to find light streaming over my shoulder; the hole was cunningly angled to catch maximum sun. More luminescence filtered in from back doors in two branching tunnels a dozen feet back. The smell of earth was mixed with something pleasant—not exactly animal, not really vegetable. No bugs, no wind. Delicate rootlets in the ceiling caressed my head, and cool, talc-like sand cushioned me beneath. It was cleaner than my own kitchen. There was nothing more. Some part of me had expected to discover a diamond on the floor. I backed out, feeling guilty; I had intruded.

Grizzlies also used the eskers, though not as consistently or cleverly. Their hibernation dens were unimpressive bear-size hollows scooped out in any south-facing spot. However, their travels were quite impressive. The Daring researchers shot dozens with drugged darts and fitted them with radio collars. Signals showed some covered at least 4,000 square miles in a few months of activity. Females in Yellowstone National Park would cover, say, 200. The barren ground grizzlies were hunting food a lot harder, so they had to go a lot farther. The researchers analyzed scat and followed bears at a distance to discover that esker berries and roots provided much nourishment. But when caribou were near, the diet could run to 90 percent meat—sometimes a calf every day. The researchers also found bear-cub bones, claws, and hair in adult scat; they were eating each other's babies. People were right to fear them.

With all this, the more biologists saw, the less they seemed to know. Up on the shoulders of the narrows lay several enormous boulders—bear rocks, the researchers called them. Before each were a dozen deep, permanent bearfoot-size tracks worn into the ground. It appeared that generations of grizzlies had approached, ritualistically placing their paws in the prints each time. The bear rock was covered with ancient lichens, except for the corner to which the footprints led, which was always scraped clean. Maybe grizzlies rubbed themselves as they claw trees in lower latitudes, to mark territories? No one was sure. The biologists also sighted musk oxen, thought exterminated from all but coastal islands. Another team studying wolverines easily spotted dozens; they are usually rare even in the best of circumstances. One was seen sitting atop a rock ledge over a caribou trail, waiting, until a caribou many times its size came along.

Then it leapt down and rode the animal bareback, gradually reposi-
tioning itself under the throat with teeth and claws. A half mile on,
the caribou dropped. The biologists did not consider this normal;
wolverines were supposed to scavenge bodies, not create them.

"A year ago, nobody thought about any of this," Fritz Mueller
said to me one night as we sat on the esker above camp watching red-
breasted Mergansers swimming around in an inlet of Yamba Lake.
"A year from now, I doubt if we will know 2 percent of what we want
to know." It seemed possible that aerial wolverine ambushes and
other unseen miracles happened all the time around Lac de Gras.
The question was whether they would keep happening if giant open-
pit mines moved in.

In June 1995, six months after BHP applied for its permits, it sub-
mitted an eight-volume Environmental Impact Statement. Company officials pointed to its height—about three feet when stacked—
and weight—sixty-four pounds—to suggest thoroughness. Cynical
environmentalists pointed out that the pages were printed on one
side only, on heavyweight paper, with plenty of white space.

BHP proposed not one mine, but five. The lakes over Panda,
Koala, Fox, Leslie, and Misery pipes, plus two others in the way, were
to be "dewatered"—that is, drained. A two-mile artificial river would
divert water so the lakes would not refill. All but Misery were situated near the now-complete Hercules airstrip. A small city would go
up there, complete with multistory dorms, an explosives factory, and
a six-story permanent processing plant covering the area of several
football fields—the largest single building in the Territories. A tank
farm for 15 million gallons of diesel fuel would run vehicles and generators for 4.4 megawatts, equivalent to Yellowknife's municipal
power grid. When it was all ready, flown-in workers on two-week
shifts would blast and dig open pits a half-mile wide each, 24 hours a
day, 365 days a year, for the next 17 to 25 years. Over a billion tons
of ore and waste rock were to be heaped into low hills, or used to fill
in one or more former lakes. Estimated produce: a coffee can of diamonds a day. Estimated profit: $8 billion to $16 billion over the
mines' life. Estimated damage: none, really.

The assessment went this way. There were 8,000 lakes on BHP's claims alone; BHP wanted only seven. Kimberlite processing produces no toxic waste. Even so, BHP promised to keep tailings from muddying the Coppermine watershed and to restore vegetation. As for the Bathurst herd, the project covered only .03 percent of its known range; the deer could walk around. As for big predators, they seemed so thinly spread that BHP hardly expected to see any. Birds might actually increase; some species could nest in mine structures. As for the eskers: Some of them would have to go. Airstrip Esker was already basically gone. The magnificent trunk esker itself conveniently ran the sixteen miles from the plant to Misery Point; BHP wanted to flatten it for trucks and mine it for side roads. They pointed out that the thirty-foot-wide roads would resemble eskers, and that perhaps when the mines were exhausted, animals would learn to use them instead. In all respects, it was presented as a temporary pinprick in a vast land. "The environmental concerns are believed to be known and mitigable with existing technology," concluded the report.

Many were unconvinced, including those who knew what was already going on. Predators, far from being thinly spread, were gravitating to camps to munch sandwiches left as bait by workers who enjoyed wildlife sightings. They liked human food so much, they began visiting the humans at home. A wolverine demolished an entire kitchen to carry away a roast beef. Workers in their bunks at Norm's Camp repeatedly heard a grizzly pawing the $7/16$-inch wallboard alongside their heads. They came outside and drove him off with flares, then rubber bullets, then a helicopter. It was well known that many animals hated low-flying aircraft—a matter of environmental concern by itself. These animals almost always came back, and soon wildlife officials at Daring were being called by radio to dart them and sling them by helicopter for release 100 miles away. They were back in days. The next step had to be killing them. And biologists now guessed there were only 1,500 grizzlies in the central tundra; shooting even a few females a year might bring an irreversible population slide. There were consistent rumors that workers had already secretly blown away several and buried them under rock piles.

Among the most persistent camp nuisances was Chuck's beloved white wolf, Snowball. BHP instituted a rule that garbage was to be incinerated, not fed to wildlife, but the incinerator did not always work, and hungry Snowball returned repeatedly to scatter leftovers and chew up any sneakers geologists left outside. One evening a pilot left the door to his Cessna 185 ajar, and Snowball shredded and shat upon the whole very expensive interior. Shortly thereafter, Snowball mysteriously disappeared following what some thought was a distant rifle shot.

Environmentalists' greatest fear, however, was not BHP itself. True, the proposal occupied only forty square miles and twenty-five years. If that was all, the wilds beyond might survive, and Lac de Gras one day could revert to its former state. But that was not all. Five pits comprised the *initial* plan. Once BHP paid off the infra-structure, it could easily build more roads to lower-grade pipes like Point Lake, or maybe find itself better ones. With 100 pipes already known, it could go on indefinitely. Monopros and the others still had camps beyond; it seemed probable that if they looked long enough, they would find something too, and if BHP were let in, it would have to be let in too. Eventually there would be a cobweb of roads and development. BHP denied any such plan; on the other hand, it had already helped identify eighteen sites for hydroelectric dams on the Coppermine and other rivers, from which wires could be strung long distances in case they needed more power some day.

The worst eventuality seemed to be an all-season road connect-ing the mines to the continental highway system. For years, in fact, ambitious northern politicians had dreamed of it: a great trans-Barrens highway bisecting the tundra from Yellowknife to Copper-mine, opening the long-untouchable metal deposits located by earlier prospectors. It had never seemed economically feasible or physically possible before, but now Lac de Gras offered wealth to pay. Then, any-one could drive in from anywhere—Edmonton, Chicago, Miami—to prospect, mine, fish, and hunt. Eventually summer tourists might drive up in recreational vehicles, and gas stations and other facilities would have to be built to ensure their survival. Underlying material for everything, of course, would have to come from eskers, cleaning them out for miles on both sides of the road.

For anyone who doubted this could happen, the lower reaches of the ice road provided a slight preview. Some Northwest Territories residents were already driving up it in moderate April weather to kill caribou within the tree line. Like old-time American buffalo hunters, some slaughtered industrially, taking only tender haunches and skins, leaving stripped, bloody, legless torsos in the snow. The slick roadside remains already had a name: "tundra seals." The Americans provided another model, too: They already had access to their part of the northern coast via the Alaska Pipeline and its parallel road to Prudhoe Bay. The road had been built in the 1970s for industry only, but public pressure opened it to all in 1991. Already, many of the feared side effects were coming to pass. As for eskers, they had already proven useful much farther south. Smaller, less dramatic ones existed as far down as Illinois and New Jersey. Much of Route 9 east out of Bangor, Maine, was actually a paved esker. By the 1990s, southerly eskers had the distinction of being the world's only endangered geologic feature: Because they were so mercilessly mined for sand and gravel, topo maps showing them were often inaccurate. They no longer existed.

"One day you have a mine. Then you have a road. Then suddenly you can stop at McDonald's along the way," declared Larry Reynolds, counsel for the Sierra Club Legal Defense Fund. A comment by Pliny, the original diamond expert, spoke to the question across 1,950 years: "To seek out gems and some little stones, we strike pits deep within the ground. Thus we pluck the very heartstrings out of her, and all to wear on our finger one gem or precious stone. How many hands are worse with digging and delving, that one joint of our finger may shine again? Surely if there were any devils or infernal spirits beneath, by this time these mines for to feed covetousness and wrath would have brought them up above ground."

As the environmentalists lobbied Ottawa officials, formed coalitions, and issued reports, the aboriginal people entered the arena in force.

Various bands had made government treaties starting in 1899, which the government said extinguished their title to traditional lands. For decades now, the Dene had said the treaties did no such thing; they had accepted yearly $5 bills, tea, and other goodies in return for not killing the Europeans—at best, for sharing the land. Up to now, the dispute focused mostly on sites well within the tree line. With the strike, it ricocheted up into the Barrens and exploded. Only the Inuit hung back from claiming Lac de Gras. They, too, had been trying to negotiate a settlement for their own lands, and just before the strike, they had concluded an agreement to take over a vast semiautonomous northern homeland, Nunavut. Its border, to take effect in 1999, more or less paralleled the ancient division between themselves and the Dene, cutting through the tundra—and just missing the diamond fields. The Inuit had to admit they had signed away their rights. They stuck to seeking assurances that the Coppermine would not carry mine wastes to them.

Among the Dene, different bands had competing claims and, within bands, disagreements about what to do. Some vowed to stop mines under any circumstances with lawsuits or civil disobedience, including blockades of the ice road. Others criticized BHP's environmental reports as shallow and incomplete, but wanted a settlement that would guarantee protection of the land and financial benefits for themselves.

The largest group was the 3,000-member Dogrib, who had expanded their range to Misery Point after the demise of the Yellowknives some seventy years before. For a generation or more, few had even passed the tree line, but memory remained. With help from local anthropologists and other allies, elders drew up computer-generated maps showing canoe routes, hunting camps, and berry-picking sites to prove their occupation of Lac de Gras. They charted ancient rock quarries and holy places, complete with long stories and the names of the spirits inhabiting them. There were hundreds of place names, all written down for the first time. Lac de Gras, they pointed out, was a translation of Ekati, or Fat Lake—named, they said, for the great caribou hunts around the Misery narrows. Yamba Lake was named for the great shaman Yamozhah, who had traveled the Barrens in a past so distant that animals and humans were the same. Yamozhah's wife was a giant beaver, said the elders. If nothing else, the sudden burst of research and documentation kept this information from dying with the old men and women who knew it. No one knew how much was gone already.

While the Dogrib were making their case, something wildly unexpected happened: The Yellowknives showed up. Anthropologists were wrong; remnant families had quietly hung on around Great Slave Lake, not extinct at all. Only about twenty elders still spoke the old dialect, and most had intermarried with the Dogrib and other groups, but diamonds gave the Yellowknives a new sense of identity. Anyone who could claim to be even part Yellowknife, about 1,000 people, banded together and claimed Lac de Gras as theirs. They efficiently generated their own canoe-route maps, oral histories, and legal arguments, overlapping with those of the Dogrib. "The fucking Dogribs haven't been up there for fifty years, and suddenly they think it's theirs," said one of the hotter-headed young Yellowknives.

Out on the diamond claims, objects from the past began appearing. Above the camp at Daring Lake (or Yamba Ti Tla Ti, Bay of Yamba Lake, as the elders identified it), biologists had walked over rounded rocks sticking from the esker for a year. Then someone noticed they formed circles. Tom Andrews, archaeologist for the territorial museum in Yellowknife, flew up. They were tent rings, once holding down caribou-skin tarps. Andrews estimated they could have been used starting 7,000 years ago by Dene or Inuit ancestors, or both. Within a short walk up and down the esker, he spotted countless hearths and chippings from stone tools. At the narrows' eastern beach, below the biggest bear rock, was a perfect white chert spearhead and other stone projectiles. Yards away were two ovals where plants grew unusually lush and green—graves? On an esker tributary jutting north into Yamba lay a bundle of long, whitened sticks—probably anchors for fur traps, hauled from the tree line. Marks made by a sharp steel ax limited their age to 200 years. On the west side of the narrows were the remains of probably the last camp—a scatter of 1940s jars and tins.

The Dene pressed BHP to hire its own archaeologists to assess matters, and BHP, sensing it had better be nice, obliged. On the esker running from the proposed main site to Misery Point lay twenty-seven ancient sites of human occupation. Some 300 others turned up elsewhere on the claims and over on Aber land. A few had already been accidentally dug up or run over. Along the shores of the targeted pipes, however, nothing was found. These sites were so bouldery that they were probably always skirted by walkers. The lakes themselves were too small for good fishing. As far as anyone knew, none had traditional names.

When I visited the Daring-Yamba sites, I could almost hear the laughter of children and the chipping of tools. In the cold and remoteness, nothing had decayed or been touched for generations.

Back in Yellowknife, I resolved to find someone still alive who remembered this place and could perhaps address its mysteries. I found Dogrib elder Suzi Rabesca. Now ninety-four, he lived with his wife outside Yellowknife in a muddy development where the boreal forest had been flattened for rows of prefabricated houses and a Catholic church. His son, Moise, was a hunting guide and entrepreneur. Moise was friendly, knowledgeable, a great talker—but admitted he had never lived on the Barrens. He was too young. "Talk to my father," he said. "Only if you have touched the land can you tell a true story."

Suzi spoke no English. When we got to the house, Moise made the introductions and translated while Suzi's wife, Sophie, her eyes blinded by age, served tea. The old man broke into a big smile when he heard I had been to Yamba and Daring lakes.

"I have never been to that exact spot, but all around it," he said. Then he began: "Between those two lakes there is a big esker"—he used the Dogrib word *wha ti*, meaning long sand—"that runs down to a narrow place. There the water is flowing through very fast. A lot of caribou cross . . ." And on with all the details of the narrows, which he supposedly never saw. I mentioned the stone projectiles on the narrows' eastern beach. "People used to wait right there in birchbark canoes to spear them," he said. "Yes, that's been a good spot for a long time." I mentioned the bear rocks. "Ah. You've been wondering? Bears hide behind those rocks and wait for a caribou to cross. When the caribou is almost on shore, the bear comes out and kills it right there in the water. Watch out!" He laughed.

Then I asked about the graves. Suzi pointed in the direction of the nearby village cemetery. "See that? You think there are a lot of people buried here, in the graveyard? Well, there are a lot more buried out there. They are everywhere. But those green spots, no. Those are not graves. That is where everybody used to dump the bones of the caribou they killed. I was a child then."

We never got around to discussing what he thought of possible diamond mines. As I left, the sightless Sophie took both my forearms and held them hard in both her hands, attentively, like a doctor assessing a patient. "Oh. You are strong. You are still growing," she pronounced, and smiled sweetly, exactly like my own Irish grandmother. Suzi took my hand and shook it for about five minutes. "Go and tell a true story," he said.

B HP courted its potentially ruinous enemies with all due vigor. The company declared itself opposed to an all-weather road from the south; for security reasons, it said, it preferred the limited access provided by aircraft and the ice road. It cracked down on wildlife encounters, prohibiting firearms and erecting electric fences to discourage predators from coming around. It explored ways to divert caribou from mine pits, including great swaths of bubble wrap on the ground. The deer were startled at first when they stepped on the wrap and the bubbles burst, but they soon figured out the wrap would not eat them and just kept coming. The esker-top road to Misery Point was dropped in favor of landfill from the already doomed Airstrip Esker, and the pipes themselves. Native guides were hired for more archaeology and wildlife studies so "traditional knowledge" could be incorporated, and elders were flown in to inspect. BHP held meetings in every community to hear concerns and ideas. It even transported a Dene delegation to the U.S. Navajo reservation, where BHP now strip-mined coal in friendly partnership with the people. This carried resonance: The Navajo are Dene who migrated south 1,000 years ago, and they speak a similar tongue. A Dene legend had predicted they would all meet again some day. Now they had. Elders did not mention that the legend also said that on the day they met, the world would end.

The Yellowknives were headed in part by a young chief, Fred Sangris, universally called Freddy. Freddy had spent years in the bush becoming a skilled hunter, but brought along books, so he was also urbane and literate in English—an unusual combination that commanded respect. He summed up the people's attitude by recalling a family story for me. One April around 1915, a tentful of Yellowknife women saw a gold prospector fresh from the south come by towing a toboggan, wearing badly ripped clothes and looking desperately cold. They shouted to him, but he failed to respond, not knowing their language. So they dragged him in by force and started stripping his clothes. The man struggled hard; he must have thought they wanted to eat him. Finally, the biggest woman sat on his chest, while the others set about sewing the rips in his clothes. They had feared he would freeze; that was what they kept trying to tell him. They fed him a hot meal and turned him loose, but were poorly repaid. Freddy named people victimized when Yellowknife was overrun by miners. These included an old woman, Liza Crookedhand, who Freddy said naively traded a gold nugget for a piece of stovepipe. An uncle of Freddy's was promised a share in a gold mine in return for land but only got $300. Where people used to hunt and fish, two dozen old mine sites were now polluted with oil, arsenic, cyanide, and radio-active tailings. People or dogs who got too near were said to lose their hair, or worse. Now the *kwet'i* wanted the Barrens.

However, Freddy was pragmatic. Having worked briefly in a gold mine—he quit after thinking about the cyanide all around him—he wanted to see if diamond mines looked any better. He got some money, went to Yellowknife airport, and flew by various connections to Kimberley, South Africa. When he got to the middle of town, he found an empty funnel a quarter-mile across, 3,330 feet deep, pene-trating countless layers of rock and ending in a lake of dead, muddy water: the Big Hole, opened 1871, abandoned 1914, toured by Chuck Fipke circa 1975. A double barbed-wire fence separated it from car dealerships, banks, skyscrapers, and old brick mine build-ings. At the lip, De Beers had a fine museum with a collection that included the De Beer family's canopied four-poster bed and three mine carts heaped with glass shards—the mock equivalent of the pipe's whole 14,504,566-carat production. Herds of tourists migrated through exhibits and on to the gift shop.

The Big Hole disturbed Freddy deeply. Still, he wondered. His people now lived in town. They needed to pay for gas, trucks, snow-mobiles, houses. If the Yellowknives ever had a copper mine, it was lost; elders had indeed spoken of a secret site in the Barrens, but they would never tell the location to Freddy or any other youngsters, for fear it would leak to Europeans. Now those elders were all dead.

Freddy's ten-year-old son, Kyle, seemed more interested in pizza than caribou steaks, and could be heard seriously discussing with his friends the question of whether Elvis Presley was truly dead. Freddy's uncle and co-chief Jonas Sangris pitched in. "Time changes everything," said Jonas. "The land is like one big stone church for us. In our hearts it feels a bit ill. But if we're going to benefit from the modern world, we have to take part. Otherwise that road just passes us by."

In January and February 1996, a federal panel held hearings in communities around the north. Environmental groups put up a ferocious fight. The World Wildlife Fund lobbied everyone from the Duke of Edinburgh to Canadian environment minister Sheila Copps for more studies. They threatened to tie up everything for years with a massive lawsuit. The Ottawa-based Canadian Arctic Resources Committee (CARC) gathered dirt on BHP's worldwide record: a whole river that caught fire near an Australian coal mine; assorted spills of oil, cyanide, tailings. At a BHP copper mine in Arizona, twenty years of ongoing lawsuits alleged overuse and pollution of nearby groundwater. In Yellowknife, at least one disgruntled former BHP employee said drills in the Barrens were leaving oil slicks on lakes and mud plumes in streams. And Chuck Fipke had already crossed paths with BHP more than once. The New Guinea jungle prospects where he had first panned for copper in the 1970s had since been bought by BHP, which found the copper Chuck had looked for. Where there had been wilderness was now the massive Ok Tedi copper mine. The native people who had been Chuck's guides were still there, and their relatives were suing: They said the mine killed everything in and around the Ok Tedi River, including fish and farmland, while they lived in the ruins and got nothing. CARC supplied funds for the Dene to fly in clan leader Alex Maun of the Wopkaimin people and his Australian lawyer to testify.

Among hundreds of others to speak and listen were Chuck and Stew. Chuck could barely contain himself. At one meeting he grabbed the microphone to say he was "personally offended" that anyone could believe his mine would cause damage. He asserted the only wildlife casualty was a single wolverine, accidentally hit by a mine truck in a snowstorm. "Geologists love wildlife," he said. "To my knowledge, not one caribou has been harmed, not one bear has been harmed." He began talking about leaving his fortune to a wildlife charity. Privately, he told me he was glad the wolverine got hit. "I don't like wolverines much. They prey on the little white foxes in the winter, and I like the little white foxes." Chuck spotted a rare white gyrfalcon near camp, but kept it to himself; aside from enjoying the secret, he did not want the environmentalists to go crazy.

Stew sat quietly and said nothing. Used to thinking in geologic time, he kept telling himself that in 100 years the mine pits would refill with water, rubble would revegetate, wildlife return. But doubts tortured him. What if wildlife did not return? "When we started this, we weren't worried about what would happen," he told me. "We had the land to ourselves. In the places where I went, I thought maybe I was the first person ever there." Now, in the Barrens, he could never feel that. "There's millions of claim posts out there," he said. "To be part of something that caused that, that really kills me. I'd give up everything I have for it never to have happened." On the other hand: Stew was still looking for diamonds out there, albeit with scaled-down, low-impact methods, including an idea to use a remote-controlled model plane for geophysical instruments.

In Ottawa, bureaucrats and high officials were enthusiastic. Mining had long been a cornerstone of Canada's economy. Canada was a big country, mostly empty; few questioned the merits of development. Furthermore, the mines had one unflinching northern constituency: the prospectors, descendants of prospectors, business owners who dealt with prospectors, and all the other prospßßector-related people who had come north over the years and stayed. They now comprised nearly 40 percent of the Territories' 64,000-some people. They were happy to see the tundra opened. To them it was the ultimate victory—the very reason they were here. During the hearings Territories Minister of Mines John Todd announced, "BHP and its partner, Dia Met, are to be congratulated for the progress made towards the development of North America's first diamond mine." The only thing Todd insisted: "The diamonds cannot come out of the ground unless there are significant benefits to northerners."

There they were: 800 mine jobs, billions in tax revenue, the chance for Yellowknife to expand past 49th Avenue. BHP offered a further deal: It would help fund regional environmental monitoring for years—which not only played to southern environmentalists but also gave even more jobs to northerners. Documents called Impact and Benefit Agreements were presented to the Dene, guaranteeing money, large job quotas, and special attention to preserving caribou. Freddy Sangris pointed out that many of his people were still illiterate; he did not want them handed shovels, but college scholarships, so they could move into management. BHP opened a college fund. Similar packages were offered to mixed-blood Métis and, for good will, the Inuit. With the decision looming, BHP stopped fighting the Wopkaimin and did public penance: It committed hundreds of millions of dollars for compensation and cleanups of its New Guinea copper mine.

The ultimate question of who owned Lac de Gras was left open, but the native people were swayed. That summer they signed the impact-benefit documents. The environmental groups saw their ground crumbling; without support from aboriginals, they could not figure out whether to fight or compromise. The World Wildlife Fund briefly filed, then withdrew, its threatened litigation.

On November 5, 1996, the federal cabinet in Ottawa granted approval to the diamond mines. On January 7, 1997, BHP obtained its final regulatory licenses. The miners rushed in.

CHAPTER 20

Ekati

That winter 1,800 heavy trucks moved up the ice road with concrete, steel, and fuel. Construction cranes poked above the low topography around the BHP airport, and the roars and beeps of heavy machinery could be heard for miles. Within months three-story dorms, truck sheds, a sewage plant, and a power plant were partway up and were being knitted together by enclosed above-ground "utilidors," so that when it was all done, no one would ever have to go outside again. Fifteen to twenty freighter planes landed each day with specialized equipment. By spring the place had its own fire department, bakery, and cell-phone network.

It was time to dewater the lakes. Aboriginals hated the idea of uselessly suffocating fish, so BHP hired them to clean out the lakes with nets first. Corporate publicists pointed out this would "allow scientists to actually identify and count every fish in a northern lake, something they have never been able to do before." The Inuit, eager for jobs, arrived while there was still ice and punched holes for nets. In July, after breakup, the Dogrib joined them with thousands of feet of gillnetting and six boats. Overall captain was Allen Niptanatiak, a trapper from what was formerly called Coppermine; both town site and Bloody Fall had now reverted to the Inuit name, Kugluktuk—Place Where the Water Falls—in preparation for the declaration of Nunavut. Residents still pulled massive salmon from the fall's roiling water, though now with fiberglass fishing poles. There was a wooden picnic table for visitors. The massacre and looting there seemed forgotten. Upon the diamond lakes, Inuit and Dene finally met in peace. They shook hands, smiled, and worked together. Only the Yellowknives stayed home; Freddy believed it would be bad luck to take part in wiping out every living thing in one spot.

With water being pumped out day and night, the fishermen ended their work in knee-deep water. The scientists counted 15,626 fish, including some venerable trout weighing over thirty pounds, but mostly smaller or less desirable specimens: grayling, chub, longnose suckers, slimy sculpins. The fish were flown to native settlements, where most were fed to dogs or left to rot.

Big loaders moved into the lake bottoms for "prestripping"—removal of twenty to sixty feet of sediment and glacial till overlying pipes. Much of the refuse went into BHP's growing system of dams and roads. Then, with the quick return of cold weather and refreezing of the ice road for winter 1997–1998, 2,100 more trucks arrived with additional building material and five fifty-ton diesel generators. BHP announced only one change: Another pipe near the central complex, the Sable, proved richer than Leslie, so it was substituted in the plan of five pipes. With Leslie put off for later, that made six.

Meantime, De Beers was quietly pressuring BHP-Dia Met to join the Central Selling Organization, its apparatus for buying up and selling independent mines' stones at set prices. BHP-Dia Met did not outright refuse to join the cartel, but it would not commit, either. The company hired an Antwerp marketing firm to consult, and BHP executives were flying to Europe for secret conferences. Projections indicated that Canada could supply 10 to 14 percent of the market—comparable to South Africa, enough to upset the order. De Beers set up a new office in Vancouver focused on marketing, apparently in hopes of swaying BHP to join up.

While considering the options, BHP-Dia Met announced in late 1997 that it had named its new place: the Ekati Diamond Mine.

Down south, scores were being evened. First Exploration, which had accused Chuck and the others of insider trading, settled before trial for a modest $6 million in Dia Met stock. Pioneer Metals settled with Stew for a lot less. Mackenzie settled his suit over the missing warrants with Chuck and Marlene on terms that remained secret. Wayne and the neighbors he had rounded up to attack both Chuck and Stew all backed off, and the suit petered out into an exchange of hostile letters between lawyers. All sides in all cases declared victory, and most avoided talking to one another ever again.

The most serious action continued, however, threatening to halt opening of the mine: Stew's suit regarding the buffer claims, on which lay the Misery Pipe. BHP executives saw huge losses looming, so they sat Chuck and Stew down in a big Vancouver boardroom for talks, flanked by platoons of lawyers. Chuck tried chiseling more, then Stew, but finally one night in 1997 they struck a deal. Chuck went out for his usual 10 P.M. dinner, came back at midnight, and together they signed the papers amid laughing and joking from relieved onlookers. BHP retained majority control of the disputed land, but Stew had won: He got most of the remainder of those parcels. Stew later claimed that Chuck caved in because Stew threatened to start a whole new round of litigation from other shareholders by documenting how much information Chuck had withheld early on. After the settlement Chuck and Stew met at planning sessions and acted civil, even friendly, toward each other—at least in front of others.

Chuck had other things to worry about now anyway. In addition to the A-Star helicopter, he had acquired a number of new objects: a Humvee military vehicle; a red Dodge Viper, whose gears he stripped while peeling out around Kelowna; a sleek speedboat for rides on the lake; nine expensive racehorses; and a twenty-two-year-old girlfriend named Tara Shaw. Marlene had always been patient, tolerant, and unflaggingly Catholic. She had long tried to save the marriage. However, that did not seem possible when it transpired that Tara Shaw delivered Chuck a baby boy, Tayler. Worse: Somehow, somewhere, Marlene got her hands on a life-insurance policy, showing Tara and Tayler as Chuck's beneficiaries. The *Edmonton Journal* reported that just after the birth, Marlene marched into Kelowna General Hospital with her two older daughters and Mark right behind her, so they could see with their own eyes. They found the long, blond mother lying in bed. She told them to get the hell out.

Marlene, now forty-eight, exiled Chuck from the lakeside house, dyed her dark hair blond, and sued for divorce. She wanted half of everything. Chuck retreated upshore to a new sixteenth-floor penthouse condominium with a fabulous water view. While the suit went forward, Marlene became a bit of a recluse, having groceries delivered and emerging mainly to walk her dog. At Christmas, Chuck sent his children expensive presents by courier, but it was said the children donated them to local charities. A few stalwart Kelowna investors defended Chuck, claiming the young woman had promised him love. In any case, he was soon seen with other women.

His newest adventures soon became legion. Asked his goal in life now, he said it was to find another diamond mine. On behalf of Dia Met he traveled to start exploration projects in Finland, Venezuela, Mauritania, and Greenland. After a trip to Yemen without finding any signs, Chuck got an inspiration: He would instead relocate Ophir, the site of King Solomon's lost gold mines. Twentieth-century research suggested it might lie buried in Yemen's searing, empty northwest desert quadrant, once ruled by the Queen of Sheba. Chuck made a deal with Yemeni President Ali Abdullah Saleh and personally went looking for the gold, accompanied by Yemeni guides garbed with traditional robes, AK-47 assault rifles on slings, and elaborate, broad curved daggers dangling in scabbards over their groins.

The diamond strike was waning—almost. All told, various companies had found a staggering 300 pipes, though only BHP's few seemed minable. Surely more lay hidden. The search promised to go on for generations. However, many juniors were out of money and had let claims lapse. Academics moved in to mop up. Pipes' xenoliths opened the first window onto the countless layers of rock below the surface and a detailed history of the cratonic depths. Radioactive isotopes dated some pipes at only 50 million years—the world's youngest. Some were loaded with the kind of fossils seen in nineteenth-century South Africa—leaves, chunks of fossil wood, turtle bones, dinoflagellates, conodonts, pollen spores, extinct fish—proving the tundra had had former incarnations as swamp, sea, and forest. In the bedrock, a few companies kept looking for diamonds. Finally, one hit.

Just below Misery Point, where BHP had staked its last claim row, Aber's geophysicists spotted an electromagnetic anomaly a few hundred yards over the boundary. It lay in the bay of a small island within Lac de Gras. Once the lake froze, Eira Thomas was put in charge of drilling it, and she hit a pipe. A bigger sample was taken, then a bigger one. There were diamonds—numerous, big, and a stunning 90-plus percent gem quality. It appeared to be the richest pipe yet—the seventh mine of Wayne Fipke's meteoric vision.

Aber and its partners formed a corporation, Diavik Diamond Mines. They wanted an operation even more radical than BHP's: diking and draining of three square miles of Lac de Gras itself. The environmental effects on the Coppermine watershed were uncertain, but now that BHP had its permits, Diavik's were virtually assured. Diavik moved quickly through the paperwork. The Thomases were rich.

Chris Jennings had his 1 percent share of Diavik—not quite the victory he had longed for, but something. Elsewhere his luck ran much as it always had. SouthernEra found nothing of note in the Barrens. Staggering under high expenses, Jennings opted to go home and explore the cheaper though civil war–racked African country of Angola, where armed factions financed themselves by selling black-market gems—so-called "blood diamonds." He was driving toward his first prospect when he narrowly missed getting killed in an ambush by soldiers of the UNITA rebel army. At the last moment, an old woman ran up the road and told him to turn back. His chief deputy, Frank Gobler, was not so lucky. In May 1998, marauders caught Gobler on the road and instantly executed him. Jennings went on to South Africa, where he discovered a series of narrow, diamond-rich kimberlite dikes that no one had yet noticed. He was closing in on 100 percent ownership when a complex property dispute arose. It turned out that De Beers was behind it. After court proceedings, De Beers emerged with a majority share of his discovery. He was bitter. "It's hard to forgive," he told me. "It's like David losing to Goliath."

De Beers did everything possible to stay ahead. Still excluded from the United States by antitrust hostilities, elsewhere it fielded 2,000 explorers on every major landmass except Antarctica. Expanding beyond kimberlites, it vacuumed seabeds off Africa's Skeleton Coast, where alluvial stones indeed lay on the bottom. It reinvestigated meteorite craters worldwide. In the Barrens, De Beers kept at it, picking up lapsed claims for re-prospecting and combing its own. It also looked back to Blackwater Lake. Chemical analyses could now distinguish indicator grains from one pipe or cluster and another; and, contrary to what everyone had long assumed, the new tests showed most grains around Blackwater had not traveled from Lac de Gras at all. Somewhere, near or far, lay something else. Similar tests on the fantastic band of garnets from the esker beach at Norm's Camp indicated they did not come from any pipe yet found.

A t the start of the twenty-first century, new kimberlites were uncovered in plains, forests, and muskegs in Alberta, Saskatchewan, Ontario, Kansas, and Montana. Near Leek Springs in the old California goldfields, a Canadian syndicate, Diadem Resources, drilled a mysterious circular breccia formation loaded with indicators and hundreds of microdiamonds—though none of minable size. No petrologist could put a name to the rock. Diadem's main consultant, by the way, was an aging Quebecois, Mousseau Tremblay. He now owned a lakeside home in Vermont. East of Lonetree, Wyoming, an outfit called Guardian Resources discovered several equally mysterious breccia pipes full of indicators—just like the ones from Tom McCandless's anthills. Possibly they are the anthills' source. McCandless himself had long given up and had taken a research position at the University of Arizona, Tucson. Over in Colorado, onetime Cominco geologist Howard Coopersmith kept at it. Finally, north of the Sloan Ranch, he proved up a minor diamond pipe. Opened in 1995 as Kelsey Lake Mine, it was more often closed than open, due to various difficulties.

Companies continued hunting the sources of the Great Lakes glacial diamonds. They made complex computer simulations of underlying rock terrains, sifted wastes from sand and gravel plants, and searched on in the boreal forests below the Keewatin. The hand-dug well on whose seventy-foot bottom tenant farmer Charles Wood had found a thumbnail-size gem in 1876 was still there, alongside the old farmhouse in Eagle, Wisconsin. But maybe not for long. Like so many places, Eagle was being devoured by development. The well now lay only one lot away from the intersection of two interstate highways.

In Arkansas the business-friendly efforts of Governor Bill Clinton brought a proposal to open Crater of Diamonds State Park to companies. It was met by a clamor from 100 outfits to get in and years of litigation from citizens who thought parkland was protected. Eventually the citizens lost and a consortium of four chosen firms tore up Crater of Diamonds with test trenches forty feet deep and drill holes to 650 feet. The results, presented in 1997, confirmed earlier tests: The pipe was unminable. Hand-diggers were allowed back into the remains. The state geologist, however, insisted the tests had been botched—perhaps sabotaged by dark forces. In the middle of all this, a local man found a yellow triangular stone of 4.25 carats at the park. It was bought by Pine Bluff jeweler Stan Kahn. Bill Clinton was inaugurated governor of Arkansas twice, and both times, Kahn loaned the diamond to Hillary Clinton to wear at the festivities, first in a necklace, then in a ring. When Bill Clinton was inaugurated as president of the United States in 1994, the First Lady had it on her left hand at the ball, in a ring designed for the occasion.

B HP scheduled the Ekati Diamond Mine opening for October 14, 1998. There was to be a big ceremony at the mine, then a gala banquet back in Yellowknife for a limited number of stockholders, executives, government dignitaries, and assorted guests. It was rumored that Prime Minister Jean Chrétien was coming.

Hugo, Ashley, and Gurney all got the coveted tickets, of course. Hugo was now BHP senior vice president in charge of world exploration. Gurney commanded such large consulting fees that he no longer remembered how many employees he had. He had taken up diving—on the Skeleton Coast, near De Beers's seabottom operation. Gurney formed a competing company and became quite good at going down and retrieving diamonds himself. Ashley got a long-delayed vacation and experienced a breakdown on an Australian beach because he suddenly had time to think about what he had seen and done in the Barren Lands. He sobbed for a long time on his wife's shoulder and wondered why he had given so much to make other people rich. He decided he did it because he was addicted to exploration, and went back the following week.

Stew Blusson's mother was now a relatively strong ninety-one. Stew knew it was unrealistic, but somehow he nurtured the hope she might be able to fly to the ceremony. Not only could he show her the mine; he had a bigger surprise. Just before the opening, it was to be announced he was giving $50 million to his and Chuck's alma mater, the University of British Columbia. It would finance a foundation for science innovation and endow Stew's version of the Nobel Prize— the Canada Prize. It was reportedly the largest single charitable gift in Canadian history. UBC officials were offering an honorary doctorate to go with his real one, but Stew shrugged; he had more money, and besides, Marilyn had kept her job as an Air Canada attendant, which carried a pension.

Stew's moment before Mother never came. Just before the bequest announcement, he got ready to head north for another diamond-prospecting expedition. He asked Mother if she would be all right. "I'll be fine," she said, and kissed his hand as he slipped away. A few days later she went out shopping, fell down, and sank into a coma. On September 11, 1998, she passed away with Stew at her bedside; she never regained consciousness. Sixty years old himself, Stew had never considered this; he had thought Mother would always be there for him. Not only did he wish it; given the way he had lived, he had always assumed he would be first to die. Now that things had changed, nature had caught up. They cleaned out her apartment, and he found the heart-shaped stone from Crater of Diamonds, still on its chain, laid away in her jewelry box. He took it home.

Among those conclusively not invited to the ceremony was Wayne Fipke. He had retreated the opposite way down the lakeshore from Chuck to a gated community called Eldorado Court. There he and his wife embarked on a five-year building plan, erecting a kind of Georgian-Italianate-Neoclassical mansion replete with curved colonnades of Corinthian pillars from Alabama, stained glass windows in the pre-Raphaelite style, Romanesque mosaic-inlaid pools with electronically choreographed water shows, and other touches that Wayne said were "a reflection of ourselves." Off and on Wayne figured to find his own diamonds, and he financed expensive ventures, including an expedition to Alaska to look at rocks that turned out not to be what his hired geologist said they were. He lost his money every time, and sometimes wondered if people were just taking advantage.

After viewing Wayne's place, I visited Chuck. I had not seen him for a couple of years. He seemed changed. Almost miraculously he had lost his stutter. He put obvious effort into speaking slowly, finishing most sentences and only occasionally wandering into tangents. It was as if the many things cluttering his head had finally been cleared out. Unfortunately, he also seemed to have lost his usual loud laugh. Marlene and Mark would not be attending the opening, and plenty of others were still mad. The divorce was proceeding; Marlene seemed virtually certain to take half the estate. Some Dia Met shareholders in fact were rooting for her to take over the whole company. Chuck had recently spent a few hours with Mark, then Wayne, trying to patch things, but with no resolution. They would barely talk to him, and even his younger children still seemed wary. When Chuck learned I was in town, he called my motel and said he would pick me up. Ten minutes later he pulled up in a nearly-destroyed old pickup with a bunch of shovels in back, but he handed me the keys: He had just managed to lose the right-hand lens from his eyeglasses and was afraid we would crash if he did the driving.

We crossed the green, rolling country outside Kelowna to visit a racehorse and her foal that Chuck had boarded on a neighbor's farm. We fed them a couple of bags of carrots, cooed at them, and petted their noses. Then we continued a few miles to Chuck's parents', bumping up a dirt road past the Fipke apple orchard. The automobile graveyard under the apple trees was still full, though the falling-down childhood house across the way was empty. His folks, now in their seventies, had fixed up a slightly better one a couple of hayfields beyond. Outside sat a new car, courtesy of Dia Met shares from Wayne. It was Sunday, and Mr. and Mrs. Fipke welcomed us in with chicken dinner and homemade wine. They were nice people, obviously proud of all their children. Wayne had just been over to help bring in hay, they said. They fussed over an honorary doctorate Chuck just got from Okanagan University, though it was just a local community college with no actual doctoral program. They expressed hope that Chuck and Wayne would work things out. Chuck said so, too. After dinner, Mrs. Fipke showed off old photos.

It was near midnight when we drove back to Kelowna. In the passenger seat of the dark cab, Chuck started a monologue.

"I did push people too hard sometimes. I know now. I pushed Mark harder than anyone else, you know. I wanted everyone to be equal. Maybe it was a mistake. I talked to him, but everything is still . . . Well, you know. It's hard when things don't work out."

We drove on in silence for a long while. Then he resumed.

"It's like if you were training for an Olympic gold medal, and were really serious, and you put all your time into that, hey? Sometimes you gain. But you can't do everything right and also win the gold medal. It had its price, hey. It had its price."

Then he brightened, as if a switch had been thrown.

"But that was *cool!* To do all that we did? It was *fun!*" And he laughed.

On October 14, 1998, several small jetloads of guests assembled at Yellowknife airport starting 6 A.M. It was snowing lightly and foggy. During the flight to Ekati, in places where the clouds broke, the Barren Lands showed through: wind-etched snow, bare rock ridges, lakes crusting with ice at the edges. Presently something else loomed. We landed. We were on a vast, perfectly flat man-made plain. You could see a mile across it through dark, drifting mists and a dusting of snow. In the distance lay a box of Pharaonic proportions, surmounted by lights that seemed to shine big as nebulae upon the near approach of space travelers. It was the central complex. Elsewhere, swarms of smaller, sharper lights flashed yellow and red, puncturing the mist—signals on the roofs of many trucks, parked and moving. Some other guests and I climbed into a schoolbus for a tour led by a megaphone-wielding guide. One was Mackenzie's teenage son. He looked weirdly like his father, partly because Mackenzie himself still looked like an overgrown teenager.

First stop was Panda Pipe, already 300 feet deep and one-third mile across. Overlying material was gone, and they were down to the start of kimberlite. The bus drove down a winding road to the floor, and we were allowed to get out and walk. Broken up from seeping water, the kimberlite resembled sodden coal. Everyone instinctively reached down, but the guide said, nicely, to stop. No one was allowed to touch the ore; they might find a gem and get tempted. In the pit's center sat the only thing that touched ore: a yellow 4,400-horsepower Komatsu hydraulic shovel with Tyrannosaurus rex–like teeth on its scoop. It was three stories high and capable of filling a 218-ton-capacity haul truck in four passes. Across the way reared the shiplike mast of a portable augur for placing explosives, the red and white maple leaf flag whipping in a powerful wind high atop. Amid spoil heaps, a few acres of undisturbed tundra lay beyond the lip, looking out of place. Mackenzie's son asked where the original exploration camp was, but the guide did not know. He told us the mine's cost—$940 million—and the time before that was paid back and profits would dwarf the exorbitant operating expenses: four years.

At the processing plant we were met by part of a fifty-member security team, headed by an ex-U.S. military colonel. They briefed us on the plant's procedures for preventing diamond theft: daily shake-downs of everything from steel-toed boots to laptop computers, voice analyzers to detect psychological stress in those leaving sensitive areas, low-intensity X-rays to spot stones secreted on or in human bodies. We then entered through a clear, one-person-size bulletproof chamber, with an outer door and an inner door. In one claustrophobic moment between the outer door's closing and the inner door's opening, you were suspended like a diamond inclusion. Beyond this was a narrow passage, and through another door the world's largest, loudest room. It was filled with a blur of pipes, ducts, girders, and roaring conveyors loaded with ore, all seen from a high, narrow catwalk. In some hidden chamber, we were told, the last 2 percent of heavies arrived via pneumatic tubes to be drawn through vacuums into X-ray sorters that would shoot the stones into locked boxes, untouched by hands.

A stage and seating were set up in the basketball gym, across from the plant cafeteria. Stew and Marilyn sat in the audience well over on stage left, Chuck well over on stage right. The prime minister did not show up. One guest was a puffy, gray-haired gentleman with a clipped accent, already wearing a souvenir Ekati Diamond Mine baseball cap—George Burne, head of De Beers's Canadian office. Things were convened by a Catholic priest. A Dogrib BHP employee sang a long prayer that was largely a nutshell history of the Barrens—migratory hunting, the Coppermine, the *kwet'i*, desertion of the land, its rediscovery. It was all in Dogrib. According to a translator, he concluded, "We pray for the safe return of our people to their families at the end of their shifts." Then came a series of toweringly boring, identical speeches by BHP leader Jerry Ellis; Dia Met's Jim Eccott; minister of Indian Affairs and Northern Development Jane Stewart; natural resources minister Ralph Goodale; Ethel Blondin-Andrew, member of Parliament from the Northwest Territories; and others. Each spoke of "historic moments," "new frontiers," "great opportunities," and "sustainable development," making nods to the "vision and determination" of Chuck and, sometimes, Stew. Chuck sat upright, expressionless; Stew slouched, looking exhausted.

For the finale there was a plaque unveiling. It listed all the dignitaries, plus "Dr. Stewart L. Blusson" and "Dr. Charles E. Fipke." Marilyn winced; Chuck had insisted on having his honorary doctorate included. Ellis and Stewart then unveiled two big vitrines filled with at least $1 million worth of diamonds. At the signal, everyone rushed over to see them. One case displayed finished Lac de Gras stones of up to 15 carats, cut in round, marquise, pear, and square shapes. In their velvet boxes, they looked like a display at Tiffany's. The other case held rough diamonds, piled in beakers, plastic bags, and spittoon-like cups and on papers. Ranging in size from coriander seeds to fingertips, they were said to weigh 9,026.97 carats—one typical day's production. Most were lumpy and grayish, and to tell you the truth, they looked like driveway gravel. George Burne came over in his Ekati cap and studied them closely with his hands behind his back.

The banquet that night was held at the Explorer Hotel, a glassy tower in New Town. Bartenders and hostesses stood at the banquet room door handing out cocktails and souvenir Ekati paperweights of polished kimberlite—presumably prescreened for diamonds. I saw Mackenzie and his tall son having a couple of beers, looking like twins. Among the crowd, Stew and Chuck were barely recognizable, dressed as they were in suits. Ashley came in with another geophysicist wearing a leather jacket. We went in with drinks in hand. Chuck was seated at a front table next to a young blond-haired woman in dark blue, identified on the guest list as Kelly Mac-Donald, no affiliation given.

I was at a table to the side, sitting between Ashley and Mackenzie's son. After salad, the same dignitaries got up and gave the same speeches, with identical additions from the mayor of Yellowknife and other locals. At least the pain this time was dulled by the predinner cocktails and beer and, now, bottles of red and white wine that the nice lady kept bringing to our table whenever one got emptied. This was often. Presently dinner was over and a very loud rock band took over. More alcohol of some kind arrived, perhaps some kind of aperitif. Mackenzie's son was slurring his words by now, and Ashley's green eyes gleamed ever brighter. Hugo was at a front table, laughing uproariously at someone's joke, while Gurney worked the room, stumping from table to table on his stiff legs. Chuck and Kelly Mac-Donald had disappeared somewhere.

Things broke up before midnight. Shortly, they reassembled, as if by migratory instinct. It started when Chuck showed up alone at the Gold Range. Chuck's kid sister Carol, who had come to the ceremony, arrived with her husband. Then one by one, the old crew filed in to join. Dynamite Dan and Brent Carr were in town without any particular plan, and they showed. Ashley came in with other BHP people. Mackenzie and his son made an appearance. Even Stew, who had never been much of a drinker, walked in. Tables were shoved together in a line so everyone could fit. The Range never changed: It was packed and deafening, and drinkers were staggering in all directions through spilled beer and broken glass. Vague possibilities for violence gusted up here and there like dust devils, then subsided, or not.

Chuck attracted a busy waitress by waving a $50 bill high in the air, and bought a double round of beer. Then Stew bought a round. Then Ashley. Passing tipplers kept falling onto everyone's laps. Most were friendly, though one threatened Brent because Brent had accidentally bumped his chair. That blew over. Nearby, a woman was clinging with both arms wrapped around a metal column as her husband tried prying her off, all the while cursing and promising her a sound beating at home. Dynamite Dan's gallantry was insulted; at great risk he inserted himself between them and drove off the man. Dan actually got away with it. Then one stranger, a towering, heavyset Yellowknife prospector-type with a beard, stumbled straight through the crowd toward Chuck and pointed a massive finger.

"Hey you! I know you!"

"You do?" asked Chuck in a small voice.

Everyone shut up.

"Yeah. You're, uh . . ." He swayed, and his eyes unfocused and refocused. "Hey! I know you!"

Chuck looked at him, silent.

"Hey! You're the diamond man!"

"Nope, not me," replied Chuck. "Must be someone else."

"No, I know you, fuck, you're the diamond man," insisted the stranger. He looked ready and suddenly quite able to throw a killing punch. "You *are!*" he screamed. Everyone around the table tensed and pushed out their chairs, though it was hard to say whose side they were going to join when the moment came.

Then the stranger held out his big paw and opened it wide for a handshake.

"I just wanna thank you," he said. "You brought this town to life again."

Chuck Fipke grinned and took the man's hand.

NOTES

Much of this book is from interviews, unpublished documents, and firsthand observation. Where the sources are published or generally available, I list them. The following abbreviations are used in the Notes:

GSC   Geological Survey of Canada

GNWT   Government of the Northwest Territories

USGS   United States Geological Survey

GSA   Geological Society of America

Prologue: Heading North

3 hundreds of companies homed in: Initial accounts of the rush include Vivian Danielson, "Discovery of Lac de Gras in NWT a Tale of Dedication and Persistence," *Northern Miner*, September 14, 1992, p. 1; Bill Richards, "Diamond Find Sparks Rush to Canada," *Wall Street Journal*, October 20, 1992, p. A4; Mary Walsh, "Great Canadian Diamond Rush," *Los Angeles Times*, October 5, 1992, p. B1; Bernard Simon, "Search for Glittering Prizes Beneath the Canadian Ice," *Financial Times*, March 15, 1993, p. 1; William Broad, "Clues Emerge to Rich Lodes of Diamonds," *New York Times*, February 15, 1994, p. D1.

3 A single discovery: See *Diamonds: Exploration, Sampling and Evaluation* (Toronto: Prospectors and Developers Association of Canada, 1993), pp. 173–184; M. Sevdermish, "The Diamond Pipeline into the Third Millennium," *Geoscience Canada*, 25, no. 2 (1998): 71–84.

6 like interstate highways: Major reviews of Barrens glacial geology include J. M. Aylsworth and W. W. Shilts, *Glacial Features Around the Keewatin Ice Divide* (Ottawa: GSC, Paper 88-24, 1989); W. W. Shilts, "Canadian Shield," in W. L. Graf, ed., *Geomorphic Systems of North America* (Boulder, CO: GSA, 1987). For eskers, see J. M. Ayslworth, "Bedforms of the Keewatin Ice Sheet," *Sedimentary Geology*, 62 (1989): 407–428; Peter U. Clark, "Subglacial

Drainage, Eskers and Deforming Beds Beneath the Laurentide and Eurasian Ice Sheets," *GSA Bulletin*, 106 (1994): 304–314. For physiography and climate, see R. J. Fulton, ed., *Quaternary Geology of Canada and Greenland* (Ottawa: GSC, 1989), pp. 177–317.

6 as one bush pilot put it: Dave Olesen, *North of Reliance* (Winocqua, WI: NorthWord Press, 1994), p. 19.

6 "We scarcely see a gang of buffaloe": Stephen Ambrose, *Undaunted Courage: Meriwether Lewis, Thomas Jefferson and the Opening of the American West* (New York: Touchstone, 1996), pp. 167, 217.

7 no-man's land: For major studies of tundra history and prehistory, see *Weledeh Yellowknives Dene: A Brief History* (Dettah, NWT: Weledeh Yellowknives Dene, 1997); C. S. MacKinnon, "An Overview of Barren Lands' History," *Musk-Ox*, 40 (1994): 13–30; Bryan C. Gordon, *People of Sunlight, People of Starlight; Barren Land Archaeology in the Northwest Territories* (Hull, Que.: Canadian Museum of Civilization, 1996); Robert McGhee, *Canadian Arctic Prehistory* (Hull, Que.: Canadian Museum of Civilization, 1990); Robert McGhee, *Ancient People of the Arctic* (Vancouver: University of British Columbia Press, 1996); Diamond Jenness, *The People of the Twilight* (Calgary: University of Calgary Press, 1959); Andrew Stewart, "Archaeology and Oral History of Inuit Land Use on the Kazan River, Nunavut: A Feature-Based Approach," *Arctic* (September 2000): 260–278.

8 the earth's top per capita diamond consumers: Early U.S. statistics from Percy Wagner, *The Diamond Fields of Southern Africa* (Johannesburg, South Africa: Transvaal Leader, 1914), p. 311. Later figures from USGS, *Mineral Industry Surveys: Gemstones* (Washington, DC: Department of Interior, 1998).

9 But that is another story: See George Kunz, "History of the Gems Found in North Carolina," *North Carolina Geological and Economic Survey Bulletin*, 12 (1907): 7–8; George Kunz, "A North Carolina Diamond," *Science*, 10 (1887): 168; USGS, *Mineral Resources of the U.S., 1886* (Washington, DC: USGS), p. 598.

9 Sotheby's of New York sold it: See Roy J. Holden, "The Punch Jones and Other Appalachian Diamonds," *Bulletin of the Virginia Polytechnic Institute, Engineering Experiment Station*, 56 (February 1944); "Punch Jones Diamond," *Gems & Gemology* (Fall 1944): 169; "Punch Jones Diamond—Monroe County, West Virginia," *Virginia Minerals*, 42, no. 4 (November 1996); Beth Macy, "A Gem of a Story," *Roanoke Times & World News*, August 5, 1990, Extra p. 1.

9 the "Lewis and Clark Diamond": See Richard Ecke, "Area Diamond Worth $80,000," *Great Falls Tribune*, September 7, 1990, p. 1; June Zeitner, "The Lewis and Clark Diamond," *Lapidary Journal* (August 1991): 79, 82, 88

10 Real diamonds of more than 2 carats: Dan Hausel, "Diamonds and Mantle Source Rocks in the Wyoming Craton with a Discussion of Other U.S. Occurrences," *Geological Survey of Wyoming Report*, 53 (1998); J. J. Brummer, "Diamonds in Canada," *Canadian Mining and Metallurgical Bulletin* (October 1978): 64–79.

11 No one knows the depth limit: For meteoritic diamonds, see "Impact Diamonds," in A. N. LeCheminant et al., eds., *Searching for Diamonds in Canada* (Ottawa: GSC, Open File 3228, 1996), pp. 183–185. For other kinds, see R. G. Berman, "Diamonds in Ultrahigh-Pressure Metamorphic Rocks" in LeCheminant et al, eds., *Searching for Diamonds*, pp. 177–180; J. G. Liou, "Into the Forbidden Zone," *Science*, February 18, 2000, pp. 1215–1216; Kenneth Collerson, "Rocks from the Mantle Transition Zone, Malaita, Southwest Pacific," *Science*, May 19, 2000, pp. 1215–1223.

13 a third of them in the 1990s: The best modern summaries of diamond natural history include Stephen Haggerty, "A Diamond Trilogy: Superplumes, Supercontinents and Supernovae," *Science*, August 6, 1999, 851–860; George E. Harlow, ed., *The Nature of Diamonds* (Cambridge: Cambridge University Press, 1998); LeCheminant et al., eds., *Searching for Diamonds*; Alfred Levinson, "Diamond Sources and Production: Past, Present and Future," *Gems & Gemology* (Winter 1992): 234–254; H. H. Helmstaedt, "Geotectonic Controls of Primary Diamond Deposits: Implications for Area Selection," *Journal of Geochemical Exploration*, 53 (1995): 125–144; Melissa Kirkley, "Age, Origin and Emplacement of Diamonds: Scientific Advances in the Last Decade," *Gems & Gemology* (Spring 1991): 2–25.

14 secret signs: In addition to the sources above, published sources on modern diamond prospecting include Warren Atkinson, "Diamond Exploration Philosophy, Practice, and Promises: A Review," in *Proceedings of the Fourth International Kimberlite Conference*, Vol. 2 (Carlton, Australia: Geological Society of Australia/Blackwell Scientific Publications, 1986), pp. 1075–1107; H. H. Helmstaedt, "Primary Diamond Deposits—What Controls Their Size, Grade, and Location?" in B. H. Whiting, ed., *Giant Ore Deposits*, Society of Economic Geologists Special Publication No. 2, pp. 13–80; C. M. H. Jennings, "The Exploration Context for Diamonds," *Journal of Geochemical Exploration*, 53 (1995): 113–124.

Chapter 1: Misery Point

23 it looked like sludge: For diamond extraction methods, see Eric Bruton, *Diamonds* (Radnor, PA: Chilton, 1978), pp. 141–155.

Chapter 2: Cap aux Diamants

25 the first real stab: See George Malcolm Thomson, *The Search for the Northwest Passage* (New York: Macmillan, 1975); Alan Cooke, *The Exploration of Northern Canada 500 to 1920: A Chronology* (Toronto: Arctic History Press, 1978); Miller Graf, *Arctic Journeys: A History of Exploration for the Northwest Passage* (New York: Peter Lang, 1978).

25 Taignoagny and Damagaya: See Thomas Costain, *The White and the Gold—The French Regime in Canada* (Toronto: Doubleday, 1954), pp. 16–37.

26 "infinite quantities of gold": H. P. Biggar, ed., *The Voyages of Jacques Cartier* (Ottawa: F. A. Acland, 1924), p. 170. Regarding native copper, see Michael Wayman, "Native Copper: Humanity's Introduction to Metallurgy?"

CIM Bulletin, August 1985, 67–77; W. W. Vernon, "New Perspectives on the Archaeometallurgy of the Old Copper Industry," *MASCA Journal*, 3, no. 5 (1985): 154–163.

26 "copper-gilt like gold": Biggar, *Voyages of Cartier*, pp. 170–171.

26 strangled with a crossbow string: Sir Arthur Helps, *The Spanish Conquest in America* (London: John Lane, 1902), pp. 377–397.

27 "they glister as it were of fire": Biggar, *Voyages of Cartier*, pp. 254–255.

27 "the King accepted certain Diamants": Ibid., p. 267.

28 "the 'unconquerable force'": Pliny, *Natural History*, Vol. 10, ed. D. E. Eicholtz (Cambridge, MA: Harvard University Press, 1962), pp. 165, 207–209, 325–327. Pliny died in the A.D. 79 eruption of Mt. Vesuvius.

29 For myths about diamonds: See George Kunz, *The Curious Lore of Precious Stones* (Philadelphia: Lippincott, 1913), pp. 41, 71–72, 152–154, 375–379; Eric Bruton, *Diamonds* (Radnor, PA: Chilton, 1978), pp. 1–15; Robert Maillard, ed., *Diamonds: Myth, Magic, and Reality* (New York: Crown, 1980), pp. 12–32, 39–41; George E. Harlow, ed., *The Nature of Diamonds* (Cambridge: Cambridge University Press, 1998), pp. 117–133, 163–164. The modern phrase "a diamond of the purest water," meaning flawless and most brilliant, may descend from the idea that the stones came from supernormal hardening of common liquid.

29 "they shall grow every year": Quoted in Kunz, *Curious Lore*, p. 72.

30 diamond could not be broken: Pliny said that only steeping a diamond in warm, fresh goat's blood would make it breakable. Medieval philosopher Albertus Magnus (1205–1280) added that this worked better if the goat consumed plenty of parsley or wine before being killed.

30 "those of bastard kinds": Thomas Nicols, *Lapidary, or, the History of Pretious Stones: With Cautions for the undeceiving of all those that deal with Pretious Stones* (Cambridge: Thomas Buck, 1652), p. 46.

31 Cap aux Diamants: George W. Brown, *Dictionary of Canadian Biography*, Vol. 1 (Toronto: University of Toronto Press, 1966), p. 169. See also John Robert Columbo, ed., *Columbo's Canadian Quotations* (Edmonton: Hurtig, 1974), p. 102; Roger Schlesinger, *Andre Thevet's North America: A Sixteenth Century View* (Montreal: McGill-Queen's University Press, 1986), p. 51.

31 Peter Martyr: Quoted in Thomson, *Northwest Passage*, p. 9.

31 "rich and wealthy in precious stones": Biggar, *Voyages of Cartier*, p. 260. The later explorer Samuel de Champlain sailed the St. Lawrence in June 1603 and observed again, "There are along the Coast of the said Quebec Diamants in the Rockes of Slate, which are better than those of Alençon [France]." This raised no recorded stir; he may have meant a species of quartz. See Arnie Bourne, *The Voyages and Explorations of Samuel de Champlain 1604–1606 Narrated by Himself* (New York: Allerton, 1922), p . 180.

32 ditches, houses, and garden walls: See D. D. Hogarth, *Mines, Minerals, Metallurgy: Martin Frobisher's Northwest Venture, 1576–1581* (Quebec: Canadian Museum of Civilization, 1994), pp. 103–104; Tryggvi J. Olesen, *Early Voyages and the Northern Approaches* (Toronto: McClelland & Stewart, 1963). For an overview of Arctic prospecting, see Richard Vaughn, *The Arctic: A History* (Gloucestershire, England: Sutton 1999).

32 unexpected coast of ice and rock: See Peter Newman, *Company of Adventurers* (New York: Penguin, 1985), pp. 42–48; Barry Lopez, *Arctic Dreams* (New York: Bantam, 1987), pp. 300–301.

32 scurvy, ice, and sheer bewilderment: Graf, *Arctic Journeys*, pp. 46–76. Arctic voyages remained insanely dangerous. Records show that from 1772 to 1852, 80 of 194 arctic whaling ships sailing from Hull, England, never returned.

33 "eat raw flesh and fish and loves it": E. E. Rich, *Hudson Bay Company 1670–1870*, Vol. 1 (Toronto: McClelland and Stewart, 1960), p. 441. This records only one brief inland trip in the seventeenth century. In June 1690, teenage apprentice Henry Kelsey was put ashore on the Bay coast at the edge of the tree line with an Indian lad to look for the Dene. They stuck near the seashore—"a brief, bitter little expedition," Kelsey called it. He was the first European to see musk oxen, but no people. See Rich, *Hudson Bay Company*, pp. 296–297; and "Henry Kelsey," *Dictionary of Canadian Biography*, pp. 307–315.

33 "they go on war parties": Quoted in E. D. Kindle, *Classification and Description of Copper Deposits, Coppermine River Area* (Ottawa: GSC, Bulletin 14, 1973), p. 4.

34 "Every Sort you see": "William Stuart," *Dictionary of Canadian Biography*, pp. 614–616.

34 This map, a version: See June Helm, "Matonabbee's Map," *Arctic Anthropology*, 26, no. 2 (1989): 28–47; Rich, *Hudson Bay Company*, p. 457.

35 Canadian Sacagawea: Elders of the Yellowknives Dene say Thanadelthur was actually named Wetsi Weko and was one of them, although there is no record that the Yellowknives met traders until later. In any case she was a remarkable leader. When a Dene elder suggested trading inferior furs to the *kwet'i*, she caught his nose, pushed him back, and called him a fool. Unafraid of Knight himself, she threatened to kill him when he took some things she felt were hers. On the day she died, the usually unsentimental Knight wrote: "She was one of a Very high Spirit and of the Firmest Resolution that I ever see in any Body in my Days and of great courage and forecast." He added that forty more like her could have taken the woodlands from the Cree. See Sylvia Van Kirk, "Thanadelthur," *The Beaver*, 304, no. 4 (1974): 1; "Thanadelthur," *Dictionary of Canadian Biography*, pp. 627–628; and Alice Johnson, "Ambassadress of Peace," *The Beaver*, 283, no. 4 (1952): 42–46.

35 people obsessed with stones: *Kwet'i*, or *theye-hotine* in some dialects, was used to denote Hudson Bay Company traders long after the stone fort ceased to exist, because it distinguished them from other traders. It was probably in the twentieth century that "stone people" shifted to its modern sense, meaning people obsessed with stones. Letter to the author from June Helm, University of Iowa, Department of Anthropology, April 30, 1996. See also June Helm, "Dogrib Oral Tradition as History: War and Peace in the 1820s," *Journal of Anthropological Research* (Spring 1981): 8–27.

35 Knight sailed to England: For more on Knight and the copper mine, see J. Geiger, *Dead Silence: The Greatest Mystery in Arctic Discovery* (London: Bloomsbury, 1993); William Barr (ed.), *Voyages to Hudson Bay in Search of a Northwest Passage* (London: Hakluyt Society, 1995); Rich, *Hudson Bay Company*, pp. 441–443; "James Knight," *Dictionary of Canadian Biography*, pp. 318–320.

Chapter 3: The Coppermine

37 Thomas Mitchell and John Longland: Alan Cooke, *The Exploration of Northern Canada 500 to 1920: A Chronology* (Toronto: Arctic History Press, 1978), p. 1744. Some sailors found what looked like wreckage of Knight's ship, but no definite evidence.

38 The last man: Samuel Hearne, *A Journey from Prince of Wales's Fort in Hudson's Bay to the Northern Ocean in the Years 1769, 1770, 1771, 1772*, ed. Richard Glover (Toronto: Macmillan of Canada, 1958), pp. lx–lxii, from which much of this chapter is drawn. See also Gordon Speck, *Samuel Hearne and the Northwest Passage* (Caldwell, ID: Caxton Printers, 1963); "Matonabbee," *Dictionary of Canadian Biography* (Toronto: University of Toronto Press, 1966), pp. 523–524; W. A. Fuller, "Hearne's Thelewy-aza-yeth," *Musk-Ox*, 27 (1980): 67–74; June Helm, "Matonabbee's Map," *Arctic Anthropology*, 26, no. 2 (1989): 28–47; E. W. Morse, "Modern Maps Throw New Light on Samuel Hearne's Route," *Cartografica*, 18, no. 4 (1981): 23–25; I. S. MacLaren, "Samuel Hearne and the Printed Word," *Polar Record*, 29 (1993): 166–167; W. A. Fuller, "Samuel Hearne's Track: Some Obscurities Clarified," *Arctic* (September 1999): 257–271.

38 Richard Norton: Speck, *Samuel Hearne*, p. 99.

38 "like a heap of pebbles": Hearne, *A Journey*, p. 112.

39 "I was pitched on": Ibid., p. lxiv.

40 "nobleness of a Turk": Ibid., pp. 35–36, 225.

40 the oldest dated earthly rocks: See S. A. Bowring, "3.96 Ga Gneisses from the Slave Province, Northwest Territories, Canada," *Geology*, 17 (November 1989): 971–975; Samuel A. Bowring, "Priscoan (4.00-4.03 Ga) Orthogenesis from Northwestern Canada," *Contributions to Mineral Petrology*, 134 (1999): 3–16.

41 Troves of silver: For overviews of Slave Province geology, see Carl Zimmer, "Ancient Continent Opens Window on the Early Earth," *Science*,

December 17, 1999, 2254–2256; Jamie Bastedo, *Shield Country* (Calgary: Arctic Institute of North America, 1994). For more technical information, see A. N. LeCheminant et al., eds., *Searching for Diamonds in Canada* (Ottawa: GSC, Open File 3228, 1996); Paul Hoffman, "United Plates of America: Birth of a Craton," *Annual Review of Earth and Planetary Science*, 16 (1988): 543–603.

41   Hundreds, maybe thousands: J. A. Pell, "Kimberlites in the Slave Craton, NWT," *Geoscience Canada*, 24, no. 2 (1997): 77–90; W. J. Davis, "A Rb-Sr Isochron Age for a Kimberlite from the Recently Discovered Lac de Gras Field," *Journal of Geology*, 105 (1997): 503–509.

42   the brief summers: See E. C. Pielou, *After the Ice Age: The Return of Life to Glaciated North America* (Chicago: University of Chicago Press, 1991).

42   fishhooks, bracelets, and spears: Some metal artifacts from the Coppermine may date to 2000 B.C. See Robert McGhee, "Excavations at Bloody Falls, NWT," *Arctic Anthropology*, 6, no. 2 (1970): 53–72; U. M. Franklin, *An Examination of Prehistoric Copper Technology and Copper Sources in Western Arctic and Subarctic North America* (Ottawa: Archeological Survey of Canada, Paper No. 101, 1981); David Morrison, "The Copper Inuit Soapstone Trade," *Arctic* (September 1991): 239–246; Michael Wayman, *Analyses of Native Copper Artifacts from a Dene Copper Workshop at Snare Lake, NWT* (Edmonton: Canadian Archeological Association Conference, 1994); A. P. McCartney, "Late Prehistoric Metal Use in the New World Arctic," in R. D. Shaw, ed., *The Late Prehistoric Development of Alaska's Native People* (Alaska Anthropological Association Monograph No. 4, 1988), pp. 57–79; David Morrison, "Thule and Historic Copper Use in the Copper Inuit Area," *American Antiquity*, 52, no. 1 (1987): 3–12; Heather Pringle, "New Respect for Metal's Role in Ancient Arctic Cultures," *Science*, August 8, 1997, 766–767; Bryan C. Gordon, *People of Sunlight, People of Starlight: Barren Land Archaeology in the Northwest Territories* (Hull, Que.: Canadian Museum of Civilization, 1996).

The Inuit also had iron, knocked off a meteorite that fell near Cape York, Greenland, some 2,000 years ago. Ten thousand basalt hammers for taking pieces were found around one mass; worked fragments were found 1,100 miles away on the Hudson Bay coast. In 1897, polar explorer Robert Peary convinced an Inuk to reveal the secret location, draped an American flag over the main piece, carved his initials in it, and hauled it away. The American Museum of Natural History still displays it. See Vagn Buchwald, "On the Use of Iron by the Eskimos in Greenland," *Materials Characterization*, 29 (1992): 139–176; A. P. McCartney, "Iron Utilization by the Thule Eskimos of Central Canada," *American Antiquity*, 38, no. 3 (1973): 328–338; John Weems, *Peary: The Explorer and the Man* (London: Eyre and Spottiswoode, 1967), pp. 142–147, 167–168; Kenn Harper, Kenn, *Give Me My Father's Body—The Life and Times of Minik, the New York Eskimo* (Frobisher Bay, NWT: Blackhead Books, 1986).

43   Inuit farther south: Just before the birth of Christ a cold climate gyration pushed the trees farther south than now; Indians retreated, and coastal people advanced. In the Middle Ages, warming readvanced the trees fifty or sixty miles, and the Indians turned the tables. When the Little Ice Age started around 1200, trees re-retreated and the northerners pushed down again. See Glen MacDonald et al., "Response of the Central Canadian Treeline to Recent Climatic Changes," *Annals of the Association of American Geographers*, 88, no. 2 (1998): 183–208; Harvey Nichols, "Historical Aspects of the Northern Canadian Treeline," *Arctic* (March 1976): 38–47; Reid Bryson et al., "Radiocarbon and Soil Evidence of Former Forest in the Southern Canadian Tundra," *Science*, January 1, 1965, pp. 46–48; Gordon, *People of Sunlight*, pp. 149–152, 217, 237–240.

44   "the capacity of migrants": Hearne, *A Journey*, pp. lxxi, 47. Others later changed "barren grounds" into "barrenground," "barren lands," "the barrens," or the French "terres stériles"; capitalized it; and wrote it onto maps. Its origins and uses are traced in Louis-Edward Hamelin, "Barren Grounds—Terres Stériles Geographie et Terminologie," in Kenneth Coates and William Morrison, eds., *For Purposes of Dominion* (North York, ON: Captus University Publications, 1989), pp. 203–207. See also Gordon, *People of Sunlight*, p. 5. *Hosi* is the term used by the now-dominant Dogrib tribe; *dechinule*, by some other Dene. Inuit, who generally live beyond the tree line, distinguish only between "where there's trees"—*napaktokanik*—and "where there's no trees"—*napaktoilgoak*.

44   "edgeways to windward": Hearne, *A Journey*, p. 3.

44   "keep up my spirits": Ibid., pp. 21–22, 44.

45   "the greatest travellers in the known world": Ibid., pp. 52, 80.

47   "nothing should be wanting": Ibid., pp. 74–75.

48   Large White Stone Lake: Ibid., p. 107.

49   "for good luck": Ibid., p. 85.

49   "the Copper-mine River": Ibid., p. 93.

50   "the friendly blow": Ibid., p. 100.

51   "more than knee-deep": Ibid., p. 112.

51   discoverer was a woman: Some details of this legend are taken from George Back, *Arctic Artist*, ed. C. Stuart Houston (Montreal: McGill-Queen's University Press, 1994), p. 54.

53   the sixty-year-old fort: An account of the fall of Prince of Wales's Fort and ensuing misfortunes is in Peter Newman, *Company of Adventurers* (New York: Penguin, 1985), pp. 364–375, 381–385.

53   iron from the stone fort: Hearne, *A Journey*, p. 116.

53   British Museum of Natural History: The copper chunk is pictured in Speck, *Samuel Hearne*, p. 223.

Chapter 4: "A Deathly Stillness"

55   the 1819–1822 foray: There are four firsthand accounts of this trip, known as Franklin's First Overland Expedition: Sir John Richardson, *Arctic*

Ordeal, ed. C. Stuart Houston (Montreal: McGill-Queen's University Press, 1984); Captain John Franklin, *Narrative of a Journey to the Shores of the Polar Sea in the Years 1819, 20, 21, and 22* (Rutland, VT: Charles F. Tuttle, 1969 [reprint]); George Back, *Arctic Artist*, ed. C. Stuart Houston (Montreal: McGill-Queen's University Press, 1994); and Robert Hood, *To the Arctic by Canoe 1819–1821*, ed. C. Stuart Houston (Montreal: McGill-Queen's University Press, 1974). Hood and Back painted watercolors of the tundra, included in the books.

56 Big Foot: See June Helm, "Akaitcho," *Arctic* (June 1983): 208–209.

56 "painful to the eye": Quoted in Kenneth Coates, *Canada's Colonies: A History of the Yukon and Northwest Territories* (Toronto: James Lorimer & Company, 1985), p. 33.

57 "the Indians have found it": Richardson, *Arctic Ordeal*, p. 312.

58 Of twenty men: Franklin and Richardson conducted a second overland expedition to chart the coast west from the Coppermine into Alaska, from 1825 to 1827. There were no casualties. In 1833–1835, Captain George Back, who made both trips, descended the Great Fish (now the Back) River in the eastern Barrens, bringing back many rocks. Akaitcho aided him on that as well. See Captain George Back, *Narrative of the Arctic Land Expedition to the Mouth of the Great Fish River, and Along the Shores of the Arctic Ocean, in the Years 1833, 1834, and 1835* (Edmonton: M. G. Hurtig, 1970 [reprint]).

58 marks from cannibalism: Today a small cult of fans still looks for the vessels. See David C. Woodman, *Unraveling the Franklin Mystery* (Montreal: McGill-Queen's University Press, 1991); Anne Keenleyside et al., "Final Days of the Franklin Expedition: New Skeletal Evidence," *Arctic* (March 1997): 36–46. For the blond Inuit story, see Coates, *Canada's Colonies*, p. 161.

59 one more lost mine: See Ted Nagle and Jordan Zinovich, *The Prospector North of Sixty* (Edmonton: Lone Pine, 1989), pp. 199–201. The other known priests were Father Alphonse Gaste, who traveled in 1868 with the Dene, and Father Emile Petitot, who in the 1860s and 1870s accompanied them several times into the tundra, sometimes to lands the Indians themselves did not know. See Donat Savoie, "Emile Petitot (1838–1916)," *Arctic* (September 1982): 446–447.

60 "a high railway embankment": Warburton Pike, *The Barren Ground of Northern Canada* (New York: Dutton, 1917), p. 68, from which this account is taken. For more on such explorers, see C. S. MacKinnon, "An Overview of Barren Lands' History," *Musk-Ox*, 40 (1994): 13–30; C. S. MacKinnon, "A History of the Thelon Game Sanctuary," *Musk-Ox*, 32 (1983): 44–60; Frank Russell, *Exploration of the Far North* (Iowa, 1898); Caspar Whitney, *On Snow Shoes to the Barren Grounds* (New York, 1896).

60 "my first shot settled him": Pike, *Barren Ground*, p. 69.

61 "a white man": Ibid., p. 134.

61 "the awful spell": Ibid., p. 117. In 1915, with World War I underway, Pike tried joining the British Army but was rejected because of his age. He then walked into the ocean off his home and plunged a knife into his heart. He was fifty-four. See R. H. Cockburn, "Warburton Pike," *Arctic* (June 1985): 152–153.

61 the "Arctic Prairies": See Ernest Thompson Seton, *The Arctic Prairies* (New York: International Universities Press, 1911); Richard Davis, "Ernest Seton," *Arctic* (June 1987): 170–171.

62 dead marmots: See George Douglas, *Lands Forlorn* (New York: Knickerbocker Press, 1914); Enid Mallory, *Coppermine: The Far North of George M. Douglas* (Peterborough, ON: Broadview Press, 1989). German geologist August Sandberg came with Douglas, ranged out from the river, and found a few bits of copper. The lesser-known Englishman David Hanbury also reached the Arctic coast in 1901, where the Inuit showed him small copper deposits on the islands. See David Hanbury, *Sport and Travel in the Northland of Canada* (New York: Macmillan, 1904).

62 spirits of the southern invaders: The killings of Radford and Street are described in Lawrence Millman, "Pursuing Justice in the Arctic," *Smithsonian* (May 1998): 74–85. The murdered missionaries were Fathers Jean-Baptiste Rouviere and Guillaume LeRoux. See Mallory, *Coppermine*, pp. 135–146; and Diamond Jenness, *The People of the Twilight* (Calgary: University of Calgary Press, 1959), p. 79.

62 under some stones: Inexplicably, Hornby is often viewed as the Barrens' most romantic figure. He hung around with traders for years, and in occasional stabs into the tundra tried trapping, writing, and prospecting. His educated accent and insectlike strength, crossed with his perpetual filth and erratic behavior, both charmed and alarmed. Before his final journey he enticed James Critchell-Bullock, late of the 18th Lancers of India, to winter in that "virgin" land, and they spent months of voluntary misery in a cave dug into an esker. Hornby's nephew kept a gripping premortem diary, found near the bodies. It is published as Edgar Christian, *Death in the Barren Ground* (Ottawa: Oberon Press, 1980). See also George Whalley, *The Legend of John Hornby* (London: John Murray, 1962); Malcolm Waldron, *Snow Man: John Hornby in the Barren Lands* (New York: Houghton Mifflin, 1931); Mallory, *Coppermine*; Nagle and Zinovich, *Prospector North of Sixty*, pp. 71–72.

62 GSC men: For more on the GSC, see Morris Zaslow, *Reading the Rocks: The Story of the Geological Survey of Canada 1842–1972* (Toronto: Macmillan, 1972); Christy Vodden, *No Stone Unturned: 150 Years of the Geological Survey of Canada* (Ottawa: GSC, 1992); Morris Zaslow, *The Opening of the Canadian North 1870–1914* (Toronto: McClelland and Stewart, 1971), pp. 77–87; Michael Clugston, "A Passion for Exploration," *Canadian Geographic* (May/June 1992): 66–77.

63 the Keewatin: In 1894 they canoed a parallel system. See J. Burr Tyrrell, *Report on the Doobaunt, Kazan and Ferguson Rivers* (Ottawa: GSC, Annual Report, Section F, 1896); J. Burr Tyrrell, "An Expedition Through the Barren Lands of Northern Canada," *Geographical Journal*, 4, no. 5 (1894): 437–449; Alex Inglis, *Northern Vagabond: The Life and Career of J. B. Tyrrell* (Toronto: McClelland and Stewart, 1978); Katharine Martyn, *J. B. Tyrrell: Explorer and*

Adventurer (Toronto: Thomas Fisher Rare Book Library, 1993). Joseph's brother James came on the 1894 trip and wrote J. W. Tyrrell, *Across the Subarctics of Canada* (Toronto: Coles, 1973 [reprint]).

64 2,500 miles altogether: At one point Canadian businessmen floated a scheme to build a transcontinental railroad from Hudson Bay to the Klondike across the Barren Lands. It was never attempted. See Pierre Berton, *Klondike: The Last Great Gold Rush, 1896–1899* (Toronto: McClelland and Stewart, 1997), pp. 120, 216–233; Arthur Arnold Dietz, *Mad Rush for Gold in Frozen North* (Los Angeles: Times-Mirror, 1914); Richard Friesen, *The Chilkoot Pass and the Great Gold Rush of 1898* (Ottawa: Parks Canada, 1981).

64 "extraordinarily rich deposits": James Bell, *Report on the Topography and Geology of Great Bear Lake and a Chain of Lakes and Streams Thence to Great Slave Lake* (Ottawa: GSC, Annual Report, Part C, 1901). Tales of early gold prospectors are from Nagle and Zinovich, *Prospector North of Sixty*, pp. 17–19, 77–90; Ray Price, *Yellowknife* (Toronto: Peter Martin Associates, 1967), pp.19–29, 56–65; Hanbury, *Sport and Travel*, p. 29.

64 "Work by day": Bertram Barker, *North of '53* (London: Methuen, 1934), pp. 3, 15.

65 in hopes of aiding development: Early Yellowknife history from Price, *Yellowknife*, pp. 65–258; NWT Chamber of Mines, *Mining: Our Northern Legacy* (Yellowknife: NWT Chamber of Mines, 1989); Mallory, *Coppermine*, pp. 220–222; Pierre Berton, *The Mysterious North* (New York: Knopf, 1956), pp. 289–321; J. O. de Wet, "Mining in the North," *The Beaver* (December 1945): 20–25.

65 Hiroshima and Nagasaki: LaBine spotted the Eldorado in 1930, even before the Yellowknife gold strike. It is said that years later, when native elders realized the connection of the uranium to the atomic bombs, they descended into deep depression. See Leslie Roberts, *The Mackenzie* (New York: Rinehart, 1949), pp. 181–188, 198–201, 227–236; Anonymous, *Historical Highlights of Canadian Mining* (Toronto: Pitt, 1973), pp. 134–137.

67 wound up a double amputee: The story of the lost prospectors is in Guy Blanchet, *Search in the North* (Toronto: Macmillan, 1960), pp. 64, 164, along with other depictions of early Barrens flying. See also Shirley Milligan, ed., *Living Explorers of the Canadian Arctic* (Yellowknife: Outcrop, 1978), pp. 255–277; Nagle and Zinovich, *Prospector North of Sixty*, pp. 44, 166–170; Roberts, *The Mackenzie*, pp. 182–187; Mallory, *Coppermine*, pp. 174–179.

67 these few planes: In July 1929, Northern Aerial staked many miles of claims on the Coppermine, as did its competitor Dominion Explorers. Staking rushes were repeated in later decades. See L. T. Burwash, *Coronation Gulf Copper Deposits* (Ottawa: Department of Interior, 1930); Peter Usher, *Economic Basis and Resource Use of the Coppermine-Holman Region* (Ottawa: Department of Northern Affairs, 1965), pp. 37–67; and E. D. Kindle, *Classification and Description of Copper Deposits, Coppermine River Area, District of Mackenzie* (Ottawa: GSC, Bulletin 214, 1972).

68 nothing of economic significance: See C. H. Stockwell, *Great Slave Lake-Coppermine River Area, Northwest Territories* (Ottawa: GSC, Annual Report, Part C, 1932), pp. 37–63. Discovery of the esker systems is reported in J. T. Wilson, "Eskers North-East of Great Slave Lake," *Transactions of the Royal Society of Canada*, Section IV (1939): 119–129; J. T. Wilson, "Further Eskers North of Great Slave Lake," *Transactions of the Royal Society of Canada*, Section IV (1945): 151–153.

69 GSC published his map: R. E. Folinsbee, *Preliminary Map: Lac de Gras* (Ottawa: GSC, Paper 47-5, 1947). He wrote up the heavy minerals in R. E. Folinsbee, "Archean Monazite in Beach Concentrates, Yellowknife Geologic Province, NWT," *Transactions of the Royal Society of Canada*, 49, series 3 (June 1955): Section 4, 7–24. Other details are from the author's interview with Folinsbee in 1999.

70 drink themselves to death: Accounts of epidemics and the decline of Barrens travel come from interviews with Dogrib elders. See also Price, *Yellowknife*, pp. 9–10; "Yellowknife," in June Helm, *Handbook of North American Indians*, Vol. 6 (Washington, DC: Smithsonian Institution, 1981), pp. 285–290; Jenness, *People of the Twilight*; Richard Condon, *A History of the Copper Eskimo* (Norman: University of Oklahoma Press, 1996); "Copper Eskimo," in Helm, *Handbook of American Indians*, Vol. 5, pp. 397–419; John Bockstoce, "Contacts Between American Whalemen and the Copper Eskimos," *Arctic* (December 1975): 298–299; Ernest Schusky, ed., *Political Organizations of Native North Americans* (Washington, DC: University Press of America, 1980), pp. 215–242; Farley Mowat, *People of the Deer* (Toronto: Seal Books, 1985); Farley Mowat, *The Desperate People* (Toronto: Seal Books, 1985); MacKinnon, *For Purposes of Dominion*, pp. 159–170.

71 "we were the timid strangers": Prentice Downes, *Sleeping Island: The Story of One Man's Travels in the Great Barren Lands of the Canadian North* (London: Jenkins, 1943), p. 118. See also "Prentice G. Downes," *Arctic* (September 1982): 448–449.

Chapter 5: Golconda

76 a huge mine: Sources on India include *Les Six Voyages de J.-B. Tavernier en Turquie, en Perse et aux Indes* (Amsterdam, 1678); Max Bauer, *Precious Stones*, Vol. I (New York: Dover, 1968 [reprint]), pp. 140–155; Sidney Ball, "Historical Notes on Gem Mining," *Economic Geology* (November 1931): 681–738; Gardner Williams, *The Diamond Mines of South Africa* (New York: B. F. Buck, 1905), pp. 11–19; Eric Bruton, *Diamonds* (Radnor, PA: Chilton, 1978), pp. 23–26; George E. Harlow, ed., *The Nature of Diamonds* (Cambridge: Cambridge University Press, 1998), pp. 73–75, 117–122, 133–134; Robert Maillard, ed., *Diamonds: Myth, Magic, and Reality* (New York: Crown, 1980), pp. 18–28, 41–46, 51–52.

77 specimens smuggled through Brazil: Sources on Brazil include John Mawe, *Travels Through the Diamond District of Brazil* (London, 1812); John Mawe, *Treatise on Diamonds and Precious Stones* (London: Longman, Hurst,

1835); Richard Burton, *Explorations of the Highlands of Brazil, with a Full Account of the Gold and Diamond Mines*, Vol. 1 (New York: Greenwood Press, 1969 [reprint]); Thomas Draper, "Diamond Mining in Brazil," *Gems & Gemology* (Winter 1949–1950): 231–242; J. M. Catharino, *Garimpo-Garimpeiro-Garimpagem* (Rio de Janeiro: Philobiblion Livros e Arte Limitada, 1986); Bauer, *Precious Stones*, pp. 155–178; Bruton, *Diamonds*, 27–28; Harlow, *Nature of Diamonds*, 75–78; Maillard, *Diamonds*, 54–65, 172–173.

78 to float off light sands: See Georgius Agricola, *De Re Metallica*, trans. Herbert Clark Hoover (New York: Dover, 1950), pp. 334–335.

79 "not that it is there": Burton, *Explorations of the Highlands*, p. 140.

80 history "cannot be traced": Quoted in ibid., pp. 138–139.

80 "as compared with other minerals": Bauer, *Precious Stones*, p. 166.

81 an argument for their origination: See Burton, *Explorations of the Highlands*, p. 139.

81 also by experiment: Early experiments on diamonds described in Anselmi Boetii de Boodt, *Gemmarum et Lapidum Historia* (Hanover: Typis Wechelianus Claudium Marnium, 1609); Sarah Maria Burnham, *Precious Stones in Nature, Art and Literature* (Boston: B. Whidden, 1886), pp. 174–184; Mawe, *Treatise*, pp. 11–29; Bauer, *Precious Stones*, pp. 113–138, 233–237; Harlow, *Nature of Diamonds*, pp. 255–257; Maillard, *Diamonds*, pp. 91, 193–196, 223; Bruton, *Diamonds*, pp. 421–422.

81 soft, black stuff—graphite: After the French Revolution of 1789, beheaded royals' diamonds were "given" to the masses, and newly expert mineralogists discovered many of the best were fakes. This only hiked the prices for real ones. "That a house with a large estate, the means of living not only at ease, but in splendour, should be . . . deemed inadequate to the purchase of a transparent crystallized stone not half the size of a hen's egg, seems almost a kind of insanity," wrote English mineralogist and jeweler John Mawe around 1820. See Mawe, *Treatise*, pp. 1–2. See also "The Colorful Career of John Mawe," *The Gemmologist* (September 1954): 157–158.

81 diamonds must form from plants: While most geologists say the carbon is primordial, variants in isotopes of some diamonds suggest multiple sources—in rare cases, possibly organic matter. In 1996, University of Edinburgh researcher Ben Harte found a diamond containing a tiny inclusion of the mineral staurolite, apparently from the crust. If staurolite got there, bacterial or plant remains could also. Some even suggest that seeds were provided by meteorites penetrating into the deep earth with intact diamonds from outer space. See Jeff Hecht, "You Can't Keep a Good Crystal Down," *New Scientist*, January 13, 1996, p. 19; Stephen Haggerty, "Superkimberlites: A Geodynamic Diamond Window to the Earth's Core," *Earth and Planetary Science Letters*, 122 (1994): 57–69.

82 compressed amber: In 1856 an American calling himself Hipponax Roset—one suspects a pen name here—argued, "Diamond has never been found in rocks, as are all other minerals, but only in gravel and mud conglomerations . . .

deep ravines on the slopes of mountains, and in cavities and water courses on the summits of, sometimes, the loftiest elevations; and hence it is believed to be the product of vegetable secretion." Hipponax Roset, *Jewelry and the Precious Stones* (Philadelphia: John Pennington & Son, 1856).

82 making synthetic diamonds: See Robert Hazen, *The Diamond Makers* (New York: Cambridge University Press, 1999); Ralph Holmes, "Synthetic and Other Man-Made Gems," *Foote Prints*, 32, no. 1 (1960): 3–16; Augusto Castellani, *Gems: Notes and Extracts* (London: Bell & Daldy, 1871), pp. 100–102; Bauer, *Precious Stones*, p. 237.

83 "nearly right angles": Lardner Vanuxem, *Geology of New York, Part III: Survey of the Third Geological District* (Albany: State of New York, 1842), p. 169. See also *First Annual Report of the Geological Survey of the State of New-York* (Assembly Document No. 161, 1837), pp. 3–7, 185–191; *Communication Relative to the Geological Survey of the State* (Assembly Document No. 275, 1839), pp. 260, 298. Vanuxem was also first to officially report on "Herkimer diamonds," very fine quartz crystals that are still found in Herkimer County, New York. See David Jensen, *Minerals of New York State* (Rochester, NY: Ward Press, 1978), pp. 11, 33–34, 85, 146.

83 gold seekers who found diamonds: See T. A. Rickard, *A History of American Mining* (New York: McGraw-Hill, 1932), pp. 18–20.

83 never found anything again: Loyd's story is from unpublished memoirs in the archives of the Georgia state geologist.

84 All were unverified: See George Kunz, "Occurrence of the Diamond in North America," *GSA Bulletin* (March 30, 1907): 692–694. Report of the Coco Creek diamonds and much other early diamond lore is found in Roy J. Holden, "The Punch Jones and Other Appalachian Diamonds," *Bulletin of the Virginia Polytechnic Institute, Engineering Experiment Station*, 56 (February 1944): 26–29. For the Vaucluse mine, see Palmer Sweet, "Diamonds in Virginia," *Virginia Minerals* (November 1996): 33–40.

84 George W. Featherstonhaugh: Featherstonhaugh surveyed the gold districts for the U.S. government in 1837, and in nearby Rutherford met Major Abram Forney, who reported finding "some very small but brilliant diamonds" in his washings. Kunz later reported that Featherstonhaugh found one himself around 1843, but his facts seem shaky. See Edmund Berkeley, *George William Featherstonhaugh: The First U.S. Government Geologist* (Tuscaloosa: University of Alabama Press, 1988), p. 200.

84 "mineral wealth of the country": C. U. Shepard, "Diamonds in North Carolina," *American Journal of Science*, 2, 2nd series (1846): 119, 253–254. Other accounts of early southern diamonds and theories include the correspondence of Assistant Geologist A. S. Furcron with Professor R. J. Holden 1943–1944 (unpublished, archives of the Georgia Geological Survey); "Mystery of Georgia's Diamonds Revived" (unpublished, archives of Georgia Geological Survey); *Mineral Resources of the U.S., 1883–84* (Washington, DC: USGS, 1885), pp. 728–733; George Kunz, *Gems & Precious Stones of North*

America (New York: Dover, 1968 [reprint]), pp. 13–24; John Sinkankas, *Gemstones of North America* (Princeton, NJ: D. Van Nostrand, 1959), pp. 32–33; Robert Cook, *Minerals of Georgia: Their Properties and Occurrences* (Atlanta: Georgia Department of Natural Resources, 1978), pp. 19–20; William Furbish, "Gold & Diamonds of North Carolina," *Rocks and Minerals* (March/April 1985): 75–76; George Kunz, "History of the Gems Found in North Carolina," *North Carolina Geological and Economic Survey Bulletin*, 12 (1907): 5–9; Dan Hausel, "Diamonds and Mantle Source Rocks in the Wyoming Craton with a Discussion of Other U.S. Occurrences," *Geological Survey of Wyoming Report*, 53 (1998): 50–59.

85 a once-famous diamond: Reports of the Dewey diamond from USGS, *Mineral Resources 1883–84*, pp. 728–729; Kunz, *Gems & Precious Stones*, pp. 16–17; Sweet, "Diamonds in Virginia," p. 35; Roset, *Jewelry and the Precious Stones*, p. 17.

86 a possible carbon source: See Kunz, *Gems & Precious Stones*, p. 18. For other early diamond-itacolumite associations, see A. J. A. Janse, "Kimberlites—Where and When," in J. E. Glover, ed., *Kimberlite Occurrence and Origin: A Basis for Conceptual Models in Exploration* (University of Western Australia, 1985); Bauer, *Precious Stones*, pp. 139, 221.

86 tiny pink garnets: See Otis Young, *Western Mining* (Norman: University of Oklahoma Press, 1970), pp. 108–110, 271.

86 east of Sacramento: The Reverend Chester Lyman, a Yale graduate gifted in sciences, authenticated it. See "Platinum and Diamonds in California," *American Journal of Science*, 8 (1849): 294. The other seminal source on early California diamonds is *Second Report of the State Mineralogist* (Sacramento: State of California, 1882), pp. 250–254. See also *Reports of the State Mineralogist* 1884 (168–171) and 1888 (106–107); *Minerals of California* (Sacramento: California State Mining Bureau, 1884), p. 164; *Minerals of California* (Sacramento: California State Mining Bureau, Bulletin 91, 1922), pp. 7–8; H. W. Turner, "Occurrence and Origin of Diamonds in California," *American Geologist*, 23 (1899): 182–185; George Kunz, *Gems, Jewelers' Materials and Ornamental Stones of California* (Sacramento: California State Mining Bureau, 1905), pp. 36–44; Eugene Blank, "Diamonds in California," *Rocks and Minerals* (December 1934): 179–182; Rudolph Kopf, "First Diamond Find in California—When and Where?" *California Geology* (July 1989): 160–162; USGS, *Mineral Resources 1883–84*, pp. 730–733; Kunz, *Gems & Precious Stones*, pp. 24–30; Bauer, *Precious Stones*, pp. 227–228; Sinkankas, *Gemstones of North America*, pp. 41–42; Hausel, "Diamonds and Mantle Source Rocks," pp. 76–82.

86 they were literally microscopic: Larger stones have been found since. In 1987, retired geologist Edgar Clark panned Hayfork Creek in Trinity County and found a 14.33-carat stone he named the Serendipity, the 17.83-carat Enigma, and the 32.99-carat Doubledipity. They are all industrials. See Rudolph Kopf, "Recent Discoveries of Large Diamonds in Trinity County, California," *Gems & Gemology* (Fall 1990): 212–219. A few have also surfaced farther upcoast, including a small stone found in 1932 in a Washington placer along the Oregon border. See Dan Hausel, "Pacific Coast Diamonds—An Unconventional Source Terrane," in A. R. Coyner, ed., *Geology and Ore Deposits of the American Cordillera* (Reno: Geological Society of Nevada, 1995), pp. 925–934.

87 "at the age of 88 years": Burr Evans, "Diamonds of Smiths Flat," *Engineering and Mining Journal*, November 4, 1916, pp. 814–815.

87 Gardner F. Williams: For more on Williams, see *Second Report of the State Mineralogist*, p. 253; Alpheus Williams, *Some Dreams Come True: Stories Leading Up to the Discovery of Copper, Diamonds and Gold in Southern Africa* (Cape Town, South Africa: Howard B. Timmins, 1948), pp. 217–221.

87 It was gold: Max Bauer noted, "The occurrence of the diamond in the United States of North America is so sparing that it has no effect whatever upon the diamond markets of the world. American stones are, however, greatly prized in the States, both for patriotic reasons and also on account of their scientific interest." Bauer, *Precious Stones*, p. 226.

87 middling-value diamonds: Kunz, "The Diamond in North America," p. 693.

88 "diamonds in rings and shirt fronts": Burton, *Explorations of the Highlands*, p. 149.

88 "in this country": *Times of London*, December 18, 1874.

89 the English Cape Colony: It is also said the first diamond seen by Europeans was in the leather bag of a Griqua sorcerer; or that in 1859 a Griqua found a small diamond along the Vaal River and sold it to a Boer. See Kunz, *Curious Lore of Precious Stones*, p. 74; and A. N. Wilson, *Diamonds from Birth to Eternity* (Santa Monica, CA: Gemological Institute of America, 1982), p. 133. Some put it even farther back, to an eighteenth-century mission map of the Orange with a notation: "Here be diamonds." See Ivor Herbert, *The Diamond Diggers* (London: Tom Stacey, 1972), p. 11. Other sources for early South Africa are W. Guybon Atherstone, "The Discovery of Diamonds at the Cape of Good Hope," *Geological Magazine* (May 1869): 208–213; Williams, *Diamond Mines of South Africa*, pp. 83–344; Oswald Doughty, *Early Diamond Days* (London: Longmans, Green, 1963); Eric Rosenthal, *River of Diamonds* (Cape Town, South Africa: Howard Timmins, n.d.); Hedley A. Chilvers, *The Story of De Beers* (London: Cassell, 1939), pp. 1–49; William Worger, *South Africa's City of Diamonds* (New Haven, CT: Yale Historical Publications, 1987), pp. 9–13; Janse, "Kimberlites—Where and When"; Stefan Kanfer, *The Last Empire: De Beers, Diamonds and the World* (New York: Farrar, Straus & Giroux, 1993); Williams, *Some Dreams Come True*, pp. 55–219; Bauer, *Precious Stones*, pp. 179–217; Bruton, *Diamonds*, pp. 28–102; Maillard, *Diamonds*, pp. 66–74.

89 "Titaniferous Iron [and] Gold": James Gregory, "Diamonds from the Cape of Good Hope," *Geological Magazine* (December 1868): 559.

89 white 83.5-carat beauty: An Anglo-Jewish settler in the region visited the Griqua chief, Nicolaas Waterboer. Almost in a whisper, Waterboer told him the truth: "I know that there are diamonds in my country. I am sorry that it is so. Some of my people have been looking for them. . . . We are not going to tell you what we know, for I know it will mean ruin." Rosenthal, *River of Diamonds*, p. 15.

90 a mistake had been made: English digger John Angove wrote, "[When] a digger was fortunate enough to find a gem, [others] were immediately made aware . . . by the great shout [of] the lucky finder; there was then a rush to the spot. [Then] an adjournment made to the nearest canteen, where he would commemorate the event by standing drinks. . . . Should he neglect doing so, he was branded as mean in the extreme, and one who was not deserving any luck." John Angove, *In the Early Days* (Kimberley, South Africa: Handel House, 1910), p. 5.

90 "while lying in bed": "The Diamond-Fields at the Cape of Good Hope," *Every Saturday*, November 19, 1870, pp. 744–745.

90 "on Sunday last": Quoted in Doughty, *Early Diamond Days*, p. 78.

91 fifteen miles from any river: Unknown until later, the first dry digging was probably actually found in July 1870 by a Boer cart driver named Bam, who picked up a diamond on the Koffiefontein ranch, sixty miles from the Vaal. The following month at the Jagersfontein ranch, even farther off, foreman Jaap de Klerk noticed a dry streambed sprinkled with small garnets. Six feet down he found a 50-carat stone. Both places were recognized a decade later as kimberlite pipes. One source says Cornelius duPlooy, owner of Bultfontein, found diamonds in the mud wall of his house, taken from the pan, in November 1869, before prospectors even arrived.

91 Hordes alit: The local paper reported that two boys, ages five and six, escaped their riverbound family and headed for the dry diggings themselves until a well-meaning soul returned them. "They protested tearfully, when found, that they could have got there the following day." Quoted in Chilvers, *Story of de Beers*, p. 14.

92 some ox yokes: Most of the original finders never profited. William Anderson, who started the rush at Dutoitspan and spearheaded the move to neighboring Bultfontein, barely missed staking actual kimberlite in both places. Richard Jackson, originator of the first De Beers rush, also missed. Fleetwood Rawstorne, who started the second De Beers rush, staked just at the edge of the pipe; when his ore ran out, he was forced to leave with just his tools. As for Erasmus Jacobs, finder of the first official diamond, it is not clear whether he received anything for it. When he was in his seventies and living without income in Kimberley, residents took pity and pooled 30 pounds for him. De Beers executives then felt shame and granted him a pension. The De Beer brothers' farm produced gems worth over 1 billion English pounds.

Chapter 6: The Great Hoax

93 "'Look at that!'": Mark Twain, *Roughing It* (Berkeley: University of California Press, 1993 [reprint]), pp. 195–196. Other North American frauds are covered in Otis Young, *Western Mining* (Norman: University of Oklahoma Press, 1970), pp. 40–51; Benjamin Fulford, "Diamonds in Canada? One of the

Oldest Scams in the Book," *Forbes* (August 1998): 1214. In Columbia County, California, some Americans in 1851 let their extra-wary Chinese customers pick a random spot to test-dig. Then a conspirator on a bluff above tossed an already dead rattlesnake onto them. The seller "saved" the terrified Chinese by shotgunning the serpent, in the process salting the hole under their noses.

94   Fraud instantly spread: South African frauds from Hedley A. Chilvers, *The Story of De Beers* (London: Cassell, 1939), p. 7; Oswald Doughty, *Early Diamond Days* (London: Longmans, Green, 1963), p. 70; Max Bauer, *Precious Stones*, Vol. I (New York: Dover, 1968 [reprint]), p. 184; John Angove, *In the Early Days* (Kimberley, South Africa: Handel House, 1910), pp. 83–86, 95–96.

94   Diamond stories about the Western deserts: Tall tales from Bruce Woodard, *Diamonds in the Salt* (Boulder, CO: Pruett Press, 1967), the best work on the Great Diamond Hoax. See also Asbury Harpending, *The Great Diamond Hoax and Other Stirring Incidents in the Life of Asbury Harpending*, ed. Glen Dawson (Norman: University of Oklahoma Press, 1958 [reprint]); Thomas Farish, *The Gold Hunters of California* (Chicago: M. A. Donohue & Co.), pp. 221–235; Henry Faul, "Century-Old Diamond Hoax Reexamined," *Geotimes* (October 1972): 23–25; Dan Hausel, "The Great Diamond Hoax of 1872," *Wyoming Geological Association 1995 Field Conference Guidebook*, pp. 13–27; Tom McCandless, "Mantle Indicator Minerals in Ant Mounds and Conglomerates of the Southern Green River Basin, Wyoming," *Wyoming Geological Association 1995 Field Conference Guidebook*, pp. 153–163; George Kunz, *Gems & Precious Stones of North America* (New York: Dover, 1968 [reprint]), pp. 36–37; T. A. Rickard, *A History of American Mining* (New York: McGraw-Hill, 1932), pp. 380–396.

95   "many surprises already": Harpending, *Great Diamond Hoax*, p. 143.

97   "the gem market of the world": Ibid., p. 158.

98   basically free extraction: "An Act to Promote the Development of Mining Resources in the United States," also known as the General Mining Law, May 10, 1872.

98   Mexicans in the states: Around 1833, Don Vincente Guerrero (later, president of Mexico) claimed Indians had brought him diamonds found inside geodes in the Sierra Madre, and that he later mined a few himself. At least one Mexican mineralogist took him seriously, but Americans scoffed. See Edmund Berkeley, *George W. Featherstonhaugh: The First U.S. Government Geologist* (Tuscaloosa: University of Alabama Press, 1988), p. 111.

99   a "real diamond field": Quoted in Woodard, *Diamonds in the Salt*, p. 53.

99   "loose pieces of brown hematite": Ibid., p. 92.

99   244-page book: M. F. Stephenson, *Geology and Mineraxzlogy of Georgia, with a Particular Description of Her Rich Diamond District* (Atlanta: Globe Publishing Co., 1871). He followed with *Diamonds and Precious Stones of Georgia* (Gainesville, GA: Eagle Job Office, 1878).

102 "the work of a sinful man": Harpending, *Great Diamond Hoax*, p. 175.

102 "out-Heroding old Herod himself": Quoted in Woodard, *Diamonds in the Salt*, p. 121.

103 "a sort of tube in the earth": Quoted in Stefan Kanfer, *The Last Empire: De Beers, Diamonds and the World* (New York: Farrar, Straus & Giroux, 1993), p. 57.

103 without killing everyone: Williams saw the Kimberley pits as "open graves" where humans worked in "imminent hazard of maiming and death." Gardner Williams, *The Diamond Mines of South Africa* (New York: B. F. Buck, 1905), pp. 274–275.

104 De Beers bought them out: This mine was named the Premier, in honor of Rhodes, who became premier of the English Cape Colony. De Beers itself nearly fell victim to a real salting scam. In 1898, two prospectors named Armstrong and Carter found diamonds on a farm called Roodedam, and Rhodes jumped to buy the claims. Williams smelled a rat when the men wanted upfront money to do preliminary tests. He disobeyed Rhodes's instructions to sign a deal and refused to even visit. It had been salted with a measly 102 carats. See Alpheus Williams, *Some Dreams Come True: Stories Leading Up to the Discovery of Copper, Diamonds and Gold in Southern Africa* (Cape Town, South Africa: Howard B. Timmins, 1948), pp. 417–424.

104 "they are very near": Quoted in Chilvers, *Story of de Beers*, p. 181.

105 Barnato's "tube in the earth": Early studies of kimberlite pipes from Theodore Reunert, *Diamonds and Gold in South Africa* (London: Edward Stanford, 1893); Henry Carvill Lewis, "On a Diamantiferous Peridotite, and the Genesis of the Diamond," *Geological Magazine* (January 1897): 22–24; Henry Carvill Lewis, *Papers and Notes on the Genesis and Matrix of the Diamond* (New York: Longmans, Green, 1897); Bauer, *Precious Stones*, pp. 187–197; Williams, *Diamond Mines of South Africa*, pp. 118–150; A. J. A. Janse, "Kimberlites—Where and When," in J. E. Glover, ed., *Kimberlite Occurrence and Origin: A Basis for Conceptual Models in Exploration* (University of Western Australia, 1985), pp. 4–7.

107 "the parent rock": T. G. Bonney, "Parent-Rock of the Diamond in South Africa," *Geological Magazine* (July 1899): 308–321.

107 no orange ones: Dozens of scientists unsuccessfully tried pinpointing diamond origins by reviving the attempt to make artificial ones, from carbon and now also kimberlite. With academies offering large cash prizes, several salted their equipment with natural stones but were caught. Ever-increasing temperatures and pressures in labs produced only explosions. The great English physicist Sir William Crookes tried wildly complex experiments and claimed success, but his results were inconclusive, his explanation incomprehensible: "I have shown that a diamond is the outcome of a series of Titanic earth convulsions, and that these precious gems undergo cycles of fiery, strange, and potent vicissitudes before they can blaze on a ring." Sir William Crookes, *Diamonds* (London: Harper & Brothers, 1909), p. xi.

107 "the light of day": Williams, *Diamond Mines of South Africa*, p. 150.

108 several small igneous knobs: The best overview of early U.S. kimberlite history, with many references, is Henry Meyer, "Kimberlites of the Continental United States," *Journal of Geology* (July 1976): 377–403. For Kentucky, see Lewis, *Genesis and Matrix of the Diamond*, pp. 10–11, 35, 53–67; J. C. Branner, "Facts and Corrections Regarding the Diamond Regions of Arkansas," *Engineering and Mining Journal*, February 13, 1909, pp. 371–372; Joseph Diller and George Kunz, "Is There a Diamond Field in Kentucky?" *Science*, 10, no. 241 (1887): 140–142; Kunz, *Gems & Precious Stones*, pp. 31–34; George Kunz, "Occurrence of the Diamond in North America," *GSA Bulletin*, March 30, 1907, p. 694; Howard Millar, *It Was Finders Keepers at America's Only Diamond Mine* (New York: Hearthstone Books, 1976), pp. 28–31; USGS, *Mineral Resources*, "Precious Stones" sections, 1886, 1889–1890, 1891, 1906, 1907, 1908. For Syracuse, see Lewis, *Genesis and Matrix of the Diamond*, pp. 58–60, 64–65; Philip Schneider, "Diamond in Central New York," *Syracuse Herald*, December 24, 1905; USGS, *Mineral Resources 1913, Part II*, p. 666; and correspondence of William Kelly, curator of mineralogy, New York State Museum, Albany.

109 "nobody knows": Branner, "Facts Regarding the Diamond Regions," 371.

109 hydrochloric acid: The next year in May, Kentuckian C. O. Helm did find a .776-carat diamond shaped like a wheat grain—but in a gravelly old field high on a hill above Cabin Fort Creek, 150 miles southwest of Isom's Creek. No one could say where it came from. See Roy J. Holden, "The Punch Jones and Other Appalachian Diamonds," *Bulletin of the Virginia Polytechnic Institute, Engineering Experiment Station*, 56 (February 1944): 25–26.

110 In nearby Ithaca: The Cascadilla dikes are still next to Central Street Bridge, obscured by dead leaves and trash. See J. H. Martens, "Igneous Rocks of Ithaca, New York, and Vicinity," *GSA Bulletin*, June 30, 1924, pp. 305–320. Apart from Syracuse, New York has three reported stones: from a rock blasted out of the Grass River near Massena around 1889; from inside a butchered chicken in Cold Spring before 1941; and from glacial drift near Plattsburgh in 1941. All are unverified. See B. Mather, "Diamonds in N. America," *Natural History Society of Maryland Bulletin*, 11, no. 6 (1941): 106–112; USGS, *Mineral Resources 1909, Part II*, p. 761.

110 "it must carry diamonds": USGS, *Mineral Resources 1906*, p. 1219. The search for California kimberlites is further described USGS, *Mineral Resources*, "Precious Stones" sections, 1907, 1908, 1909, 1910, 1911, 1912, 1913, 1914, 1915; Winfield Scott, "A Diamond Quest in California," *Scientific American* (May 1926): 312–313; John Sinkankas, *Gemstones of North America* (Princeton, NJ: D. Van Nostrand, 1959), p. 41; Dan Hausel, "Pacific Coast Diamonds—An Unconventional Source Terrane," in A. R. Coyner, ed., *Geology and Ore Deposits of the American Cordillera* (Reno: Geological Society of Nevada, 1995), p. 930.

110 a somewhat ill-fated one: Kunz wrote the USGS's "Precious Stones" reports from 1882 to 1905 and the U.S. Bureau of the Census gem reports from 1905 to 1929. Among his many other pursuits, he spearheaded an effort to convert the United States to the metric system and promoted the planting of trees in Iceland, as well as in New York's Central Park. Information on him can be found in George Kunz, "American Travels of a Gem Collector," *Saturday Evening Post*, November 26, 1927; December 10, 1927; January 21, 1928; March 16, 1928; March 31, 1928; and May 5, 1928; Joseph Purtell, *The Tiffany Touch* (New York: Random House, 1971), pp. 61–80; Paul Kerr, "Memorial of George Frederick Kunz," *American Mineralogist* (March 1933): 91–94; Herbert Whitlock, "Memorial and Bibliography of George Frederick Kunz," *GSA Bulletin* (April 30, 1933): 377–394; "George Frederick Kunz," *Natural History* (September 1932): 554; Kunz, *Gems & Precious Stones*; "Dr. George F. Kunz," *New York Times*, June 30, 1932; numerous other *New York Times* articles, 1888–1932; Lawrence Conklin, *Notes and Commentaries on Letters to George F. Kunz* (New Canaan, CT: Privately printed, 1986). Caches of his papers are at the USGS Rare Book Library, Reston, Va.; the American Museum of Natural History; and the New-York Historical Society.

112 "poorest of all": Kunz, "American Travels," November 26, 1927, p. 85.

113 his 1886 USGS report: USGS, *Mineral Resources 1886*, p. 598.

113 rings mounted with uncut industrials: See Kunz, *Gems & Precious Stones*, pp. 37–38; George Kunz, "History of the Gems Found in North Carolina," *North Carolina Geological and Economic Survey Bulletin*, 12 (1907): 1–3; Bauer, *Precious Stones*, p. 228.

114 Arizona's Canyon Diablo meteorite: Probably the first known meteoritic diamonds were from fragments in Novy Urej, Russia, found in the 1880s. Most agreed they were created on impact, with no connection to kimberlite. However, in 1890 chemist A. Meydenbauer suggested that kimberlite pipes were bored by meteorites bearing diamonds made in space and that alluvial diamonds rained from the sky when meteors disintegrated in midair. While most geologists doubted it, they noted kimberlite often did resemble meteorites for unknown reasons.

115 "very peculiar circumstances": Kunz, *Gems & Precious Stones*, p. 35. Other Great Lakes material is from USGS, *Mineral Resources*, "Precious Stones" sections, 1883–1884, 1893, 1894, 1895, 1896, 1898, 1899, 1900, 1902, 1904, 1905; "The Diamond Mine of Wisconsin," *Scientific American*, Supplement No. 583 (1887): 9307–9308; George Kunz, "On the Occurrence of Diamonds in Wisconsin," *GSA Bulletin*, August 7, 1891, pp. 638–639; George Kunz, "Diamonds in Wisconsin," *Engineering and Mining Journal*, 1 (1890): 686; Rudolph Kopf, "Newspaper Articles That Refer to Diamond Prospecting and Recovery in Wisconsin, 1887–1889" (unpublished, Wisconsin Geological Survey archives, 1987); correspondence and unidentified news clippings on Plum Creek diamonds, 1906–1907 (unpublished, Wisconsin Geological Survey archives); William H. Hobbs, "On a Recent Diamond Find in Wisconsin," *American Geologist* (July 1894): 31–35; William H. Hobbs, "The Diamond Field of the Great Lakes," *Journal of Geology* (May–June 1899): 375–388; W. S. Blatchley, *Gold and Diamonds in Indiana* (Bloomington: Indiana Department of Geology, 1903) (expanded version republished in 1963 and again as John Hill, *Gold and Diamonds in Indiana: An Update* [Indiana Geological Survey circular 12, 1994]); Oliver Farrington, "Correlation of Distribution of Copper and Diamonds in Glacial Drift of the Great Lakes Region," *Science*, May 8, 1908, p. 729; Neil

Clark, "Wisconsin's Mysterious Diamond Hoard," *Saturday Evening Post*, January 7, 1950, pp. 28, 97; Edwin Olson, "History of Diamonds in Wisconsin," *Gems & Gemology* (Spring 1953): 284–285; Christopher Gunn, *Provenance of Diamonds in Glacial Drift of the Great Lakes Region, North America* (unpublished master's thesis, University of Western Ontario, London, 1967); Christopher Gunn, "Relevance of the Great Lakes Discoveries to Canadian Diamond Prospecting," *Canadian Mining Journal* (July 1968): 39–42; Esther Middlewood, "Tracking the Dowagiac Diamond," *Michigan Natural Resources Magazine* (March–April 1976): 20–21; Amy Silvers, "When Eagle Pinned Its Hopes on Diamonds," *Milwaukee Journal*, November 9, 1978, pp. 1, 10; W. F. Cannon, "Potential for Diamond-Bearing Kimberlite in Northern Michigan and Wisconsin" (Washington, DC: USGS, Circular 842, 1981); Michael Hansen, "Diamonds from Ohio," *Ohio Geology* (Fall 1982): 1–3; Michael Hansen, "Additional Notes on Ohio Diamonds," *Ohio Geology* (Winter 1985): 1–4; Burton Westman, "Indiana Gold and Diamonds: Are They Indigenous?" (unpublished, Indiana Geological Survey archives, circa 1980).

115 "he soon may be": Unpublished document, Indiana Geological Survey archives, 1888. The first known Indiana diamond was found by gold panner Peter Davis—a greenish 3-carat stone from a tributary to Little Indian Creek.

116 "the ancestral home": Hobbs, "Diamond Field," p. 383. While Hobbs was the moving force, first mention of a "northern" diamond mine belongs to Professor E. T. Cox, who first noted the Indiana stones, saying: "These are interesting facts, and point to the existence of a true diamond field, somewhere in the beds of crystalline rock to the north." But it is not clear how far north he meant. *Report of the Geological Survey of Indiana*, 1875, p. 294.

117 "barren of minerals of any kind": Robert Bell, "Occurrence of Diamonds in the Drift of Some Northern States," *Engineering and Mining Journal* (November 3, 1906): 819. See also *Economic Geology of the Rainy Lake Region* (Ottawa: GSC, Annual Report 1887–88, Vol. 3), p. 180f.

117 "I think we ought to look": Archibald Blue, "Are There Diamonds in Ontario?" *Report of the [Ontario] Bureau of Mines 1900*, pp. 119–124. Reports of other early Canadian exploration from USGS, *Mineral Resources*, "Precious Stones" sections, 1902, 1904, 1905, 1906.

118 "gem expert of Tiffany & Co.": "Diamonds Discovered in Wisconsin," clipping from *Chicago American 1906* (undated), and related unpublished correspondence from the Wisconsin Geological Survey archives.

118 its biggest customer: Statistics combined from USGS, *Mineral Resources*, 1887, pp. 560–571; 1893, pp. 687–689; 1905, pp. 986–987. Kunz gave the official figure of U.S. diamond production from 1880 from 1905 as $1,460; I have approximated pre-1880 finds, plus a few diamonds Kunz obviously left out. See also George E. Harlow, ed., *The Nature of Diamonds* (Cambridge: Cambridge University Press, 1998), pp. 105, 186–188, 208–213; Robert Maillard, ed., *Diamonds: Myth, Magic, and Reality* (New York: Crown, 1980), pp. 51–53; Percy Wagner, *The Diamond Fields of Southern Africa* (Johannesburg, South Africa: Transvaal Leader, 1914), pp. 241–242.

119 "Wealth may still lie": Kunz, "American Travels," November 26, 1927, p. 87.

119 "Ivory is my life": Quoted in Purtell, *Tiffany Touch*, pp. 73–74.

119 Kimberley, Nevada: Philip Earl, "Tonopah's Great Diamond Rush," *California Mining Journal* (January 1983): 36–37.

Chapter 7: Crater of Diamonds

121 a real diamond pipe surfaced: See J. C. Branner, "Peridotite of Pike County, Arkansas," *American Journal of Science*, 38 (1889): 50–59; J. C. Branner, "Facts and Corrections Regarding the Diamond Regions of Arkansas," *Engineering and Mining Journal*, February 13, 1909, pp. 371–372.

122 "extraction [should be entirely] successful": George Kunz, "Diamonds in Arkansas," *American Institute of Mining Engineers Bulletin*, 20 (1908): 191–192. Other accounts of Arkansas include USGS, *Mineral Resources*, "Precious Stones" sections, 1906, 1907, 1908, 1909, 1910, 1911, 1912, 1913, 1914, 1915, 1916, 1919, 1920, 1921; George Kunz, "Note on the Forms of Arkansas Diamonds," *American Journal of Science*, 24 (1907): 275–276; A. H. Purdue, "A New Discovery of Peridotite in Arkansas," *American Geologist* (August–September 1908): 525–528; Hugh Miser, "Diamond-Bearing Peridotite in Pike County, Arkansas," *Economic Geology* (December 1922): 662–674; Jim Ferguson, *Minerals in Arkansas* (Little Rock: Arkansas Department of Mines, Manufactures and Agriculture, 1922), pp. 55–56; George Branner, *Outlines of Arkansas' Mineral Resources* (Little Rock: Arkansas Geological Survey, 1927), pp. 125–133; Eugene Blank, "Diamonds in Arkansas," *Rocks and Minerals* (January 1935): 7–10; and (February 1935): 23–26; Philip Henson, "Arkansas' Diamond Fields," *Gems & Gemology* (Fall 1940): 109–112; J. R. Thoenen, "Investigation of the Prairie Creek Diamond Area," *Earth Science Digest*, 4, no. 6 (1950): 3–8; Martin Gross, "Why We Have No Diamonds," *True* (September 1959): 52–102; Harold Begell, "Blue Ground + Yellow Dirt = Diamonds," *Rocks and Minerals* (May 1971): 306–307; Howard Millar, *It Was Finders Keepers at America's Only Diamond Mine* (New York: Hearthstone Books, 1976); Burton Westman, "If Diamonds Are Not Vigorously Sought, It Is Not Logical to Argue That Diamonds Are Rare," *California Mining Journal* (July 1982): 26–31; (August 1982): 66–70; and (September 1982): unpaged; Albert Kidwell, "Famous Mineral Localities: Murfreesboro, Arkansas," *Mineralogical Record* (November–December 1990): 545–555.

124 "The only way": John Fuller, "Diamond Mine in Pike County, Arkansas," *Engineering and Mining Journal*, January 16, 1909, pp. 152–155.

128 others headed up: Hudson Bay accounts from *Mineral Industry of the British Empire: Gemstones* (London: Imperial Institute, 1933), p. 58; Gilean Douglas, "Diamond Rush in Canada," *Canadian Mining Journal* (January 1952): 63–64.

128 the size of a robin's egg: This first Canadian diamond was called the Peterborough, for the locality where it was found. See George Kunz, "Diamonds in North America," *GSA Bulletin* (March 31, 1931): 221–222.

128 at least that far north: In 1911 the GSC announced discovery of kimberlitelike rocks in the British Columbia mountains, sixty miles southwest of Kelowna, and recovered microscopic crystals Kunz thought were diamonds. Almost forty years later, better technology proved they were periclase crystals a labworker had made by heating the sample. At the time it was enough to set off a brief staking rush and bolster theories that U.S. stones came from the north. A similar mistake was made in Ontario a few years later. See Charles Camsell, "New Diamond Locality in the Tulameen District, B.C.," *Economic Geology* (September 1911): 604–611; A. N. LeCheminant et al., eds., *Searching for Diamonds in Canada* (Ottawa: GSC, Open File 3228, 1996), p. 171; *Ontario Bureau of Mines 23rd Annual Report* (Toronto: Bureau of Mines, 1914), p. 47; Joseph Brunet, *Diamonds in Canada* (Ottawa, unpublished manuscript, 1967).

128 "in the Glacial Period": Kunz, "Diamonds in North America," p. 222.

128 "I have no doubt": Quoted in "Dr. Kunz Tells British Association of Discoveries Made in Glacial Drift," *New York Times*, August 13, 1924, p. 18.

128 "far from these sites": George Kunz, "American Travels of a Gem Collector," *Saturday Evening Post*, January 21, 1928, p. 33.

129 "Why not return those": Kunz, "Diamonds in North America," p. 222.

130 lonely twos or threes: Many South Africans still work small claims. See Jon Manchip White, *The Land God Made in Anger* (New York: Rand McNally, 1969); "The Diamond Diggers," *The Gemmologist* (December 1947): 345–353; "The Last Diamond Rush," *The Gemmologist* (January 1952): 15: "All Quiet at the Bloemhof Diggings," *The Gemmologist* (January 1953): 13; "Not Many in Latest Diamond Rush," *The Gemmologist* (February 1954): 38–39.

131 a beautiful autobiography: Fred Cornell, *The Glamour of Prospecting* (London: T. Fisher Unwin, 1920; Cape Town, South Africa: David Philip, 1986 [reprint]), p. 191. Cornell also depicted diamond prospectors in a book of short stories, *Rip van Winkle of the Kalahari* (1915).

131 forked divining rods: See Georgius Agricola, *De Re Metallica*, trans. Herbert Clark Hoover (New York: Dover, 1950), pp. 38–41; "The Diamond Diviner," *The Gemmologist* (September 1947): 254; "Finding Gems by Divining Rod," *The Gemmologist* (October 1948): 275; "Diamond Find in Lost River," *The Gemmologist* (September 1952): 188–189.

131 "lack of scientific knowledge": Alpheus Williams, *Some Dreams Come True: Stories Leading Up to the Discovery of Copper, Diamonds and Gold in Southern Africa* (Cape Town, South Africa: Howard B. Timmins, 1948), p. 435.

131 "A new era": Quoted in Timothy Green, *The World of Diamonds* (New York: William Morrow, 1981), p. 52.

132 *The Genesis of the Diamond*: Alpheus Williams, *The Genesis of the Diamond*, Vol. 1 (London: Ernest Benn, 1932), p. 373. About the only other major works useful to prospectors before the 1970s were Percy Wagner, *The Diamond Fields of Southern Africa* (Johannesburg, South Africa: Transvaal Leader, 1914); Djalma Guimaraes, *A Margem de Os Satelites do Diamante*

(Estado de Minas Gerais, Brazil: Servicio Geologico, Monografia 2, 1934); Edmond Bruet, *Le Diamant* (Paris: Payot, 1952).

133 U.S. government retaliated: See Edward Jay Epstein, *The Rise and Fall of Diamonds* (New York: Simon & Schuster, 1982), pp. 88–96.

133 "Sir Ernest Oppenheimer": Quoted in ibid., p. 92.

134 "to keep their production off the market": *U.S. v. de Beers Consolidated Mines Ltd. et al.* (U.S. SDNY, Civil Action 29-446, January 29, 1945). The charges were dropped from 1947 to 1949.

135 Canadian John Williamson: Sources on Williamson include John Gawaine, *The Diamond Seeker* (Johannesburg, South Africa: Macmillan, 1976); Timothy Green, "The John Williamson Story," *Diamond International* (November–December 1995): 110–111; H. F. Burgess, *Diamonds Unlimited* (London: The Adventurers Club, 1960); Robert Maillard, ed., *Diamonds: Myth, Magic, and Reality* (New York: Crown, 1980), pp. 101–103; Stefan Kanfer, *The Last Empire: De Beers, Diamonds and the World* (New York: Farrar, Straus & Giroux, 1993), pp. 242–246, 268–269; W. G. Lundstrum, "The Richest Four Miles on Earth," *The Gemmologist* (July 1952): 125–131. A scientific account of his pipe appears in C. B. Edwards, "Kimberlites in Tanganyika with Special Reference to the Mwadui Occurrence," *Economic Geology*, 61 (1966): 537–554.

136 *sekhama* mines: See G. M. Stockley, *Report on the Geology of Basutoland* (Maseru: Basutoland Government, 1947), p. 83; USGS, *Mineral Resources*, 1895, pp. 900–901; Eric Bruton, *Diamonds* (Radnor, PA: Chilton, 1978), p. 95; A. N. Wilson, *Diamonds from Birth to Eternity* (Santa Monica, CA: Gemological Institute of America, 1982), p. 203; Epstein, *Rise and Fall of Diamonds*, pp. 36–39.

137 the Russians were looking: Russian history from Jiri Strnad, "Discovery of Diamonds in Siberia and Other Northern Regions: Explorational, Historical and Personal Notes," *Earth Sciences History*, 10, no. 2 (1991): 227–246; P. J. Wylie, *Ultramafic and Related Rocks* (New York: John Wiley, 1967), pp. 251–261; V. M. Barygin, "Prospecting for Kimberlite Pipes from the Air," *Mining Magazine* (London) (August 1962): 73–78; "Diamond Deposits of Yakutia" (book review), *Economic Geology*, 55 (1960): 1569; "Secret of the Yakut Diamonds" (book review), *Economic Geology*, 54 (1959): 1326; "The Diamonds of Siberia" (book review), *Economic Geology*, 53 (1958): 220–221; C. F. Davidson, "The Diamond Fields of Yakutia," *Mining Magazine* (London) (December 1957): 329–338; Kanfer, *The Last Empire*, pp. 268–270; Warren Atkinson, "Diamond Exploration Philosophy, Practice, and Promises: A Review," in *Proceedings of the Fourth International Kimberlite Conference*, Vol. 2 (Carlton, Australia: Geological Society of Australia/Blackwell Scientific Publications, 1986), pp. 1084–1095; Wilson, *Diamonds from Birth to Eternity*, pp. 334–343; Green, *World of Diamonds*, pp. 100–111.

139 American geologist Noel Stearn: Noel Stearn, *Practical Geogmagnetic Exploration with the Hotchkiss Superdip*, American Institute of Mining and Metallurgical Engineers, Technical Publication 370, 1930.

142 many camps now moved: See Bruton, *Diamonds*, pp. 156–166.

144 "I own the only diamond mine": Millar, *Finders Keepers*, p. 98. He had a next-door competitor on another part of the pipe, variously called the Arkansas Diamond Mine, or the Big Mine.

Chapter 8: De Beers in America

145 more barren kimberlitic rocks: See George Hall, "Igneous Rock Areas in the Norris Region, Tennessee," *Journal of Geology*, 52 (1944): 424–430; Palmer Sweet, "Diamonds in Virginia," *Virginia Minerals* (November 1996): 37–39; J. E. Lamar, *Gold and Diamond Possibilities in Illinois* (unpublished, Illinois State Geological Survey, 1968); Henry Meyer, "Kimberlites of the Continental United States," *Journal of Geology* (July 1976): 377–403; Dan Hausel, "Diamonds and Mantle Source Rocks in the Wyoming Craton with a Discussion of Other U.S. Occurrences," *Geological Survey of Wyoming Report*, 53 (1998): 76–82.

145 near the Great Lakes: These were single small crystals near Peru, Indiana, and Salem, Ohio, some 300 miles apart. In the 1950s a University of Wisconsin mineralogy student brought in diamonds from a man supposedly mining them near Eagle, but they contained saw marks and a polished facet—apparent leftovers from a previous salting or part of a new one. See Frank Wade, "Another Diamond Found in Indiana," *Gems and Gemology* (Winter 1949–1950): 249–250; Michael Hansen, "Additional Notes on Ohio Diamonds," *Ohio Geology* (Winter 1985): 1–4; Arthur Vierthaler, "Wisconsin Diamonds," *Wisconsin Academy Review* (Spring 1958): 53–55; Arthur Vierthaler, "Wisconsin Diamonds," *Gems and Gemology* (Fall 1961): 210–215; "Mine Your Back Yard for Riches," *True* (March 1958).

146 a "dykelet" of kimberlite: See J. Satterly, *Geology of Michaud Township*, Vol. 55, Part 4 (Toronto: Ontario Department of Mines, Annual Report, 1949); K. D. Watson, "Kimberlite at Bachelor Lake, Quebec," *American Mineralogist*, 40 (1955): 565–579.

146 "few white men have set foot": Quoted in "Secret Rich Diamond Lode," *The Gemmologist* (February 1948): 36. See also D. S. M. Field, "The Question of Diamonds in Canada," *Journal of Gemmology* (July 1949): 103–111; D. S. M. Field, "Diamond Pipes in Canada," *Canadian Mining Journal* (July 1950): 54–57; "La Course aux Diamants," *L'Echo* (Val d'Or, Quebec), April 19, 1950; Robert Webster, *Gems: Their Sources and Identification* (Oxford: Butterworth Heinemann, 1994), p. 24.

146 a "wall of blue clay": Accounts of western Canada from "Canada," *The Gemmologist* (August 1948): 212; "Diamond Find in Canada?" *The Gemmologist* (September 1948): 219; Field, "Question of Diamonds in Canada," p. 111; Jiri Strnad, "Discovery of Diamonds in Siberia and Other Northern Regions: Explorational, Historical and Personal Notes," *Earth Sciences History*, 10, no. 2 (1991): 242–243; M. R. Gent, "Diamond Exploration in Saskatchewan," *CIM Bulletin*, 956 (1992): 64–71; David Duval, *New Frontiers in Diamonds* (London: Rosendale Press, 1996), pp. 65–66; Barry Price, *History of Diamond Exploration in Western Canada* (Vancouver, unpublished manuscript, Petra Gem Exploration, 1999).

146 the world's largest kimberlite: See V. Ben Meen, "The Study of Chubb Crater, Ungava, Quebec, 1951," in *National Geographic Society Research Reports 1890–1954* (Washington, DC: National Geographic Society, n.d.), p. 213; V. Ben Meen, "Solving the Riddle of the Chubb Crater," *National Geographic* (January 1952): 1–32; Lex Schrag, *Crater Country* (Toronto: Ryerson Press, 1958), pp. 1–5.

147 deep rocks containing earthly diamonds: H. P. Schwarcz, "Origin of Diamonds in Drift of the North Central United States," *Journal of Geology*, 73 (1965): 657–663. See also Christopher Gunn, "Origin of Diamonds in Drift of the North Central U.S.: A Discussion," *Journal of Geology*, 75 (1967): 232–233.

151 nothing serious, but annoying: Narratives of this period in eastern Canada include J. J. Brummer, "Discovery of Kimberlites in the Kirkland Lake Area Northern Ontario, Canada," *Exploration and Mining Geology*, 1, no. 4 (1992): 339–370; I. E. Reed, "The Search for Kimberlite in the James Bay Lowlands of Ontario," *CIM Bulletin* (March 1991): 132–139; W. J. Wolfe, *Heavy Mineral Indicators in Alluvial and Esker Gravels, Moose River Basin* (Toronto: Ontario Division of Mines Geoscience Report 126, 1975); Hulbert Lee, *Investigation of Eskers for Mineral Exploration* (Ottawa: GSC, Paper 65-14, 1965).

153 the ants were mining: The process by which living creatures rearrange particles is called bioturbation. Lamont was also aided by deflation—when winds in a dry environment remove light matter like quartz grains, leaving heavies like garnets in a layer near the surface. Orapa was discovered in 1967, the nearby Letlhakane mine shortly after. The Jwaneng, in another cluster, was tracked down by De Beers in 1973. See Eric Bruton, *Diamonds* (Radnor, PA: Chilton, 1978), pp. 159–160; Robert Maillard, ed., *Diamonds: Myth, Magic, and Reality* (New York: Crown, 1980), pp. 101–102; Timothy Green, *The World of Diamonds* (New York: William Morrow, 1981), pp. 93–98; A. N. Wilson, *Diamonds from Birth to Eternity* (Santa Monica, CA: Gemological Institute of America, 1982), pp. 225–231.

154 the apartheid system: It must be noted that Harry Oppenheimer spoke publicly against apartheid many times. When he took over in 1957, he worked steadily to change it, encouraging black labor unions and higher education and gradually bringing blacks into management—thus garnering enduring hostility from other whites. When he died in August 2000 at age ninety-one, people of all races showed up at his funeral, at which he was eulogized by Cyril Ramaphosa, past secretary general of the African National Congress.

Chapter 9: Broken Skull River

160 skulless skeletons of two prospectors: According to legends persisting even into the 1930s, somewhere in the recesses was a "poisonous Shangri-La" sheathed in tropical mist, inhabited by prehistoric animals and an evil queen who presided over visitors' deaths. See Pierre Berton, *The Mysterious North* (New York: Knopf, 1956), pp. 11–12, 24–26, 59–63.

161 shortened to twenty-five years: See C. S. Lord, "Operation Keewatin, 1952: A Geological Reconnaissance by Helicopter," *Canadian Mining and*

Metallurgical Bulletin (April 1953): 224–233; Michael Clugston, "A Passion for Exploration," *Canadian Geographic* (May/June 1992): 66–77; Morris Zaslow, *Reading the Rocks: The Story of the Geological Survey of Canada 1842–1972* (Toronto: Macmillan, 1972), pp. 433–437.

165 Chuck grew up in Kelowna: Some details on Fipke's early life are taken from Vernon Frolick, *Fire into Ice: Charles Fipke & the Great Diamond Hunt* (Vancouver: Raincoast Books, 1999).

166 *The Diamond Seeker*: Heinz Heidgen, *The Diamond Seeker: The Story of John Williamson* (London: Blackie & Son, 1959)—not to be confused with the later, juicier book of the same title by John Gawaine (Johannesburg, South Africa: Macmillan, 1976).

Chapter 10: Deliverance

174 everyone wanted back in: See David Gold, "Natural and Synthetic Diamonds and the North American Outlook," *Earth and Mineral Sciences* (February 1968): 37–43; F. M. Lampietti, "Prospecting for Diamonds—Some Current Aspects," *Mining Magazine* (August 1978): 117–123; John Bristow, "Current Status of Diamond Exploration in the USA," *California Mining Journal* (November 1982): 60–69; David Gold, "A Diamond Exploration Philosophy for the 1980s," *Earth and Mineral Sciences* (Summer 1984): 37–45.

174 interest had increased: For kimberlites of modern times, see K. D. Watson, "Kimberlites of Eastern North America," in P. J. Wyllie, ed., *Ultramafic and Related Rocks* (New York: John Wiley, 1967), pp. 312–323; W. F. Cannon, *Potential for Diamond-Bearing Kimberlite in Northern Michigan and Wisconsin* (Washington, DC, USGS Circular 842, 1981); Elaine McGee, "Potential for Diamond in Kimberlites from Michigan and Montana," *Economic Geology*, 83 (1988): 428–432; Elaine McGee, "The Lake Ellen Kimberlite, Michigan, U.S.A.," in J. Kornprobst (ed.), *Kimberlites and Related Rocks*. Proceedings of the Third International Kimberlite Conference (Amsterdam: Elsevier, 1984), pp. 143–154; Henry Meyer, "Kimberlites of the Continental United States," *Journal of Geology* (July 1976): 377–403.

174 Antigo, Wisconsin: Shortly after, a ninth-grade science teacher took his class to a cornfield near Akron, Ohio, for a lesson on minerals, and student Jeni Croft picked up a small bluish-gray diamond. See Renee Bendheim, "Diamonds in Them Thar Glacial Deposits?" *SUR/VIEW (Wisconsin Geological Survey)* (December 1984); Bill Sloat, "Elusive Summit Diamond Perpetuates Jinx," *Akron Beacon Journal*, August 13, 1984, p. C2; "Junior High Student Finds 1st Ohio Diamond of This Century," *Cincinnati Enquirer*, August 15, 1984. The first Alaska alluvial diamonds were also found at this time. See R. B. Forbes, *Preliminary Evaluation of Alluvial Diamond Discoveries in Placer Gravels of Crooked Creek* (Alaska Department of Natural Resources, Investigation 87-1, 1995 [reprint]); "Diamonds in Crooked Creek," *Alaska Magazine* (June 1987).

175 the Great Lakes: See Paul Hayes, "Diamond Hunters Treasure the North," *Milwaukee Journal*, September 23, 1984, p. 1; "Dow Continuing Exploration for Diamonds in Michigan," *Skillings' Mining Review*, June 20, 1987.

175 without a search warrant: *U.S. v. de Beers Industrial Diamond Division Ltd. et al.* (U.S. SDNY, Criminal Action 74, CR1151, December 10, 1974). See also Daniel Nossiter, "Justice vs. de Beers," *Barron's*, March 30, 1980, pp. 17–21. Sibeka was mentioned in a 1994 U.S. indictment against De Beers.

185 competitors had the same idea: See Dale Buss, "Industry Vies with Prospectors for Rights to Strike It Rich at Crater of Diamonds Park," *Wall Street Journal*, July 30, 1981, p. 27. Later efforts by companies to take over the park can be found in "Clinton Mining Support Bearing Fruit?" *Engineering & Mining Journal* (June 1993): 14–17; Michael Taylor, "Field of Dreams," *Audubon* (July–August, 1993): 72–74; Doug Thompson, "Who's Minding the Mine?" *Arkansas Democrat-Gazette*, March 6, 1994, pp. 1G–2G; *History and Summary of the Diamond Evaluation Program* (unpublished, archives of Crater of Diamonds State Park, September 30, 1998).

192 reclassified as lamproites: For lamproites, see B. H. Scott, "A New Look at Prairie Creek, Arkansas," in Kornprobst, ed., *Kimberlites and Related Rocks*, pp. 255–283; Michael Waldman, *Geology and Petrography of Twin Knobs #1 Lamproite* (Boulder, CO: GSA, Special Paper 215, 1987), pp. 205–216; W. L. Griffin, "Indicator Minerals from Prairie Creek and Twin Knobs Lamproites," in H. O. A. Meyer, *Proceedings of the Fifth International Kimberlite Conference*, Vol. 2 (Rio de Janeiro: Companhia de Pesquisa de Recuursos Minerais, 1991), pp. 302–311; T. E. McCandless, "Macrodiamonds and Microdiamonds from Murfreesboro Lamproites," in ibid., pp. 78–97; R. H. Mitchell, "Kimberlites and Lamproites: Primary Sources of Diamond," *Geoscience Canada*, 18, no. 1 (1991): 1–16.

Chapter 11: The Secret of the Anthill

195 a dozen true kimberlites: For Colorado pipes, see John Chronic, "Lower Paleozoic Rocks in Diatremes, Southern Wyoming and Northern Colorado," *GSA Bulletin* (January 1969): 149–156; M. E. McCallum, "Mineralogy of Sloan Diatreme, a Kimberlite Pipe in Colorado," *American Mineralogist* (September–October 1971): 1735–1749; M. E. McCallum, "Diamonds in an Upper Mantle Peridotite Nodule from Kimberlite in Southern Wyoming," *Science* (April 1976): 253–256; M. E. McCallum, *Diamond in State-Line Kimberlite Diatremes* (Geological Survey of Wyoming, Report 12, 1976); Dan Hausel, *Exploration for Diamond-Bearing Kimberlite in Colorado and Wyoming* (Geological Survey of Wyoming Report 19, 1979); Dan Hausel, *Geology, Diamond Testing Procedures and Economic Potential of the Colorado-Wyoming Kimberlite Province* (Geological Survey of Wyoming Report 31, 1985); Dan Hausel, "Diamond-Bearing Kimberlite Pipes in Wyoming and Colorado," *Rocks and Minerals* (September–October 1983): 241–244; Donley Collins, "History of the Colorado-Wyoming State Line Diatremes," *Rocks and Minerals* (January–February 1984): 35–37; "Diamonds," *Wyoming Geo-Notes*, 47 (1995): 34–39; James Cappa, *Alkalic Igneous Rocks of Colorado and Their Associated Ore Deposits* (Colorado Geological Survey, 1998).

203 contents scattered everywhere: McCandless later wrote about it. See T. E. McCandless, "Mantle Indicator Minerals in Ant Mounds and Conglomerates of the Southern Green River Basin, Wyoming," in *Wyoming Geological*

Association Conference Fieldbook (1995), pp. 153–163; T. E. McCandless, "Detrital Mantle Indicator Minerals in Southwestern Wyoming U.S.A.: Evaluation of Diamond Exploration Significance," *Exploration and Mining Geology*, 5, no. 1 (1995): 33–44; T. E. McCandless, "Kimberlite Xenocryst Wear in High-Energy Fluvial Systems: Experimental Studies," *Journal of Geochemical Exploration*, 37 (1990): 323–331.

204 inclusion or indicator grain: Until that time the study of inclusions had lagged so badly that in the 1950s celebrated Swiss gemologist E. Gübelin took magnified pictures of them, then whiled away hours dreaming up objects they looked like: a witch on a broom, a Buddhist shrine, Egyptian pyramids, or, apropos of the time, a mushroom cloud. See L. G. A. Trillwood, "Inclusions in Diamond," *The Gemmologist* (January 1934): 186–189; E. Gübelin, "Inclusions in Diamond," *The Gemmologist* (December 1951): 241–243; E. Gübelin, "Inclusion Fantasy," *The Gemmologist* (September 1954): 170–171. For the history of inclusion study, see also Robert Maillard, ed., *Diamonds: Myth, Magic, and Reality* (New York: Crown, 1980), pp. 196, 221–226; A. N. Wilson, *Diamonds from Birth to Eternity* (Santa Monica, CA: Gemological Institute of America, 1982), pp. 28–31, 146–147, 416–417, 434–436.

204 with or near diamonds: The seminal modern papers on inclusions are Henry Meyer, "Mineral Inclusions in Diamonds," *Carnegie Institution of Washington Yearbook* (1966): 446–450; Henry Meyer, "Chrome Pyrope: An Inclusion in Natural Diamond," *Science* (June 1968): 1446–1447; Henry Meyer, "Composition and Origin of Crystalline Inclusions in Natural Diamonds," *Geochimica et Cosmochimica Acta*, 36 (1972): 1255–1273; J. J. Gurney, "Silicate and Oxide Inclusions in Diamonds from the Finsch Kimberlite Pipe," in F. R. Boyd, ed., *Proceedings of the Second Annual International Kimberlite Conference*, Vol. 1 (Washington, DC: American Geophysical Union, 1979), pp. 1–15; Hsiao-ming Tsai, "Mineral Inclusions in Diamond," in ibid., pp. 16–26; F. R. Boyd, "Low-Calcium Garnets: Keys to Craton Structure and Diamond Crystallization," *Carnegie Institution of Washington Yearbook* (1981): 261–267.

204 diamond-rich pipe: This work was supplemented by lab re-creations of roughly similar garnets from ground-up chemicals, using pressures and heats calculated to be roughly in the diamond stability field. This was around the same time that others made laboratory diamonds themselves starting in 1953, led by the Swedes, then a U.S. General Electric team. See George Kennedy, "The Genesis of Diamond Deposits," *Economic Geology*, 63 (1968): 495–503; and J. B. Dawson, "Recent Researches on Kimberlite and Diamond Geology," *Economic Geology*, 63 (1968): 504–511.

204 Smithsonian's brand-new microprobe: See J. J. Gurney, "Discovery of Garnets Closely Related to Diamonds in the Finsch Pipe, South Africa," *Contributions to Mineralogy and Petrology*, 39 (1973): 103–116; G. S. Switzer, "Composition of Garnet Xenocrysts from Three Kimberlite Pipes," *Smithsonian Contributions to Earth Sciences*, 19 (1975): 1–21.

205 wrote a thesis: P. J. Lawless, *Some Aspects of the Geochemistry of Kimberlite Xenocrysts* (unpublished master's thesis, University of Cape Town, Cape Town, South Africa, October 1974).

206 an earlier researcher: J. B. Dawson, "Statistical Classification of Garnets from Kimberlite and Associated Xenoliths," *Journal of Geology*, 83 (1975): 589–607.

206 vaporized, gone: This was not strictly Gurney's work, but he figured out how to apply it. See Stephen Haggerty, "Redox State of Earth's Upper Mantle from Kimberlitic Ilmenites," *Nature*, May 26, 1983, pp. 295–300; Stephen Haggerty, "Redox State of the Continental Lithosphere," in Martin Menzies, ed., *Continental Mantle* (Oxford: Clarendon Press, 1990), pp. 87–105.

206 a secret diamond-detecting weapon: The system was partially published long after. See J. J. Gurney, "Interpretation of the Major Element Compositions of Mantle Minerals in Diamond Exploration," *Journal of Geochemical Exploration*, 53 (1995): 293–309; J. J. Gurney, "A Review of the Use and Application of Mantle Mineral Geochemistry in Diamond Exploration," *Pure and Applied Chemistry*, 65 (1993): 2423–2442; J. J. Gurney, "Kimberlite Garnet, Chromite and Ilmenite Compositions: Applications to Exploration," paper 21 in *Proceedings of the Canadian Institute of Mining* (1991).

206 The now-accepted theory of diamonds: The theory was generally accepted by the mid-1980s, when researchers in conjunction with Gurney studied garnet inclusions in diamond and found them to be billions of years older than the kimberlite from which they came. Geologists have found about 100 recorded diamondiferous eclogite xenoliths. But only a handful of diamondiferous garnet harzburgites have been discovered, starting in the 1970s in Russia. See S. H. Richardson and J. J. Gurney, "Origin of Diamonds in Old Enriched Mantle," *Nature*, July 19, 1984, pp. 198–202; F. R. Boyd and J. J. Gurney, "Evidence for a 150–200-Km Thick Archaean Lithosphere from Diamond Inclusion Thermobarometry," *Nature*, May 30, 1985, pp. 387–389; Stephen Haggerty, "Diamond Genesis in a Multiply-Constrained Model," *Nature*, March 6, 1986, pp. 34–38; F. R. Boyd and J. J. Gurney, "Diamonds and the South African Lithosphere," *Science*, April 25, 1986, pp. 472–477; Stephen Richardson, "Latter-Day Origin of Diamonds of Eclogitic Paragenesis," *Nature*, August 14, 1986, pp. 623–626.

Chapter 12: A Nunatak

209 a few other pipelike formations: See D. A. Grieve, "Diatreme Breccias in the Southern Rocky Mountains," in *British Columbia Fieldwork 1980–81* (British Columbia Ministry of Mines, 1981), pp. 97–103; D. A. Grieve, "Petrology and Chemistry of the Cross Kimberlite," in *Geology in British Columbia* (British Columbia Ministry of Mines, 1985).

213 line of high peaks: See H. T. Dummett, "Diamond Exploration Geochemistry in the North American Cordillera," in *Geoexpo/86*. Proceedings of the Geological Association of Canada (Vancouver: University of British Columbia, 1986), pp. 168–176; M. E. McCallum, "Lamproitic(?) Diatremes in the Golden Area of the Rocky Mountain Fold and Thrust Belt," in *Proceedings of the Fifth International Kimberlite Conference* (Rio de Janeiro: Companhia de Pesquisa Minerais, 1991).

Chapter 13: Blackwater Lake

225 Mountain Diatreme: See Colin Godwin, "Geology of the Mountain Diatreme Kimberlite, North-Central Mackenzie Mountains," in James Morin, ed., "Mineral Deposits of the Northern Cordillera," *Canadian Institute of Mining & Metallurgy*, Special Issue 37 (1986): 298–310; O. J. Ijewliw, "Diatreme Breccias in the Cordillera," in A. N. LeCheminant et al., eds., *Searching for Diamonds in Canada* (Ottawa: GSC, Open File 3228, 1996), pp. 91–95.

228 Somerset Island: See Roger Mitchell, "Kimberlite from Somerset Island, District of Franklin, N.W.T.," *Canadian Journal of Earth Sciences*, 10 (1973): 384–393; B. A. Kjarsgaard, "Somerset Island Kimberlite Field, District of Franklin, N.W.T.," in LeCheminant et al., eds., *Searching for Diamonds*, pp. 61–65.

230 taken up by plants: Biogeochemistry is the technical word for this. See P. Gregory, "Geochemical Prospecting for Kimberlites," *Quarterly of the Colorado School of Mines*, 64 (1969): 265–305; C. E. Dunn, "Biogeochemical Studies of Kimberlites," in LeCheminant et al., eds., *Searching for Diamonds*, pp. 219–223; Donna Collins, "A Colorado-Wyoming Border Diatreme and a Possible Potential Kimberlite Indicator Plant," *Mountain Geologist* (April 1984:) 68–71.

236 they forced Howard out: See Susan Fraker, "Brawl in the Family at Superior Oil," *Fortune*, May 30, 1981, pp. 70–73; Susan Fraker, "Jousting over Superior's Oil Riches," *Fortune*, January 9, 1984, pp. 84–86.

Chapter 14: Tree Line

242 shown by new GSC maps: See *Pleistocene Geology of Arctic Canada* (Ottawa: GSC, Paper 60-10, 1960); Weston Blake, *Notes on Glacial Geology, Northeastern District of Mackenzie* (Ottawa: GSC, Paper 63-28, 1963); G. M. Wright, *Geology of the the Southeastern Barren Grounds* (Ottawa: GSC, Memoir 350, 1967); R. J. Allan, *Reconnaissance Geochemistry of a 36,000-Square-Mile Area of the Northwestern Canadian Shield* (Ottawa: GSC, Paper 72-50, 1972).

250 Glacial Lake McConnell: See B. G. Craig, *Glacial Lake McConnell and the Surficial Geology of Parts of Slave River and Redstone River Map-Areas* (Ottawa: GSC, Bulletin 122, 1963).

253 "Clifford's Rule": See Tom N. Clifford, "Tectono-Metallogenic Units and Metallogenic Provinces of Africa," *Earth and Planetary Science Letters*, 1 (1966): 421–434. For a historical perspective, see also Martin Menzies, ed., *Continental Mantle* (Oxford: Clarendon Press, 1990), 1–52.

257 maps, recently published: Assembly of such maps is described in Gerald McGrath, ed., *Mapping a Northern Land: The Survey of Canada 1947–1994* (Montreal: McGill-Queen's University Press, Montreal, 1998).

Chapter 15: The Sandman

259 The town had changed little: For later history of Yellowknife and the northern search for minerals, see John Hamilton, *Arctic Revolution: Social*

Change in the Northwest Territories 1935–1994 (Toronto: Dundurn Press, 1994); *Mining: Our Northern Legacy* (Yellowknife: NWT Chamber of Mines, 1989); W. W. Nassichuk, "Forty Years of Northern Non-Renewable Resource Development," *Arctic* (December 1987): 274–284; W. A. Padgham, *Mineral Deposits of the Slave Province, Northwest Territories* (Ottawa: GSC, Open File 2168, 1990); Edith Iglauer, *Denison's Ice Road* (New York: Dutton, 1975).

262 "Glacial Map of Canada": V. K. Prest, *Glacial Map of Canada* (Ottawa: GSC, 1967). A far more detailed map of the Barrens came out a few years later; see J. M. Aylsworth and W. W. Shilts, *Glacial Features Around the Keewatin Ice Divide* (Ottawa: GSC, Map 24, 1987).

268 an obscure Australian publication: J. J. Gurney, "A Correlation Between Garnets and Diamonds," in J. E. Glover, ed., *Kimberlite Occurrences and Origin* (University of Western Australia, Publication 8, 1984), pp. 143–166. See also N. V. Sobolev, "On Mineralogical Criterions of Diamond Content in Kimberlites," *Geology and Geophysics* (Academy of Sciences of the USSR, Siberian Branch) (March 1971) [in Russian].

268 a GSC scientific grant: Fipke was required to write a report to help other diamond prospectors, but it came out long after it could be used by immediate competitors. See C. E. Fipke, *Diamond Exploration Techniques Emphasising Indicator Mineral Geochemistry and Canadian Examples* (Ottawa: GSC, Bulletin 423, 1995).

269 near the old penitentiary: See John de Mont, "A Diamond Rush Excites Saskatchewan," *Maclean's*, November 7, 1988, p. 48.

Chapter 16: The Barren Lands

286 Government rules: *Acquiring Mining Rights in the Northwest Territories* (Yellowknife: Department of Northern and Indian Affairs, 1993).

307 "Canadian claim blocks": "BHP-Utah Enters Joint Venture for Canadian Diamond Exploration," *Northern Miner*, September 20, 1990.

308 It showed several roots: Stephen P. Grand, "Tomographic Inversion for Shear Velocity Beneath the North American Plate," *Journal of Geophysical Research*, December 10, 1987, pp. 14,065–14,090. See also Thomas H. Jordan, "Structure and Formation of the Continental Tectosphere," *Journal of Petrology*, Special Lithosphere Issue (1988): 11–37; Shawna Vogel, *Naked Earth: The New Geophysics* (New York: Dutton, 1995), pp. 45–61.

308 dating of the Acasta gneisses: Bowring and his partners left the coordinates vague to keep away curiosity seekers and competing scientists, but Yellowknife gold prospector Walt Humphries managed to get them from a government geologist. He registered a claim, flew up with an Otterload of dynamite, and blasted ten tons to trinket-size bits for sale to Japanese tourists and others who might wish to own a sample of the world's oldest rock—ultimate proof of the word *kwet'i*. The pure scientists were appalled because part of their study site had been blown away, but it was legal.

Chapter 17: The Great Staking Rush

328 160 tons of kimberlite: See Virginia Heffernan, "Diamond Sample Shipped Out, Land Scramble Goes On," *Northern Miner*, March 30, 1992. For an overview of the rush, see *Diamonds and the Northwest Territories, Canada* (Yellowknife: GNWT Department of Energy, Mines and Petroleum, 1993).

Chapter 18: The Corridor of Hope

344 "the best bet in the world": See "Experts Compare Dia Met with South Africa," *Northern Miner*, June 1, 1992, p. 1; Paul Luke, "Lac de Gras Diamond Field 'Best in the World,'" *Province Money Newspaper*, May 14, 1993, p. 1; Matthew Hart, "Filthy Rich," *Toronto Life* (May 1996): 73–76.

345 "geologist Charles Fipke": Virginia Danielson, "Charles Fipke—1992 Mining Man of the Year," *Northern Miner*, December 28, 1992, p. 1.

346 "best mushroom I ever ate": Maryanne McNellis, "The Great Canadian Diamond Rush," *Financial Post Magazine* (October 1993): 18–38.

346 "exploration in this country": "Closer and Closer," *Canadian Mining Journal* (February 1994): 11.

346 *Fire into Ice*: Vernon Frolick, *Fire into Ice: Charles Fipke & the Great Diamond Hunt* (Vancouver: Raincoast Books, 1999). This more or less official biography of Fipke contains many details of Fipke's family history and early career, as well as narration of the search for diamonds.

347 "with muscular forearms": Neil Ulman, "Call of the Wild," *Wall Street Journal*, May 27, 1994, p. A4.

349 the fund filed a huge lawsuit: *Equity Investments Corp. v. Charles Fipke et al.* (Supreme Court of British Columbia A934048, October 25, 1993).

355 U.S. magazines asked me to cover: See Kevin Krajick, "The Great Canadian Diamond Rush," *Discover* (December 1994): 70–79; Kevin Krajick, "The Rich Barrens," *Audubon* (January–February 1995): 18–21; Kevin Krajick, "Digging Frozen Carats," *Newsweek International*, August 21, 1995, pp. 36–37; Kevin Krajick, "An Esker Runs Through It," *Natural History* (May 1996): 28–37.

Chapter 19: An Esker Runs Through It

359 The Dene had a saying: See Ed Hall, ed., *People & Caribou in the Northwest Territories* (Yellowknife: GNWT Department of Renewable Resources, 1989).

359 they starved to death: See D. C. Heard, "Distribution of Wolf Dens on Migratory Caribou Ranges in the Northwest Territories," *Canadian Journal of Zoology*, 70, no. 8 (1992): 1504–1510; T. Mark Williams, *Summer Diet and Behaviour of Wolves Denning on Barren-Ground Caribou Range in the Northwest Territories* (unpublished master's thesis, University of Alberta, Edmonton, Alberta, Canada, 1990).

360 the only good place to dig: Fritz Mueller, *Tundra Esker Systems and Denning by Grizzly Bears, Wolves, Foxes, and Ground Squirrels in the Central Arctic,*

NWT (Yellowknife: GNWT Department of Renewable Resources, File Report 115, 1995).

360 700 years old: L. David Mech, "Possible Use of Wolf Den over Several Centuries," *Canadian Field Naturalist*, 104 (1990): 484–485.

361 Grizzlies also used the eskers: See Philip McLoughlin, *Spatial Organization and Habitat Selection of Barren-Ground Grizzly Bears in the Central Arctic* (unpublished Ph.D. dissertation, University of Saskatchewan, Saskatoon, Canada, 2000); Robert Gau, *Food Habits, Body Condition and Habitat of the Barren-Ground Grizzly Bear* (unpublished master's thesis, University of Saskatchewan, Saskatoon, Canada, 1998); Wayne Lynch, "King of the Barrens," *Canadian Geographic* (May/June 1992): 26–34.

361 studying wolverines: See John Lee, *Ecology of the Wolverine on the Central Arctic Barrens—Progress Report* (Yellowknife: GNWT Department of Renewable Resources, Manuscript Report 75, 1993).

363 "mitigable with existing technology": *NWT Diamonds Project Environmental Impact Statement Summary* (Yellowknife: BHP Diamonds Inc., 1995), p. 53.

364 on both sides of the road: Aspects of the environmental debate are carried in James Raffan, "Diamonds & Coppermine," *Equinox* (December–January 1998): 32–43; Jamieson Findlay, "Blue Ground," *Canadian Geographic* (January–February 1998): 49–58.

365 The road had been built: See Kevin Krajick, "Road to Ruin?" *Audubon* (July–August 1997): 120.

365 "brought them up above ground": Pliny, *Natural History*, Vol. 1 (London: Philemon Holland Translation, 1661), p. 30.

366 research and documentation: See Thomas Andrews, "The Idaa Trail," in George Nicholas, ed., *At a Crossroads: Archaeology and First Peoples in Canada* (Burnaby, BC: Archaeology Press, Simon Fraser University, 1997); Thomas Andrews, "On Yamozhah's Trail: Dogrib Sacred Sites and the Anthropology of Travel," in Jill Oakes, ed., *Sacred Lands: Claims and Conflicts* (Edmonton: Canadian Circumpolar Institute, 1996).

372 "North America's first diamond mine": Quoted in Bill Braden, "Gov't Embraces Mammoth Diamond Project," *The Yellowknifer*, February 25, 1994, p. 1.

Chapter 20: Ekati

377 *Edmonton Journal* reported: Ric Dolphin, "The King of Diamonds," *Edmonton Journal*, January 12, 1997, pp. E1, E4–E5.

378 Chuck made a deal: See Priscilla Ross, "Going for Gold," *The Middle East* (June 1997): 29–30.

378 Academics moved in: Major papers on Lac de Gras include J. A. Carlson, "Recent Canadian Kimberlite Discoveries," in *Proceedings of the Seventh International Kimberlite Conference, 1998* (Cape Town, South Africa: University of Cape Town, 1999); G. W. Berg, "Leslie Kimberlite of Lac de Gras, NWT," in

Proceedings of the Seventh International Kimberlite Conference, 1998; M. G. Kopylova, "Upper-Mantle Stratigraphy of the Slave Craton, Canada," *Geology* (April 1998): 315–318; J. A. Pell, "Kimberlites in the Slave Craton, NWT," *Geoscience Canada*, 24, no. 2 (1997): 77–90; W. J. Davis, "A Rb-Sr Isochron Age for a Kimberlite from the Recently Discovered Lac de Gras Field," *Journal of Geology*, 105 (1997): 503–509; J. A. Carlson, "Geology and Exploration of Kimberlites on the BHP/Dia Met Claims," in *Proceedings of the Sixth International Kimberlite Conference, 1995* (New York: Allerton Press), pp. 98–100.

378 swamp, sea, and forest: See Harrison Cookenboo, "Remnants of Paleozoic Cover on the Archean Canadian Shield," *Geology* (May 1998): 391–394; W. W. Nassichuk, "Fossils from Diamondiferous Kimberlites at Lac de Gras, NWT," in A. N. LeCheminant et al., eds., *Searching for Diamonds in Canada* (Ottawa: GSC, Open File 3228, 1996), pp. 43–48; L. D. Stasiuk, "Thermal History and Petrology of Wood and Other Organic Inclusions in Kimberlite Pipes at Lac de Gras," *GSC Current Research* (1995-B): 1–10; W. W. Nassichuk, "Cretaceous and Tertiary Fossils Discovered in Kimberlites at Lac de Gras," ibid., pp. 110–114.

379 Diavik moved quickly: Diavik was scheduled to open in 2003.

379 majority share of his discovery: See Matthew Hart, "The Saga of SouthernEra Resources," *Financial Post Magazine* (June 1998): 22–28.

380 start of the twenty-first century: For an excellent account of modern-day prospecting, as well as old diamond finds and source documents that have only recently come to light, see John Sinkankas, *Gemstones of North America*, Vol. III (Tucson, AZ: Geoscience Press, 1997), pp. 109–134.

380 Kelsey Lake Mine: See Jack Murphy, "Colorado Diamonds: An Update," *Rocks & Minerals* (September–October 2000): 350–354; Howard Coopersmith, "Kelsey Lake: First Diamond Mine in North America," *Mining Engineering* (April 1997): 30–33; Michael Pollak, "A Diamond Industry in the Rough," *New York Times*, August 1, 1996, pp. B1, 48.

380 the Great Lakes glacial diamonds: See J. M. Memmi, "Application of Diamond Exploration Geoscientific Information System Technology for Integrated Exploration in the North-Central U.S.," *International Journal of Remote Sensing*, 18, no. 7 (1997): 1439–1464; Nelson Shaffer, "Heavy Minerals in Fine-Grained Waste Materials from Sand and Gravel Plants of Indiana," *New Mexico Bureau of Mines Bulletin*, 154 (1996): 215–222.

381 Crater of Diamonds: See Doug Thompson, "Who's Minding the Mine?" *Arkansas Democrat-Gazette*, March 6, 1994, pp. 1G–2G. Other information, press releases from the superintendent of Crater of Diamonds State Park.

INDEX